NEUROMETHODS ☐ 27

Voltammetric Methods in Brain Systems

NEUROMETHODS

Program Editors: Alan A. Boulton and Glen B. Baker

27. **Voltammetric Methods in Brain Systems**
    Edited by *Alan A. Boulton, Glen B. Baker, and Ralph N. Adams,* 1995
26. **Patch-Clamp Applications and Protocols**
    Edited by *Alan A. Boulton, Glen B. Baker, and Wolfgang Walz,* 1995
25. **Neurotrophic Factors**
    Edited by *Alan A. Boulton, Glen B. Baker, and Franz Hefti,* 1993
24. **Animal Models of Drug Addiction**
    Edited by *Alan A. Boulton, Glen B. Baker, and Peter Wu,* 1992
23. **Practical Cell Culture Techniques**
    Edited by *Alan A. Boulton, Glen B. Baker, and Wolfgang Walz,* 1992
22. **Animal Models of Neurological Disease, II: Metabolic Encephalopathies and the Epilepsies**
    Edited by *Alan A. Boulton, Glen B. Baker, and Roger F. Butterworth,* 1992
21. **Animal Models of Neurological Disease, I: Neurodegenerative Disease**
    Edited by *Alan A. Boulton, Glen B. Baker, and Roger F. Butterworth,* 1992
20. **Intracellular Messengers**
    Edited by *Alan A. Boulton, Glen B. Baker, and Colin W. Taylor,* 1992
19. **Animal Models in Psychiatry, II**
    Edited by *Alan A. Boulton, Glen B. Baker, and Mathew T. Martin-Iverson,* 1991
18. **Animal Models in Psychiatry, I**
    Edited by *Alan A. Boulton, Glen B. Baker, and Mathew T. Martin-Iverson,* 1991
17. **Neuropsychology**
    Edited by *Alan A. Boulton, Glen B. Baker, and Merrill Hiscock,* 1990
16. **Molecular Neurobiological Techniques**
    Edited by *Alan A. Boulton, Glen B. Baker, and Anthony T. Campagnoni,* 1990
15. **Neurophysiological Techniques: Applications to Neural Systems**
    Edited by *Alan A. Boulton, Glen B. Baker, and Case H. Vanderwolf,* 1990
14. **Neurophysiological Techniques: Basic Methods and Concepts**
    Edited by *Alan A. Boulton, Glen B. Baker, and Case H. Vanderwolf,* 1990
13. **Psychopharmacology**
    Edited by *Alan A. Boulton, Glen B. Baker, and Andrew J. Greenshaw,* 1989
12. **Drugs as Tools in Neurotransmitter Research**
    Edited by *Alan A. Boulton, Glen B. Baker, and Augusto V. Juorio,* 1989
11. **Carbohydrates and Energy Metabolism**
    Edited by *Alan A. Boulton, Glen B. Baker, and Roger F. Butterworth,* 1989
10. **Analysis of Psychiatric Drugs**
    Edited by *Alan A. Boulton, Glen B. Baker, and Ronald T. Coutts,* 1988
9. **The Neuronal Microenvironment**
    Edited by *Alan A. Boulton, Glen B. Baker, and Wolfgang Walz,* 1988

NEUROMETHODS ☐ 27

Voltammetric Methods in Brain Systems

Edited by

Alan A. Boulton

University of Saskatchewan, Saskatoon, Canada

Glen B. Baker

University of Alberta, Edmonton, Canada

and

Ralph N. Adams

University of Kansas, Lawrence, KS

Humana Press Totowa, New Jersey

999 Riverview Drive, Suite 208
Totowa, New Jersey 07512

This publication is printed on acid-free paper. ∞
ANSI Z39.48-1984 (American National Standards Institute)
Permanence of Paper for Printed Library Materials.

Printed in the United States of America.

ISBN 0-89603-312-0
ISSN 0893-2336

Preface to the Series

When the President of Humana Press first suggested that a series on methods in the neurosciences might be useful, one of us (AAB) was quite skeptical; only after discussions with GBB and some searching both of memory and library shelves did it seem that perhaps the publisher was right. Although some excellent methods books had recently appeared, notably in neuroanatomy, it was a fact that there was a dearth in this particular field, a fact attested to by the alacrity and enthusiasm with which most of the contributors to this series accepted our invitations and suggested additional topics and areas. After a somewhat hesitant start, essentially in the neurochemistry section, the series has grown and will encompass neurochemistry, neuropsychiatry, neurology, neuropathology, neurogenetics, neuroethology, molecular neurobiology, animal models of nervous disease, and no doubt many more "neuros." Although we have tried to include adequate methodological detail and in many cases detailed protocols, we have also tried to include wherever possible a short introductory review of the methods and/or related substances, comparisons with other methods, and the relationship of the substances being analyzed to neurological and psychiatric disorders. Recognizing our own limitations, we have invited a guest editor to join with us on most volumes in order to ensure complete coverage of the field. These editors will add their specialized knowledge and competencies. We anticipate that this series will fill a gap; we can only hope that it will be filled appropriately and with the right amount of expertise with respect to each method, substance or group of substances, and area treated.

Alan A. Boulton
Glen B. Baker

Preface

Efforts to measure neurotransmitters and related species in living brain tissue by faradic electrochemistry began in earnest in the early 1970s. During the ensuing years, several monographs and various symposia on the subject have appeared. The present volume of *Neuromethods, Voltammetric Methods in Brain Systems,* can be considered different from previous offerings in several respects. The methodology of in vivo electrochemistry has now reached a stable level of maturity. It is no longer necessary to convince interested readers that the techniques can measure, with specified degrees of reliability, certain neurotransmitters and metabolites in the extracellular fluid space. Moreover, the basic approaches of electroanalytical measurements are more widely appreciated by neuroscientists and one need not explain in tutorial fashion the details of oxidative electrochemistry. Nevertheless, in keeping with the nature of the *Neuromethods* series, this is a "how to" volume and readers will find ample detail on the preparation of electrodes, practical experimental details, and the interpretation of results. But the emphasis is strongly on the application of voltammetric methods.

The first two chapters deal with fundamentals inherent in employing in vivo electrochemical measures. The first article offers an authoritative account of our understanding of the surface states of carbon and carbon fiber electrodes. It provides a rationale for the various pretreatments that have allowed the "tuning" of electrode properties to obtain sensitivity and fast response times. The second chapter covers the unusual properties of the extracellular fluid space in which the measurements are made and offers an appreciation of mass transport and active reuptake in such a medium.

The remaining chapters cover the entire spectrum of applications for tissue, brain slices, transplant media, cell cultures, and single cells, with an emphasis on fast measurements that reveal release–reuptake information on transmitter systems. One chapter is devoted to the measurement of ascorbate. The latter is often considered a major interferent in the detection of the biogenic amines, but is now of great interest with regard to the coupling of ascorbate to glutamate reuptake.

Finally, the last chapter is not an *in situ* technique, but introduces a new electrochemical approach to neurochemical studies, the rotated disk electrode. This technique employs synaptosomes and tissue homogenates. It enables one to do very fast and sophisticated measurements on the kinetics of catecholamine transporter systems, a subject of special interest to the authors of this volume.

It is a distinct pleasure to have served as guest editor of *Voltammetric Methods in Brain Systems,* and to have been associated for so many years with all of the ranking experts, and their colleagues, represented in its pages who have brought in vivo electrochemistry to its present status. It is gratifying to see their work now reported in such a manner as to render the book a valuable and important tool for all neuroscientists employing, or seeking to employ, these powerful techniques.

Ralph N. Adams

Contents

The Measurement of Brain Ascorbate and Its Link with Excitatory Amino Acid Neurotransmission
Robert D. O'Neill

Voltammetric and Amperometric Probes for Single-Cell Analysis
Andrew G. Ewing, Ta-Kung Chen, and Guangyao Chen

Contributors

GUANGYAO CHEN • *Department of Chemistry, Pennsylvania State University, University Park, PA*

TA-KUNG CHEN • *Department of Chemistry, Pennsylvania State University, University Park, PA*

COLIN DAVIDSON • *Anaesthetics Unit (Neurotransmission Laboratory), London Hospital Medical College, Royal London Hospital, London, UK*

ANDREW G. EWING • *Department of Chemistry, Pennsylvania State University, University Park, PA*

PAUL A. GARRIS • *Department of Chemistry, The University of North Carolina at Chapel Hill, NC*

GREG A. GERHARDT • *Departments of Psychiatry and Pharmacology, Neuroscience Training Program, and the Rocky Mountain Center for Sensor Technology at the University of Colorado Health Sciences Center, Denver, CO*

F. GONON • *INSERM, Pierre-Bénite, France*

CHRISTINE M. JORM • *Anaesthetics Unit (Neurotransmission Laboratory), London Hospital Medical College, Royal London Hospital, London, UK*

RICHARD L. MCCREERY • *Department of Chemistry, The Ohio State University, Columbus, OH*

SUSAN M. MEIERGERD • *Department of Chemistry, Washington State University, Pullman, WA*

CHARLES NICHOLSON • *Departments of Neurosurgery, Physiology, and Biophysics, New York University Medical Center, New York, NY*

ROBERT D. O'NEILL • *Chemistry Department, University College, Dublin, Ireland*

PETER PALIJ • *Anaesthetics Unit (Neurotransmission Laboratory), London Hospital Medical College, Royal London Hospital, London, UK*

PAUL E. M. PHILLIPS • *Anaesthetics Unit (Neurotransmission Laboratory), London Hospital Medical College, Royal London Hospital, London, UK*

MARGARET E. RICE • *Departments of Neurosurgery, Physiology, and Biophysics, New York University Medical Center, New York, NY*

JAMES O. SCHENK • *Departments of Chemistry, Biochemistry, and Biophysics, and Programs in Pharmacology/Toxicology and Neuroscience, Washington State University, Pullman, WA*

JONATHAN A. STAMFORD • *Anaesthetics Unit (Neurotransmission Laboratory), London Hospital Medical College, Royal London Hospital, London, UK*

R. MARK WIGHTMAN • *Department of Chemistry, The University of North Carolina at Chapel Hill, NC*

Carbon Electrode Surface Chemistry

Optimization of Bioanalytical Performance

Richard L. McCreery

1. Introduction

The utility of electroanalytical probes for in vivo monitoring of neurotransmitters is derived principally from the small size and high sensitivity of microelectrodes. The high sensitivity of voltammetry compared to optical or magnetic resonance techniques derives from the ability to measure small currents, since a relatively easily measured 1 pA corresponds to $\sim 10^{-17}$ mol/s of an electroactive molecule. An analytical probe with dimensions of a few micrometers has major advantages in spatial and temporal resolution over a technique requiring extraction or preconcentration of the target molecule from tissue. The ability of a microvoltammetric electrode to detect neurotransmitters in vivo depends not only on sensitivity (signal per unit of neurotransmitter concentration), but also on discrimination of a small analytical signal on top of an often large background from electrode capacitance and/or interfering species in the tissue. Although voltammetry is extraordinarily sensitive, it is not particularly selective and is often hampered by background currents.

The success of techniques for in vivo voltammetry thus depends on methods for maintaining or enhancing sensitivity while suppressing interferences. These characteristics are in turn controlled by electrode surface chemistry, since the

From: *Neuromethods, Vol. 27: Voltammetric Methods in Brain Systems*
Eds: A. Boulton, G. Baker, and R. N. Adams © 1995 Humana Press, Inc.

measurement inherently involves electron transfer across an interface between a solid and a solution. Beginning with the inception of in vivo voltammetry for neurotransmitter monitoring by the Adams group in 1973 (Kissinger et al., 1973; McCreery et al., 1974; Adams, 1990), advances in the field have correlated with new electrode pretreatments. For example, the early demonstration of the discrimination of dopamine (DA) and ascorbic acid (AA) in vivo by Gonon (Gonon et al., 1980,1983), depended on an electrochemical pretreatment that enhanced DA response and suppressed ascorbate. Carbon electrode properties have been reviewed on several occasions (Wightman et al., 1984; Kinoshita, 1988; McCreery, 1991), as have their use for in vivo voltammetry (Kawagoe et al., 1983; Feng et al., 1987; Justice, 1987; Stamford et al., 1988; Adams, 1990; Stamford, 1990).

Given the nearly exclusive use of carbon fiber microelectrodes for in vivo voltammetry, one inevitably is faced with questions about how carbon surface chemistry controls analytical response. In particular, how can electrode preparation be modified to suit a particular application? This chapter discusses the surface chemistry of carbon electrodes, with particular attention to factors affecting analytical response for catecholomines and related biological components. The first section addresses the surface phenomena that affect voltammetric measurements, including electron-transfer kinetics, background current, and adsorption. The second section involves carbon electrode surfaces in particular, with attention to how surface chemistry affects sensitivity, selectivity, and so forth. Finally, an overview of the commonly used carbon fiber pretreatments, and their effects on surface structure and reactivity is presented.

2. Electrode Phenomena Affecting Analytical Utility

Before considering carbon fiber electrodes themselves, it is useful to discuss the factors that determine any electrode's analytical utility. For the case of in vivo voltammetry of neurotransmitters, the most important issues are selectivity, sensitivity, detection limits, stability, and response time.

Sensitivity is the slope of a plot of response (usually current) vs the desired quantity (usually concentration in the fluid near the electrode). Sensitivity will be greater when there are more electrons *(n)* involved in the redox reaction of a given analyte molecule, but *n* is nearly always 1 or 2, so there is not much room for improvement. A more practical means to increase sensitivity is preconcentration of the analyte on or near the electrode by adsorption or ion exchange. For voltammetry at relatively high scan rates (>10 V/s), the sensitivity to adsorbed analyte can be greatly enhanced compared to a solution species, since peak current is linear with scan rate for adsorbed species, whereas current varies with the square root of scan rate for a solution species. To a first approximation, sensitivity is directly proportional to electrode area, so enhanced sensitivity becomes particularly important for small in vivo probes.

Selectivity determines the ability to measure a neurotransmitter in the presence of possibly large concentrations of interferences, such as AA and NADH. Since voltammetric peaks are fairly broad relative to the available potential range, overlap of peaks from different components is often a problem. However, there are several means to enhance selectivity above that derived from differences in voltammetric peak potential. A fast heterogeneous electron-transfer rate constant (k^o cm/s) yields a sharper peak, whereas a small k^o can shift a peak outside the available potential range. Thus, a high k^o can result in selectivity of one component over an interferent with smaller k^o. Figure 1A shows theoretical voltammograms for two different electron-transfer rate constants (k^o cm/s). The voltammogram for the larger k^o value is more analytically useful, since it is sharper and larger and can be observed at higher scan rates. As k^o decreases (or the scan rate increases), the response is broadened and shifted, often so much that it is off scale. As noted earlier, adsorption of the analyte to the electrode can significantly enhance the signal and also provide selectivity over a nonadsorbed species. Figure 1B is a

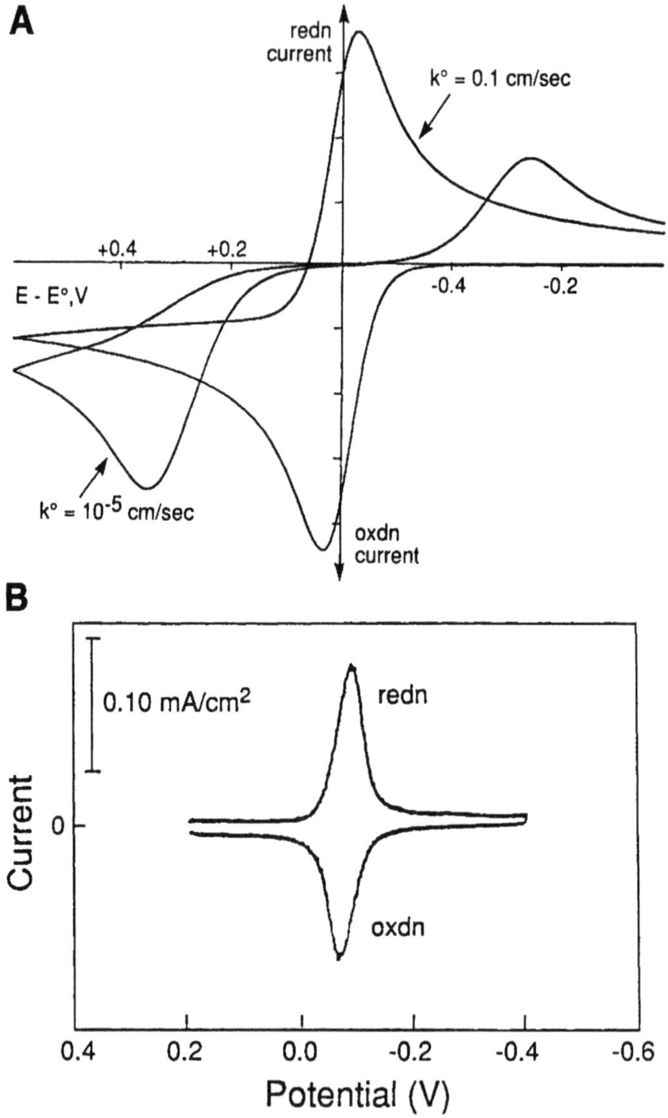

Fig. 1. **(A)** Calculated voltammograms for two different electron-transfer rate constants, assuming a chemically reversible, 1-electron redox reaction. **(B)** Voltammogram of $10^{-5}M$ anthraquinone 2-6 disulfonate on glassy carbon in $0.1M\ HClO_4$. The response is dominated by the adsorbed species, leading to sharp, symmetric peaks. Adapted with permission from McDermott et al. (1992).

voltammogram of anthraquinone 2,6 disulfonate adsorbed to glassy carbon. The sharp peaks and preconcentration afforded by adsorption permit quite low concentrations to be observed,

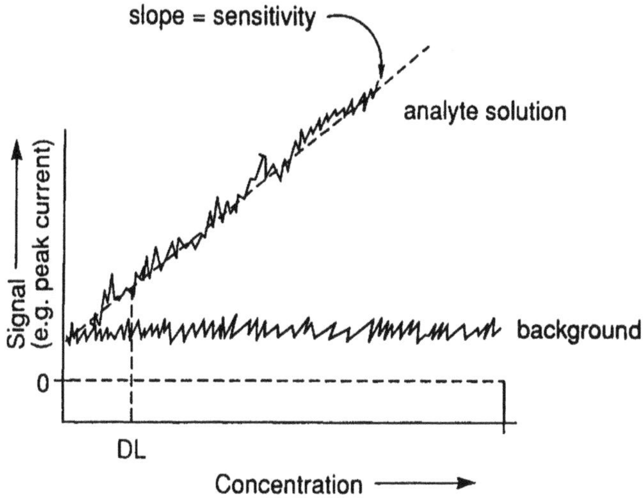

Fig. 2. Hypothetical calibration curves for in vivo voltammetry, plotting peak current vs analyte concentration. The DL is generally defined as the concentration yielding a signal above background equal to three times the standard deviation of the background.

$10^{-5}M$ in this case. A related selectivity enhancement technique is an electrostatic "gate" on the electrode surface. A good example is Nafion, a sulfonated fluoropolymer that can be coated onto electrodes to form a thin film (Kristensen et al., 1987; Gerhardt et al., 1989; Rice et al., 1989). Cations, such as DA and norepinephrine (NE), can pass through the Nafion film and be detected at the electrode, whereas anions like AA and 3,4-dihydroxyphenylacetic acid (DOPAC) are repelled by the SO_3^- groups.

Although often confused with sensitivity, the detection limit (DL) is the more important parameter for in vivo analysis. The DL is determined by the size of the analytical signal compared to the background signal, which consists of noise, capacitance, and interferences. As shown in Fig. 2, the sensitivity determines the slope of the current vs concentration plot, whereas the DL depends on the size of the signal compared to background at low analyte concentrations. A common convention is to consider the DL to be the analyte concentration yielding a signal above the background equal to three times the standard deviation of the background signal determined from a blank. In more qualitative terms, the DL is the concentration where a definite response is observable over the variation in the background. For in vivo voltammetry, the lowest DLs occur when high sensitivity is combined with low background. Even if a pretreatment procedure increases sensitivity, it will not improve the DL if it also significantly increases the background.

Stability is a particular concern for in vivo voltammetry, since monitoring in live animals may require a constant (or at least known) response over hours or days in an environment that would normally be considered hostile to a voltammetric probe. Any electrode is prone to adsorption of molecules in solution, and this adsorption can eventually lead to total loss of activity. Any means to protect the electrode from adsorption and deactivation will extend useful analytical life.

Finally, in vivo voltammetric probes vary in response time, often for reasons involving surface chemistry. Neurochemical events of interest have time scales ranging from a few milliseconds (synaptic vesicle release) to many minutes (e.g., drug response). In some cases, there is a trade-off between response time and another analytical variable, such as sensitivity. For example, if preconcentration of the analyte on the electrode is required, response time can be slow.

In general, the design and preparation of the in vivo electrode and associated electronics will determine the sensitivity, selectivity, detection limits, stability, and lifetime of the in vivo voltammetric technique. There are often trade-offs among these variables, so a particular procedure may optimize one factor at the expense of another. In the next section, we will consider carbon electrode surface chemistry in order to understand its effects on analytical performance.

3. Surface Chemistry of Carbon Electrodes

Carbon differs from metals in both its electronic properties and its surface chemistry. The form of carbon relevant to in vivo voltammetry is the sp^2 hybridized form in a variety of manifestations, including graphite, glassy carbon, and carbon fibers (McCreery, 1991). Schematic drawings of highly ordered graphite, glassy carbon, and carbon fiber are shown

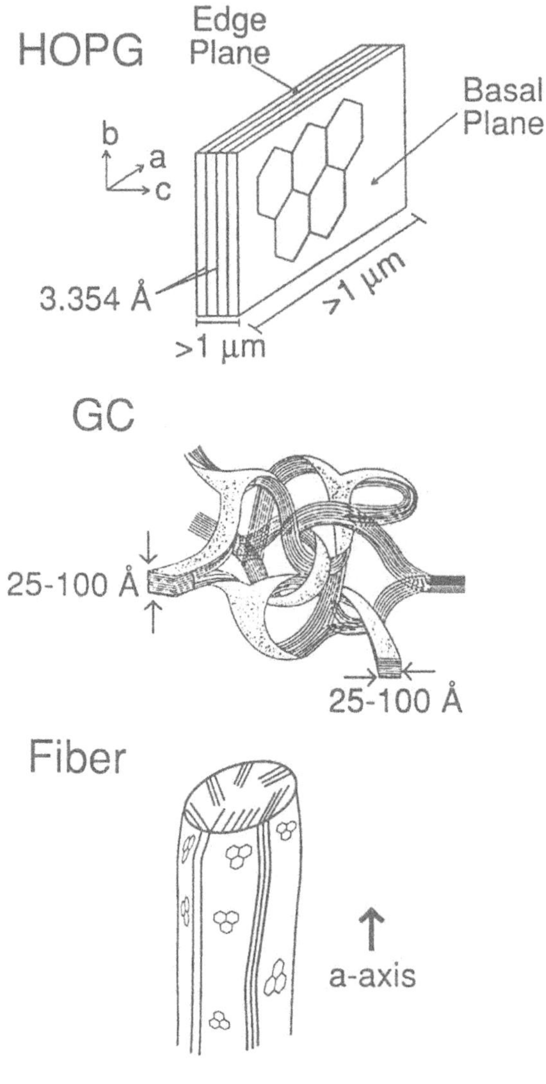

Fig. 3. Three forms of sp^2 hydridized carbon. The parallel layers in GC are similar to those in HOPG, but are much smaller. Fiber morphology differs significantly with fiber type, but the a axis (in plane) of the graphitic domains generally is oriented along the fiber axis.

in Fig. 3. The term "graphite" is often used generically, but usually means a layered material with an interlayer spacing of 3.354 Å. The size of the layers along the "a" axis in Fig. 3 can be large (>10 µm) for specialized materials, such as highly ordered pyrolytic graphite (HOPG) or quite small (~100 Å) for polycrystalline graphite (pencil lead, lubricants). Glassy carbon (GC) is made by heat-treating a polymer, such as polyacrylonitrile, and large graphitic domains are prevented from forming by the original polymeric backbone. Thus, GC has domains that are limited to 50–100 Å, but the material is isotropic at larger distance scales. In contrast, graphite is nonisotropic at the crystallite level, and the anisotropy can extend to bulk crystals (several centimeters).

Most carbon fibers are made by a process similar to that leading to GC, starting with polyacrylonitrile or petroleum pitch, and then heat treating (Dresselhaus, 1988). However, the curing process is designed to orient preferentially the a axis (parallel to the graphite plane) along the fiber axis. Since the major application of carbon fibers is structural composites, this orientation is desirable to improve tensile strength. Fiber types differ significantly in the degree of orientation, the crystallite size, and tensile strength, and these parameters are important to electrochemical applications. Even the fiber cross-section depends on the preparation procedure, with at least three generic types shown in Fig. 4. Unfortunately, the relationship between fiber structure and electrochemical activity has not been established in detail as yet, in part because of the large number of fiber types and manufacturers. Consequently, most users of carbon fibers for electrochemistry use a single source (and even batch) of fiber, which has been characterized for its particular application.

Another major difference between carbon and metal electrodes is the nature of the surface oxides. Metals form oxides and hydroxides easily and reversibly in aqueous solution, often resulting in a background current in voltammetry. Carbon forms a richer variety of surface oxides (Fig. 5), and these surface groups are difficult to remove. Although electrochemical formation of oxides on carbon is usually only observed at extreme potentials, surface oxides can have a variety of effects on electron-transfer, and adsorption of catecholamines and interferences. We will see that manipulation of surface oxides is quite important for certain in vivo applications.

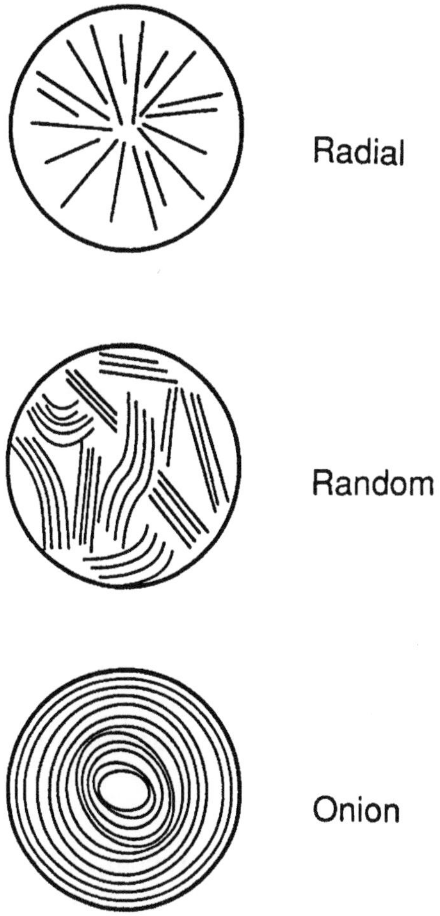

Fig. 4. Commonly encountered carbon fiber cross-sections. Lines represent the edges of graphitic planes.

The effects of carbon microstructure and oxides on electrochemical properties have been discussed in detail (Kinoshita, 1988; McCreery, 1991), but will be summarized here when relevant to in vivo voltammetry. The most basic issue deals with bulk structural effects on electrode kinetics and voltammetric background. The basal plane of a perfect crystal of graphite has substantially different electronic properties from disordered (i.e., polycrystalline) graphite. As shown in Fig. 6, this results in a severe anisotropy in electrochemical properties of the basal plane compared to the edge

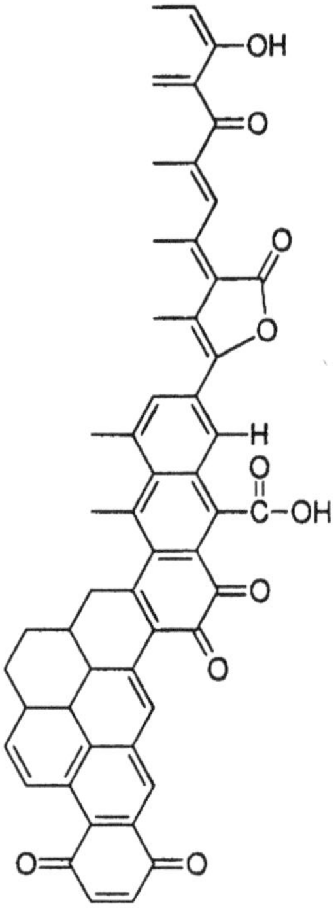

Fig. 5. Representative, but not exhaustive, structures occurring on oxidized carbon surfaces.

(Kneten et al., 1992; McDermott et al., 1992; Cline et al., 1994). The electron-transfer rate is much faster for a variety of systems on edge plane, and the capacitance on basal plane is anomalously low (1 μF/cm^2 instead of ~70 μF/cm^2 for edge plane). Furthermore, adsorption is very weak on basal plane and is below the DL for voltammetry on low-defect HOPG basal plane.

When a graphite crystal is disordered by defects or by a smaller crystallite size, the unusual electronic properties are masked by "defect" or "localized" electronic states, and the material behaves more like a metal (Cline et al., 1994). Glassy carbon and most carbon fibers are quite disordered, and exhibit more typical capacitance (~20–50 μF/cm^2) and faster electron-transfer rates. In fact, clean GC or HOPG edge plane can exhibit the highest k^o values yet reported for $Fe(CN)_6^{-3/-4}$ and similar inorganic redox systems (Rice et al., 1990; Kneten

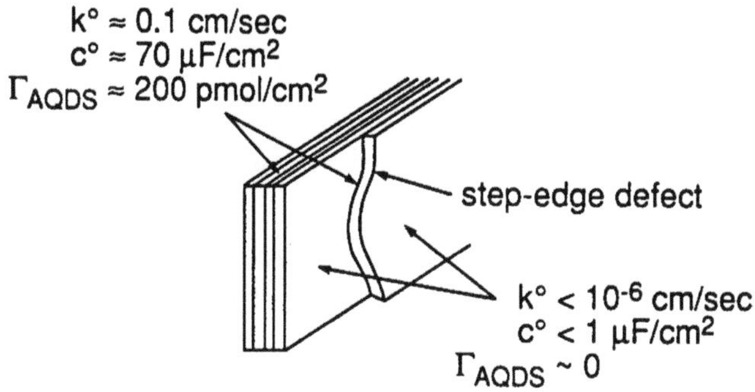

Fig. 6. Electrochemical properties of graphitic "edge plane" compared to perfect basal plane. k^o is electron-transfer rate for $Fe(CN)_6^{-3/-4}$ in 1*M* KCl, C^o is capacitance in 0.1*M* $HClO_4$, and Γ_{AQDS} is the amount of AQDS adsorbed. An edge plane defect on an otherwise perfect basal plane has properties similar to that of the crystal's edge.

Table 1
Electrochemical Properties of Several Carbon Surfaces[a]

	Γ_{AQDS}[b] pmol/cm²	C^o[c] F/cm²	$k^o_{Fe(CN)_6}$ cm/s	k^o, $Eu^{+2/+3}$ cm/s
GC-20, polished	186	33	0.002	2×10^{-4}
GC-20				
laser-activated[d]	336	34	≥0.5	2×10^{-4}
GC-20, fractured[e]	533	75	0.5	8×10^{-5}
GC-20, ECP[f]	—	>100	—	4×10^{-2}
HOPG				
basal plane	<1	<1	10^{-5}	$<3 \times 10^{-6}$
HOPG, edge	>200	~70	~.1	—

[a]Data from Rice et al., 1990; Kneten et al., 1992; McDermott et al., 1992, 1993a; Cline et al., 1994.
[b]Adsorption of anthraquinone 2,6 disulfonate, 1 monolayer ≈ 135 pmol/cm².
[c]Capacitance in 1*M* KCl.
[d]Three 25 MW/cm² laser pulses applied.
[e]GC-20 surface freshly exposed in solution, without polishing.
[f]Oxidized by an *in situ* 0–2.2 V scan (50 mV/s) in 1*M* H_2SO_4.

et al., 1992). Electrochemical properties of several carbon surfaces are listed in Table 1. Note that the capacitance and k^o for $Fe(CN)_6^{-3/-4}$ are largest for the disordered, clean surfaces (e.g., GC-20). It is useful at this point to discuss the term "site," which is often applied to regions on the carbon surface where electron transfer or adsorption occur. The electronic effects just described are related to the band structure and conductivity of the bulk material (Cline et al., 1994), and are not necessarily related to functional groups, such as oxides. Although the electronic disturbance may be localized near a defect, it does not require a specific chemical group. Although electron transfer for simple redox processes (e.g., $Ru[NH^3]_6^{+2/+3}$) and adsorption (e.g., of phenanthrenequinone) depend on the electronic perturbation associated with disorder, they are not dependent on specific surface chemical interactions. This distinction is shown in Fig. 7.

Compared to HOPG, all but a few carbon fibers are disordered in this electronic sense. Such materials will support fast electron-transfer for $Fe(CN)_6^{-3/-4}$ and strong adsorption of many materials, whether or not any oxidative pretreatment is involved. However, disorder is necessarily accompanied by exposure of edge sites to oxygen and water, and some degree of oxidation occurs, resulting in the variety of functional groups shown in Fig. 5. The nature and extent of such surface oxides are a strong function of the sample history, and it is difficult to generalize about which groups are most prevalent. Typical observed oxygen/carbon ratios for several carbon surfaces are listed in Table 2.

The presence of oxide functional groups raises the possibility of specific chemical interactions between the surface and a molecule in solution. Such interactions can catalyze redox reactions, promote or inhibit adsorption, or affect background. An example is the catalysis of $Eu^{+2/+3}$ on oxidized carbon surfaces, an effect that is reversed when oxides are occupied by organosilane reagents. As shown in the last column of Table 1, $Eu^{+2/+3}$ is quite slow (small k^o) on unoxidized surfaces, but fast following oxidation. This catalysis involves the more familiar definition of "site." The surface is disordered with or without the oxide present, but the oxide provides a specific chemical interaction that accelerates electron transfer (McDermott et al., 1993a). A prominent biological example is the catalysis of cytochrome c electron transfer by surface oxides on graphite (Armstrong et al., 1989).

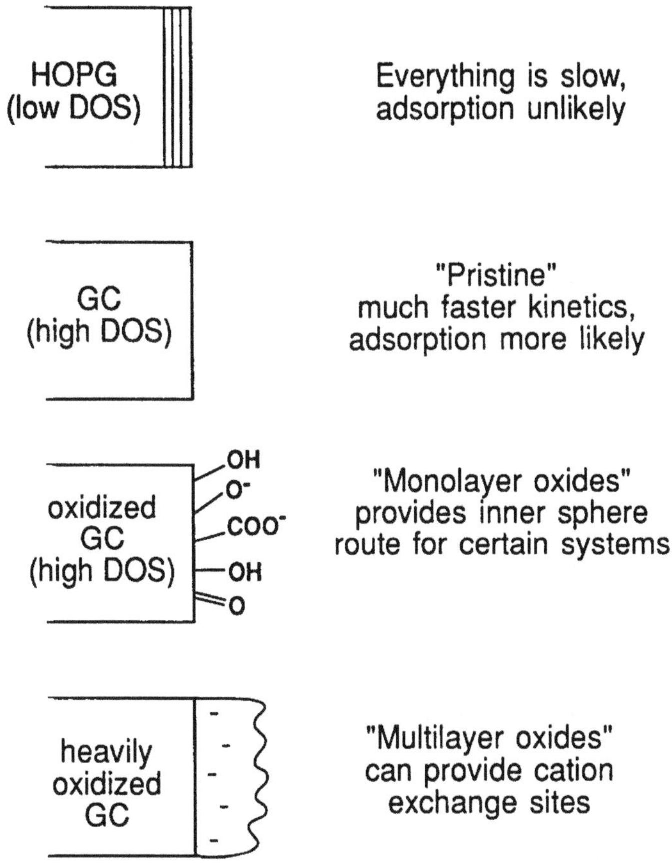

Fig. 7. Schematic pictures of several carbon surfaces. "DOS" refers to the density of electronic states and is higher for disordered materials. The basal plane of HOPG (top) has both a low DOS and a lack of oxide functional groups. Fast electron transfer and adsorption both require a high DOS (Cline et al., 1994).

Surface oxides can also greatly increase background current owing in part to interactions between charged surface groups and electrolyte ions. This effect can increase the voltammetric background by a factor of 10 or more, resulting in a less favorable DL. An example will be discussed later for the case of dopamine on GC.

In addition to the electronic properties of bulk carbon and specific chemical effects of oxides, an additional factor that can alter electrochemical behavior is the presence of

Table 2
Surface O/C Ratio on Various Carbons

Material	Pretreatment	O/C ratio[a]	References
HOPG, basal plane	None	<.01	Schrader, 1975
GC-20	Fractured in UHV[b]	<.01	
GC-20	UHV-heat treatment	0.06	Fagan et al., 1985
GC-20	Polished	0.12–0.18	Cabaniss et al., 1985; Hu et al., 1985; Kamau et al., 1985
GC-20	Polished, then ECP[c]	0.22–32	Cabaniss et al., 1985; Engstrom and Strasser, 1984
E-120 Carbon fiber	None	0.01	Sherwood and Xie, 1990
E-120 Carbon fiber	ECP in HNO_3, 1.5 V	0.076	Sherwood and Xie, 1990
E-120 Carbon fiber	ECP in HNO_3, 2.0 V	0.42	Sherwood and Xie, 1990
E-120 Carbon fiber	ECP in HNO_3, 3.0 V	0.35	Sherwood and Xie, 1990

[a]Atomic ratio of surface oxygen to surface carbon, determined by X-ray photoelectron spectroscopy.
[b]UHV, $<10^{-7}$ Torr.
[c]Electrochemical pretreatment.

"thick" films on the carbon surface. "Thick" in this context means much thicker than a monomolecular layer of oxide or adsorbate. Such films can vary greatly in composition and properties, but several are relevant to in vivo voltammetry. First, a thick layer of oxidized carbon can be formed by anodization and is often referred to as "electrochemical graphitic oxide" (EGO) (Besenhard and Fritz, 1983). EGO is not a stoichiometrically well-defined material, but it is a hydrated, anionic film that is capable of exchanging cations in aqueous solutions. Second, the Nafion film mentioned earlier can be applied as an alcoholic solution to a carbon electrode (Kristensen et al., 1987). The resulting film is permeable to cations, and the sulfonates can easily act as cation-exchange sites.

4. Generic Carbon Surface Types

Given this overview of carbon-surface properties, we can now turn to how various preparation procedures control surface chemistry, and ultimately how these factors control electroanalytical performance. The practical objective here is to modify the preparation procedure to maximize the desired analytical property, such as sensitivity, speed of response, selectivity, and so on. In considering these issues, we will discuss three general types of carbon surfaces. First, "pristine" will refer to a disordered carbon surface without oxides or impurities. Second, "monolayer oxides" will refer to a carbon surface with oxygen-containing surface functional groups present, but at a level less than or equal to a monolayer. "Multilayer oxides" will refer to a surface with a "thick" film of oxides, usually generated electrochemically. Although these designations are somewhat idealized, they are useful for describing electrode behavior.

4.1. Pristine Carbon Surface

Owing to the propensity of oxygen (and many other species) to chemisorb on carbon, a clean, oxide-free surface is difficult to achieve. Strictly speaking, such surfaces exist only in ultrahigh vacuum (UHV) after extensive heat treatment (700–900°C) and cannot be prepared by conventional polishing in air. However, carbon surfaces with very low O/C ratio have been evaluated electrochemically and provide useful insight into the behavior of more common surfaces. Fagan et al. (1985) used a 700°C UHV treatment to reduce the O/C ratio of a GC electrode below 6%, and then transferred the electrode into an electrochemical cell. Our lab fractured a GC rod in solution, thus exposing an intially oxide-free surface to the solution directly (Rice et al., 1990; McDermott et al., 1993). Both the vacuum heat-treated and fractured surfaces exhibit high electron-transfer reactivity, with nearly reversible behavior toward $Fe(CN)_6^{-3/-4}$, DA, and AA. The voltammogram of a mixture of DA (10 m) and AA (1 m*M*) in Fig. 8 was obtained on a fractured GC surface (Allred and McCreery,

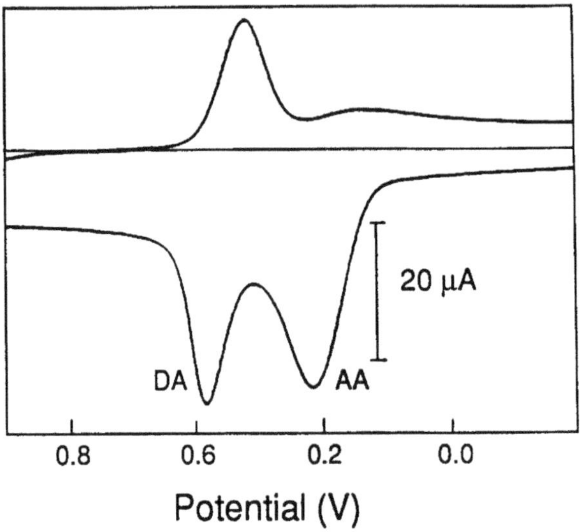

Fig. 8. Voltammogram of 10 μ*M* DA and 1.0 m*M* AA in 0.1*M* $HClO_4$ on a fractured GC surface. This surface is a reasonable approximation of the "pristine" surface shown in Fig. 7. Adapted with permission from Allred and McCreery (1992).

1992). Note that the peak potentials are near the thermodynamic E° values, indicating fast electron-transfer.

In addition, Fig. 8 demonstrates adsorption of DA on the fractured surface, leading to a much higher sensitivity. Comparable peak heights are observed for DA and AA, even though AA is 100× more concentrated in solution. This results from the preconcentration of DA on the fractured surface through adsorption. In the absence (or near absence) of surface oxides, the adsorption of DA (a cation), 4-methylcatechol (neutral), and DOPAC (an anion) at pH 7.0 are approximately equal, indicating that the adsorption mechanism in this case is not electrostatic (Allred and McCreery, 1992). We conclude that the "pristine" GC surface adsorbs through dispersion interactions with the phenyl ring of the catechols, and this interaction is much weaker for ascorbate or metal ions.

As noted earlier, the reactivity of the pristine surface depends on disorder, and basal plane HOPG is quite unreactive. For carbon fibers, this HOPG-like low reactivity would be observed only in very ordered regions, such as the cylindrical exterior (not the disk-shaped end) of a very ordered fiber. Most carbon fibers are sufficiently disordered to behave electronically like GC, and their disk-shaped ends will always be disordered in this sense. We should also note that the disordered surface must be quite clean to have the high reactivity and adsorption behavior of the "pristine" surface. Although careful polishing can lead to reactive GC surfaces (Hu et al., 1985; Thornton et al., 1985), conventional polishing with alumina slurries usually deactivates the surface, particularly if commercial alumina slurries containing organic deagglomerating agents are used. Without unusual care, conventionally polished surfaces will show poor electron transfer and adsorption reactivity toward DA and AA. Carbon fibers are often cut (Strein and Ewing, 1992) or very lightly polished without an alumina slurry (Kawagoe et al., 1993). The resulting surfaces appear to approximate the pristine surface, since the carbon fiber is essentially fractured.

To summarize the behavior of the pristine surface, it exhibits fast electron-transfer to catechols, AA, and many metal ions, provided the redox reactions of interest do not require surface functional groups. Adsorption of catechols occurs, but without selectivity for adsorbate charge. The pristine GC (or fiber) surface is a good model for electronically disordered carbon without apparent oxide involvement. Finally, the pristine surface exhibits strong voltammetric selectivity for catechols over ascorbate owing to selective adsorption through the phenyl ring.

4.2. Submonolayer Oxidized Carbon Surfaces

Exposure of a pristine carbon surface to even trace amounts of oxygen results in a submonolayer coverage of the oxygen-containing functional groups in Fig. 5. Exposure to water also leads to surface oxides, but at a much slower rate (and may in fact be the result of trace dissolved oxygen). Oxides can also be formed by electrochemical oxidation, perhaps unintentionally, during voltammetry. In this section on monolayer (and submonolayer) oxidized surfaces, we are not

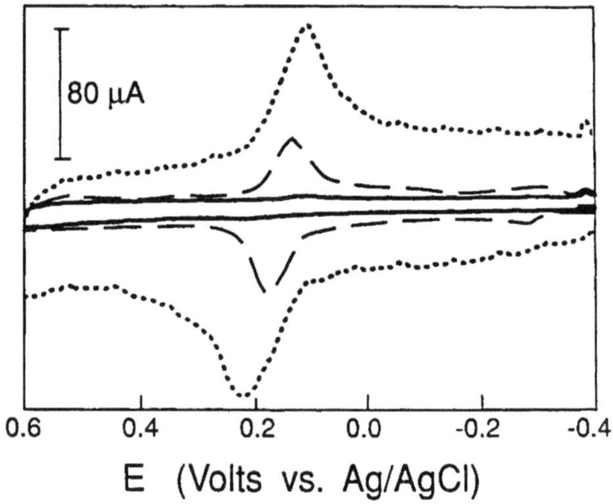

Fig. 9. Voltammetry of dopamine (10*M*) at pH 7.0 on three carbon surfaces. Solid line is on polished GC. The DA response is negligible because polishing debris has prevented DA adsorption, and the response from DA in solution is too weak. The dashed line is fractured GC, demonstrating that DA adsorption on the clean surface yields a good voltammetric response. Dotted line is the same surface after very mild electrochemical pretreatment, showing both increased DA response and background current. Adapted from Allred and McCreery, 1992.

addressing the thick EGO films resulting from intentional oxidation, but are limiting the discussion to carbon surfaces with ca. 5–15 atom % of oxygen.

Submonolayer coverage of oxides results in several important properties, compared to the pristine surfaces. First, the voltammetric background can increase, since oxides can increase double-layer capacitance through electrostatic attraction to ions in solution. Second, oxides are neutral or anionic (never cationic), so they enhance adsorption of cationic species. Both the increased background and the enhanced response for DA are shown in Fig. 9. DOPAC (an anion at pH 7) response is suppressed by oxides, whereas 4-methylcatechol (neutral) shows little change. A third effect of submonolayer oxides is the possibility of "redox" mediation, in which a redox active oxide (e.g., a quinone) can mediate electron-transfer to a solution species. An example is the quinone-catalyzed oxidation of

NADH by a modified carbon surface (Tse and Kuwana, 1978). The kinetics of direct electron-transfer from NADH to the carbon are quite slow, and the quinone/catechol surface group can act as an effective catalyst. A fourth effect of submonolayer oxides was noted earlier: binding of redox species in solution to accelerate electron-transfer (McDermott et al., 1993b). Finally, mild oxidation of the carbon can desorb impurities and thereby clean the surface (Engstrom and Strasser, 1984).

The oxide effects of particular importance to in vivo voltammetry include the increased selectivity and sensitivity toward cationic species. Since common interferences (AA and DOPAC) are anionic at pH 7, their response is suppressed. Furthermore, the adsorption of cationic analytes (DA and NE) is enhanced on the anionic oxide surface, increasing their response. As the voltammetric scan rate increases, the peak current for adsorbed species increases linearly, whereas that for solution species increases with the square root of the scan rate. This is a very useful feature at fast scan rates (10–300 V/s), where the response to adsorbed species is dominant and the measurement is insensitive to even high concentrations of AA or DOPAC (Baur et al., 1988). As shown in Fig. 8, the pristine surface adsorbs catechols over AA, but it does not discriminate between anions and cations.

Nafion modification of carbon fiber surfaces has been used to improve selectivity and stability. Starting with a polished surface with submonolayer oxide coverage, the electrode is dipped in a solution of Nafion in mixed alcohols. After removal and drying, the exposed carbon disk is coated with a Nafion film several hundred angstroms thick (Kristensen et al., 1987; Rice et al., 1989). The carbon surface is presumably unaltered in this process, and retains the activity expected for the submonolayer oxidized surface. However, the relatively thick Nafion layer both repels anions and preconcentrates cations. Furthermore, the carbon surface is at least partially protected from adsorption of large molecules from the extracellular fluid, resulting in a longer analytical lifetime.

The beveled and Nafion-coated microdisk electrodes used by the Wightman group (Kristensen et al., 1987; Kawagoe et al., 1993) demonstrate the effectiveness of the modified submonolayer oxidized surface. Light polishing on a diamond wheel yields a reactive, but slightly oxidized surface with good selectivity for DA and NE. This selectivity is enhanced, and the lifetime extended, by the Nafion coating. Note also that the Nafion layer is quite thin, so mass transport through the film is fast (a few milliseconds), and the electrode response time is not degraded significantly.

4.3. Multilayer Oxide Films on Carbon

Intentional oxidation of sp^2 carbon electrodes has proven to be a useful surface modification with often beneficial effects on electrode performance (Engstrom and Strasser, 1984; Cabaniss et al., 1985; Kepley and Bard, 1988; Nagaoka et al., 1988). There is a large variety of paradigms used for electrochemical pretreatment (ECP) of carbon electrodes, with both the applied potential (or current) and solution composition varying from lab to lab. Although these variations lead to differences in performance for different pretreatments, some generalizations about ECP are available:

1. ECP leads to multilayer oxide films, in some cases thick enough to be visible to the eye. A measurement of growth rate by Kepley and Bard (1988), using 0–2.2 V sweeps in 1*M* H_2SO_4, yielded an average growth rate of 4 nm/cycle.
2. Film formation depends on the identity of the anion in the solution, probably because of intercalation. Insertion of the anion between the graphite layers may be an intermediate step during anodization. NO_3^-, HSO_4^-, and ClO_4^- are more aggressive intercalators than phosphate or organic anions (Alsmeyer and McCreery, 1992).
3. The EGO film is semitransparent and nonconductive, but does contain particles of sp^2 carbon. These may be very small crystallites of sp^2 material, but they do not appear to be in electrical contact with the surface.
4. The film is unstable in base (pH > 13) and appears to hydrolyze to soluble products (Anjo et al., 1989).

5. The film is anionic and hydrated at pH 7.0, and can exchange cations.
6. ECP removes impurities probably resulting from gas evolution and loss of surface carbon (Engstrom and Strasser, 1984).

A working model for an EGO film is useful, based on the model surfaces of Fig. 7. A thin EGO film behaves similarly to the "monolayer oxide" film of Fig. 7. The oxidation process cleans the surface of previously adsorbed impurities and forms a variety of oxides, some of which are anionic. These act as sites for cation adsorption and anion repulsion, but the film is too thin to impede mass transport significantly. A thick EGO film will have a high capacity for cation exchange, and therefore higher sensitivity for cations than a pristine surface or a thin EGO film. However, time is required to load the film (and mass transport through the film is retarded), so the response time is slow. Although the background current for oxidized surfaces is higher than for the "pristine" surfaces (as shown in Fig. 9), the preconcentration for cations leads to better detection limits.

The many ECP paradigms are divided by some authors into "mild" and "strong" ECP (McCreery, 1991). "Mild" ECP usually involves potential steps or sweeps from 0 to 1.5–2.2 V and has been shown to enhance electron-transfer rates and DA adsorption significantly. At least part of this improvement is the result of the removal of surface impurities resulting from polishing and/or handling. "Strong" ECP involves more aggressive potentials of 2.5–2.8 V, and results in a thicker EGO film. At severe potentials (≥3.0 V), the EGO film is insulating, and activity is suppressed completely. Kovach et al. (1986) concluded that "strong" ECP yielded an insulating EGO film interspersed with reactive microcracks covering a small fraction (a few percent) of the surface. The insulating film could adsorb cations, but electron transfer occurred at the cracks. Sherwood et al. (Koslowski and Sherwood, 1985; Sherwood, 1985; Sherwood and Xie, 1990), have discussed the variation in oxide functional group densities with pretreatment potential and oxidation medium.

Table 3
Electroanalytical Properties of Various Carbon Surfaces

	Ordered basal plane	Pristine, disordered	Monolayer oxide	Multilayer oxide
Capacitance, µF/cm	<1	20–50	~50–70	>100
Electron transfer kinetics for $Fe(CN)_6^{-3/-4}$ cm/s	~10^{-6}	>0.5	~0.1	Slow[a]
Selectivity	Poor	Moderate[b]	Good[c]	Excellent[c]
Response time		Excellent	Good	Moderate-slow
Sensitivity	Poor	Moderate	Moderate	Excellent

[a]Anionic $Fe(CN)_6^{-3/-4}$ is excluded by the film.
[b]Selective for catechols over nonoromatic analytes, but not for cations over anions. Cation selectivity can be provided by Nafion film with minimal increase in response time.
[c]Selective for cations and aromatics.

5. Summary

Figure 7 and Table 3 present models and properties for carbon electrode surfaces. At the risk of oversimplification, these models incorporate the major factors affecting voltammetric performance. The "pristine" surface represents a limit that can probably not be achieved in aqueous solution. However, significant evidence does indicate that an oxide-free surface is reactive toward simple redox systems, and oxides are not necessarily required for adsorption and fast electron transfer. Vacuum-heat-treated and fractured GC surfaces and freshly cut fiber ends approximate the pristine surface. Oxides affect electrode behavior by increasing the voltammetric background, providing selectivity for cations over anions, and often increasing sensitivity by preconcentrating the analyte. Typical polished carbon electrodes behave like the monolayer oxide surface, but with varying levels of adsorbed impurities. Multilayer oxides can exhibit quite high sensitivity for cations, but usually with an increase in response time. The Nafion-coated carbon fibers combine the properties of the monolayer oxide surface with the cation selectivity of the Nafion film.

Acknowledgments

The author thanks the many students who contributed to this work and whose papers were cited. Valuable conversations with Ralph Adams, Andrew Ewing, Ted Kuwana, and Mark Wightman are also acknowledged, as is financial support from the National Science Foundation, and the Air Force Office of Scientific Research.

References

Adams R. N. (1990) In vivo electrochemical measurements in the CNS. *Prog. Neurobiol.* **35,** 297–311.

Allred C. D. and McCreery R. L. (1992) Adsorption of catechols on fractured glassy carbon electrode surfaces, *Anal. Chem.* **64,** 444–448.

Alsmeyer D. C. and McCreery R. L. (1992) In situ Raman monitoring of electrochemical graphite intercalation and lattice damage in mild aqueous acids. *Anal. Chem.* **64,** 1528–1533.

Anjo D. M., Kahr M., Khodabakhsh M., Nowinski S., and Wanger M. (1992) Electrochemical activation of carbon electrodes in base: minimization of dopamine adsorption and electrode cappacitance. *Anal. Chem.* **61,** 2603–2608.

Armstrong F. A., Bond A. M., Hill H. A. O., Oliver B. N. I., and Psalti I. S. M. (1989) Electrochemistry of cytochrome c, plastocyanin, and ferredoxin at edge- and basal-plane graphite electrodes interpreted via a model based on electron transfer at electroactive sites of microscopic dimensions in size. *J. Am. Chem. Soc.* **111,** 9185–9189.

Baur J. E., Kristensen E. W., May L. J., Wiedemann D. J., and Wightman R. M. (1988) Fast-scan voltammetry of biogenic amines. *Anal. Chem.* **60,** 1268–1272.

Besenhard J. O. and Fritz H. P. (1983) The electrochemistry of black carbons. *Agnew Chem. Int. Ed. Engl.* **22,** 950–975.

Cabaniss G. E., Diamantis A. A., Murphy W. R. Jr., Linton R. W., and Meyer T. J. (1985) Electrocatalysis of proton-coupled electron-transfer reactions at glassy carbon electrodes. *J. Am. Chem. Soc.* **107,** 1845–1853.

Cline K. K., McDermott M. T., and McCreery R. L. (1994) Anomalously slow electron-transfer at ordered graphite electrodes: influence of electronic factors and reactive sites. *J. Phys. Chem.* **98,** 5314–5319.

Dresselhaus M. S. (1988) *Graphite Fibers and Filaments.* Springer-Verlag, Berlin.

Engstrom R. C. and Strasser V. A. (1984) Characterization of electrochemically pretreated glassy carbon electrodes. *Anal. Chem.* **56,** 136–141.

Fagan D. T., Hu I. F., and Kuwana T. (1985) Vacuum heat treatment for activation of glassy carbon electrodes. *Anal. Chem.* **57,** 2759–2763.

Feng J.-X., Brazell M., Renner K., Kasser R., and Adams R. N. (1987) Electrochemical pretreatment of carbon fibers for in vivo electrochemistry: effects on sensitvity and response time. *Anal. Chem.* **59,** 1863–1867.

Gerhardt G. A., Oke A. F., Nagy G., Moghaddam B., and Adams R. N. (1989) Nafion-coated electrodes with high selectivity for CNS electrochemitry. *Brain Res.* **290,** 390–395.

Gonon F., Buda M., Desimoni G., and Pujol J. F. (1983) Catecholamine metabolism in the rat locus coeruleus as studied by in vivo differential pulse voltammetry. II. pharmacological and behavioral study. *Brain Res.* **273,** 207–216.

Gonon F., Buda M., Cespuglio R., Jouvet M., and Pujol J. F. (1980) In vivo electro-chemical detection of catechols in the neostriatum of anesthetized rats: dopamine or DOPAC? *Nature* **286,** 902–904.

Hu I. F., Karweik D. H., and Kuwana T. (1985) Activation and deactivation of glassy carbon electrodes. *J. Electroanal. Chem.* **188,** 59–72.

Justice J. B. (ed.) (1987) *Voltammetry in The Neurosciences.* Humana, Clifton, NJ.

Kamau G. N., Willis W. S., and Rusling J. F. (1985) Electrochemical and electrochemically pretreated glassy carbon electrodes. *Anal. Chem.* **57,** 545.

Kawagoe K. T., Zimmerman J., and Wightman R. M. (1993) Principles of voltammetry and microelectrode surface states. *J. Neurosci. Meth.* **48,** 225–240.

Kepley L. J. and Bard A. J. (1988) Ellipsometric, electrochemical, and elemental characterization of the surface phase produced on glassy carbon electrodes by electrochemical activation. *Anal. Chem.* **60,** 1459–1467.

Kinoshita K. (1988) *Carbon: Electrochemical and Physicochemical Properties.* Wiley, New York.

Kissinger P. T., Hart J. B., and Adams R. N. (1973) voltammetry in brain tissue—a new neuro-physiological measurement. *Brain Res.* **55,** 209–213.

Kneten K. K. and McCreery R. L. (1992) Effects of redox system structure on electron-transfer kinetics at ordered graphite and glassy carbon electrodes. *Anal. Chem.* **64,** 2518–2524.

Kovach P. M., Deakin M. R., and Wightman R. M. (1986) Electrochemistry at partially blocked carbon-fiber microcylinder electrodes. *J. Phys. Chem.* **90,** 4612–4617.

Kozlowski C. and Sherwood P. M. A. (1985) X-Ray photoelectron spectroscopic studies of carbon-fibre surfaces, Part 5. *J. Chem. Soc. Farad. Trans.* I, **81,** 2745.

Kristensen E. W., Kuhr W. G., and Wightman R. M. (1987) Temporal characterization of perfluorinated ion exchange coated microvoltammetric electrodes for in vivo use. *Anal. Chem.* **59,** 1752–1757.

McCreery R. L. (1991) Carbon electrodes: structural effects on electron transfer kinetics, in *Electroanalytical Chemistry,* vol. 17 (Bard A., ed.), Dekker, New York, pp. 221–373.

McCreery R. L., Dreiling R., and Adams R. N. (1974) Voltammetry in brain tissue: quantitative studies of drug interactions. *Brain Res.* **73,** 23–33.

McDermott C. A., McDermott M. T., and McCreery R. L. (1993a) Scanning tunneling microscopy of carbon surfaces: relationships between electrode kinetics, capacitance, and morphology for glassy carbon electrodes. *Anal. Chem.* **65,** 937–944.

McDermott C. A., Kneten K. R., and McCreery R. L. (1993b) Electron-transfer kinetics of aquated $Fe^{+3/+2}$, $Eu^{+3/+2}$ and $V^{+3/+2}$ at carbon electrodes. *J. Electrochem. Soc.* **140,** 2593–2599.

McDermott M. T., Kneten K., and McCreery R. L. (1992) Anthraquinonedisulfonate adsorption, electrons-transfer kinetics, and capacitance on ordered graphite electrodes: the important role of surface defects. *J. Phys. Chem.* **96,** 3124–3130.

Nagaoka T., Fukunaga T., Yoshino T., Watanabe I., Nakayama T., and Okazaki S. (1988) Uptake of ions into electrochemically treated glassy carbons. *Anal. Chem.* **60,** 2766–2769.

Rice R. J., Pontikos N., and McCreery R. L. (1990) Qualitative correlations of heterogeneous electron-transfer kinetics with surface properties of glassy carbon electrodes. *J. Am. Chem. Soc.* **112,** 4617–4622.

Rice M. E. and Nicholson C. (1989) Measurement of nanomolar dopamine diffusion using low-noise perfluorinated lonomer coated carbon fiber microelectrodes and high-speed. cyclic voltammetry. *Anal. Chem.* **61,** 1805–1810.

Schrader M. E. (1975) Ultrahigh vacuum techniques in the measurment of contact angles. IV, water on graphite. *J. Phys. Chem.* **79,** 2508–2515.

Sherwood P. M. A. (1985) Photoelectron spectroscopic studies of electrode and related surfaces. *Chem. Soc. Rev.* **14,** 1–44.

Sherwood P. M. A. and Xie Y. (1990) X-ray photoelectron spectroscopic studies of carbon fiber. Part XIV: electrochemical treatment of pitch-based fibers and the surface and bulk structure changes monitored by XPS, XRD, and SEM. *Applied Spectros.* **44,** 1621.

Stamford J. A. (1990) Fast cyclic voltammetry: measuring transmitter release in "real time." *J. Neurosci. Meth.* **34,** 67–72.

Stamford J. A., Kruk Z. L., and Millar J. (1988) Stimulated limbic and striatal dopamine release measured by fast cyclic voltammetry: anatomical, electrochemical and pharmacological characterization. *Brain Res.* **454,** 282–288.

Strein T. G. and Ewing A. G. (1992) Characterization of submicron-sized carbon electrodes insulated with a phenol-allylphenol copolymers. *Anal. Chem.* **64,** 1368–1373.

Thornton D. C., Corby K. T., Spendel V. A., Jordan J., Robbat A., Rutstrom D. J., Gross M., and Ritzler G. (1985) Pretreatment and validation procedure for glassy carbon voltammetric indicator electrodes. *Anal. Chem.* **57,** 150–155.

Tse D. C. S. and Kuwana T. (1978) Electrocatalysis of dihydronicotinamide adenosine diphosphate with quinones and modified quinone electrodes. *Anal. Chem.* **50,** 1315–1318.

Wightman R. M., Deakin M. R., Kovach P. M., Kuhr P. M., and Stutts K. J. (1984) Methods to improve electrochemical reversibility at carbon electrodes. *J. Electrochem. Soc.* **131,** 1578–1583.

Diffusion and Ion Shifts in the Brain Extracellular Microenvironment and Their Relevance for Voltammetric Measurements

The Brain Is Not a Beaker:
In Vivo vs In Vitro Voltammetry

Margaret E. Rice and Charles Nicholson

1. Overview of Brain Extracellular Microenvironment

1.1. Structural Characteristics and Their Influence on Mass Transport

The space between cells of the brain is not just a void separating membranes, but a complex milieu that plays an essential role in many brain functions. The term brain extracellular microenvironment (BEM) will be used here to designate the environment encountered by a molecule when that molecule is in the space between cells of the brain. A similar term (brain cell microenvironment) was coined by Schmitt and Samson (1969). As our ideas have matured, it has become necessary to distinguish the extracellular from the intracellular microenvironment so the term BEM is now more appropriate. Several publications have refined and developed the concepts that are now embodied in the BEM; *see* espe-

From: *Neuromethods, Vol. 27: Voltammetric Methods in Brain Systems*
Eds: A. Boulton, G. Baker, and R. N. Adams

cially Nicholson (1979,1980), Schmitt (1984), Cserr (1986), and Fuxe and Agnati (1991).

The BEM consists of the extracellular space, the fluid that fills it, and the long-chain glycosaminoglycans and proteoglycans that are tethered to membranes, but extend into the space. The abundance and role of the latter molecules in the extracellular space remain unclear, but a recent review may be consulted for some current concepts (Margolis and Margolis, 1993). The extracellular space is the narrow interstitium between cell membranes. In the words of Kuffler and Potter (1964): "these spaces...in three dimensions...resemble the water phase of a foam." The BEM is a major channel for the movement of metabolic substrates, such as glucose, and for the transmission of chemical signals between cells. In addition, it provides essential isolation between cellular elements.

1.1.1. Volume Fraction (α)

The extracellular volume fraction of the brain is defined as the volume that is extracellular space expressed as a fraction of total tissue volume for some small region, frequently called a Representative Elementary Volume (REV) (Fig. 1). Extracellular volume fraction is denoted by α. This is a dimensionless quantity and is often expressed as a percentage. This quantity can be measured morphometrically (Bondareff and Pysh, 1968; Cragg, 1979; Rees et al., 1982), by radiotracer analysis using sucrose or inulin (Levin et al., 1970), or more recently by the TMA-method (Nicholson and Phillips, 1981; Nicholson, 1992). Methods based on impedance have also been used, as have determinations based on the chloride space, but both these methods have technical uncertainties. In the past there was disagreement about the size of the volume fraction (*see* Van Harreveld, 1972), but all the methods now agree that the volume fraction typically amounts to 0.2 (20%) of the total brain volume. The value is this large because every area of cellular membrane in the brain is associated with an "atmosphere" of extracellular space. Table 1 lists some data recently obtained with the TMA-method in different brain regions and in a mammal (rat), a reptile (turtle), and an invertebrate (cuttlefish).

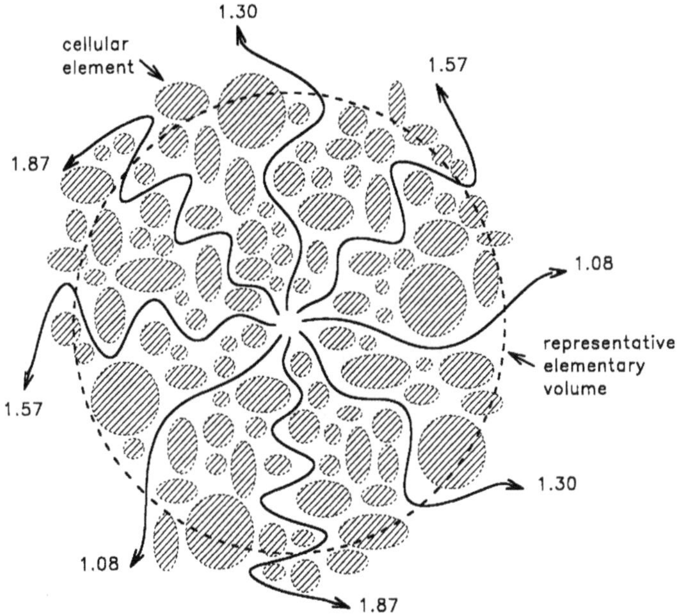

Fig. 1. Geometry of the brain cell microenvironment. Extracellular volume fraction is the ratio of the interstitial volume (i.e., space that is not cellular elements) within a small REV to the whole of that volume, which is the region inside the dotted circle. The volume fraction of brain tissue is about 0.2 (20%). Tortuosity is approximately the increase in path length for a diffusing particle imposed by the obstructions of cellular elements. In the figure, lines are depicted with the tortuosities, relative to straight line, indicated by the numbers. The tortuosity of brain tissue is about 1.6. In fact, the concept of tortuosity is more complex than this.

1.1.2. Tortuosity (λ)

Tortuosity is a measure of the extent to which the cellular obstructions, such as the bodies or processes of cells, hinder the movement of substances. It cannot be defined as easily as volume fraction, because it is necessary to compare the mean distance a particle has to travel between two points in the presence of the obstructions with the mean distance in the absence of the obstructions (Fig. 1). Tortuosity is denoted by the dimensionless parameter λ. As a simple example, if the obstructions consisted of spheres of radius r, fairly close together, but not touching, the tortuosity would be approximately the ratio of the distance around the sphere from one

Table 1
Values of Volume Fraction (α) and Tortuosity (λ) Obtained with the TMA Method

Animal	Region	α	λ	Ref.
Rat	Neocortex	0.18	1.57	Cserr et al., 1991
	Neocortex	0.18	1.40	Lundbæk and Hansen, 1992
	Neocortex	0.21	1.63	Lemenkühler et al., 1993
	Hippocampus, CA1	0.12	1.67	McBain et al., 1990
	Hippocampus, CA3	0.18	1.83	McBain et al., 1990
	Neostriatum	0.21	1.54	Rice and Nicholson, 1991
	Spinal cord, dorsal horn	0.20	1.62	Syková et al., 1994
Turtle	Cerebellum, molecular layer	0.31	1.44, 1.95, 1.58	Rice et al., 1993
	Cerebellum, granular layer	0.21	1.77	Rice et al., 1993
Cuttlefish	Optical lobe	0.29	1.86	Nicholson et al., 1995
	Vertical lobe	0.10	1.65	Nicholson et al., 1995

side to the other (πr) divided by the straight path through the sphere ($2r$), i.e., $\pi/2 = 1.57$. A medium without obstructions would have a tortuosity of 1. The actual values measured in brain tissue are about 1.6 (Table 1).

Tortuosity was introduced into biological diffusion measurements as far back as 1948 by Harris and Burn, and has been measured extensively with radiotracer methods (Fenstermacher and Patlak, 1975) and the TMA-method (*see* Table 1 for some references). Discussions of tortuosity can be found in Nicholson and Phillips (1981) and Mathias (1983).

1.1.3. Anisotropy and Heterogeneity in Different Regions

Different brain regions show different anatomical structures, so that an obvious question is: To what extent are such differences seen in the volume fraction and tortuosity? Table 1 shows that the volume fraction is quite similar among brain regions. Two cases have been found where the volume fraction is about half that of the average; these are the CA1 region of the rat hippocampus (McBain et al., 1990) and the vertical lobe of the cuttlefish (Nicholson et al., 1994). In contrast, the adjacent CA3 region of the hippocampus shows an almost typical volume fraction and the optical lobe of the cuttlefish shows a larger than normal value (*see* Table 1).

An instance where the volume fraction is larger than the average is in the molecular layer of the cerebellum where it has recently been estimated at 31% (Rice et al., 1993; *see also* Table 1). Another case where the volume fraction is large is in the neonatal rat cortex, just after birth, where the volume fraction is about 40% (Bondareff and Pysh, 1968; Lehmenkühler et al., 1993).

Tortuosity generally does not vary much from region to region, lying in the range 1.4–2.0, but an interesting phenomenon has recently been revealed in the cerebellum of the turtle (Rice et al., 1993), where three different values of the tortuosity were estimated, each associated with one direction of the intrinsic rectangular axes of this highly organized brain region. Thus, a value of 1.44 was associated with the axis running along the parallel fibers, 1.95 at right angles to these fibers in the plane of the surface and 1.58 associated with the vertical axis. This is an example of anisotropy and confirms that tortuosity is actually a second-order tensor (Nicholson and Phillips, 1981). Anisotropic diffusion is also anticipated in white matter tracts, based on earlier impedance measurements (Ranck and BeMent, 1965).

1.1.4. Hindered Diffusion

The typical probe substances used to measure volume fraction are small molecules that do not readily enter cells. The most commonly used compound today is TMA^+ (tetramethylammonium), a cation with a mol wt of 74.1. Obviously, the spaces between cells are of finite width, and it may be expected that when the size of a molecule exceeds some limit, the molecules will no longer freely pass through the extracellular space. This has been explored using dextrans, labeled with fluorescent tags, in the range 3000–70,000 mol wt in the cortex of the rat (Nicholson and Tao, 1993). It was found that

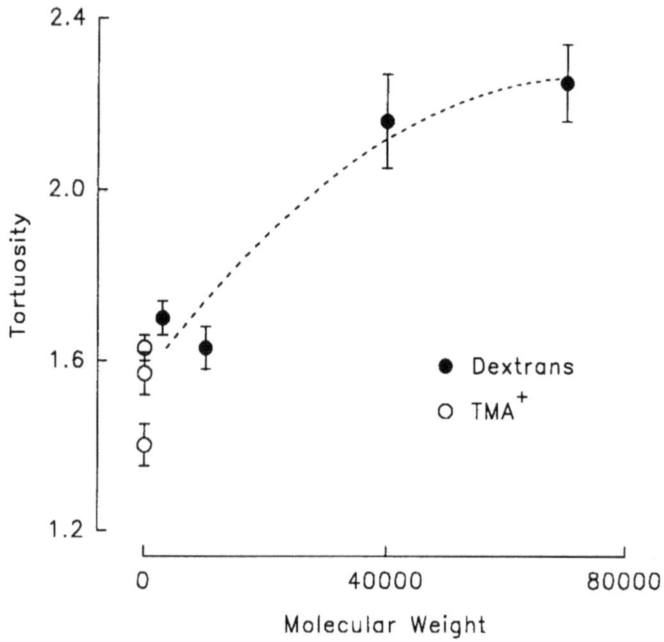

Fig. 2. Hindered diffusion of high-mol-wt molecules in the cerebral cortex. The three measurements of tortuosity (λ) for TMA^+ (mol wt 74.1) in the rat neocortex, detailed in Table 1, are plotted as open circles. The values of tortuosity measured with integrative optical imaging for dextrans of 3000, 10,000, 40,000, and 70,000 mol wt are shown as closed circles. A second-order regression curve is plotted through the dextran points. Error bars represent standard errors of the mean. Note that the small dextrans are subject to a similar tortuosity as TMA^+, but the two larger dextrans are significantly more hindered. *See* Nicholson and Tao (1993) for details.

up to 40,000 mol wt, the molecules were subject to a similar tortuosity (1.6–1.7) as the much smaller TMA ions. At 40,000 and 70,000 mol wt, however, the dextrans showed tortuosities of about 2.2 (Fig. 2). One interpretation of these data is that the larger dextrans are excluded from some regions of the BEM. Such exclusion would likely apply to other molecules of similar size, including growth factors, for example.

1.1.5. Implications of Volume Fraction and Tortuosity for the Movement of Substances in the BEM

Although a substance remains free to move in the BEM, it is primarily subject to diffusion. Interstitial fluid flow, if present, is a slow process, perhaps on the order of 10 µm/min (based on data in Cserr and Knopf, 1992). For comparison, a direct measurement of interstitial fluid flow in ear tissue of the rabbit obtained a value of 36 µm/min (Chary and Jain, 1989). On a microscopic level, diffusion of a single molecule represents the net imbalance of collisions with water molecules (the concentration of which, at about 50 mol/L, enormously exceeds that of any other substance). The result is that the molecule executes a random walk in the solvent. Volume fraction affects the initial concentration of a released substance. If a synapse or other site releases a given number of molecules into a restricted extracellular space, then the initial concentration will be higher by the reciprocal of the volume fraction. Expressed in quantitative and macroscopic terms, if we denote volume fraction by α and the source by Q (mol/s), then the apparent source in the restricted medium will be Q/α. Since α is about 0.2 in brain tissue, the apparent source strength will be five times greater than in a free solution.

As the particles execute a random walk, they will strike the cellular obstructions from time to time so that the number of steps required to move a given distance (measured as a straight line from the initial point to the final point) from the source will be increased. Macroscopically, this is expressed by the tortuosity λ and an apparent diffusion coefficient $D^* = D/\lambda^2$, where D (cm^2/s) is the diffusion coefficient (at an appropriate temperature) in a free medium.

1.1.6. Definitions of Extracellular Concentration

Concentration needs careful definition in tissues that have multiple compartments (Nicholson, 1995b). This is particularly true in studies involving uptake of substances, where comparisons inevitably arise between the kinetic parameters measured in classical studies on slices and homogenates using radiotracers, and the parameters appearing when in vivo voltammetry is employed. For a substance in the extracellular space, two definitions of concentration arise:

1. The amount of substance found in the extracellular space expressed as a concentration with respect to the whole

volume of tissue, i.e., the sum of extracellular and intracellular volumes in an appropriate REV. This definition frequently arises in working with radiolabels and tissue homogenates through the use of wet-weight or protein content as a reference parameter.

2. The amount of substance found in the extracellular space expressed as a concentration with respect to only the extracellular space. This is the quantity generally measured by an ion-selective microelectrode (ISM) or carbon fiber microelectrode (CFM) and is the quantity that will be used in this chapter, denoted by C (mol/L).

Since the two concentrations described above differ by the volume fraction, substantial errors may arise if the different definitions enter implicitly into an analysis. Similar considerations apply to intracellular concentrations.

1.1.7. Measurement of Diffusion Parameters in Brain Tissue

Diffusion of molecules from a point-source in a complex, i.e., restricted, medium, can be expressed by the equation

$$\frac{\partial C}{\partial t} = \left(\frac{D}{\lambda^2}\right)\frac{1}{r^2}\frac{\partial}{\partial r}\left(r^2\frac{\partial C}{\partial r}\right) + \frac{Q}{\alpha}\left(\frac{\delta(r)}{r^2}\right)[H(t) - H(t-d)] \tag{1}$$

Here t is time (s) and r (cm) is distance from the source, which is represented in this example as the function $Q[\delta(r)/r^2]H(t-d)$, where $\delta(r)$ is a Dirac delta function and $H(t)$, $H(t-d)$ are Heaviside step functions that define the duration of the source pulse. This means that a pulse of substance of duration d (s) is emitted from a very small region, approximating a point, at the origin of the spherically symmetric coordinate system. This type of source can be precisely realized in the form of an iontophoresis microelectrode that delivers a cationic or anionic substance to the tissue from a microelectrode with a typical tip diameter of 1–4 µm. Depending on the substance, the molecule can be detected with an ISM or CFM located at a distance r (cm) from the source. This is the basis of the TMA-method (Nicholson and Phillips, 1981; Nicholson 1992).

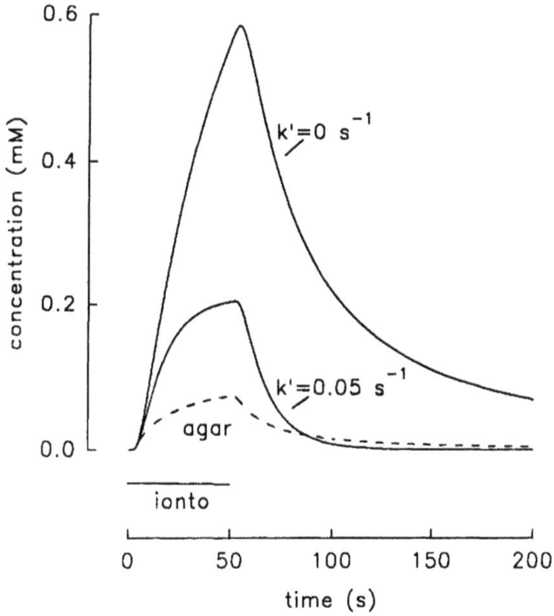

Fig. 3. Curves illustrating TMA^+ iontophoresis curves computed from average results. The lowest curve shows the diffusion curve in a dilute agar gel (free medium) with $\alpha = 1$, $\lambda = 1$. The two upper curves are for the concentrations produced by the same source, in brain tissue with $\alpha = 0.21$, $\lambda = 1.54$ for linear uptake ($k' = 0.05/s$) and no uptake ($k' = 0$). The other parameters for all curves were: distance from source, 100 μm; iontophoretic current, 100 nA for 50 s (indicated by horizontal bar); transport number of iontophoretic electrode, $n = 0.4$; $D = 1.3 \times 10^{-5}$ cm^2/s (TMA^+ at 37°C).

It is evident from Eq. (1) that the tortuosity and volume fraction enter the equation at different places. The effect of these two parameters is made evident by solving Eq. (1) for the case of brain tissue assuming that the boundaries are far from the source point. In this case, writing "erfc" for the complementary error function:

$$C = \frac{Q\lambda^2}{4\pi D\alpha r}\left\{\operatorname{erfc}\left(\frac{r\lambda}{2(Dt)^{\frac{1}{2}}}\right) - H(t-d)\operatorname{erfc}\left(\frac{r\lambda}{2\left[D(t-d)\right]^{\frac{1}{2}}}\right)\right\} \tag{2}$$

The tortuosity affects the shape of the curve as a function of both space and time, because it is a component of the argument of the error function. In contrast, the amplitude is determined by both tortuosity and volume fraction (Fig. 3).

To describe a free medium, the values $\alpha = 1$ and $\lambda = 1$ would be inserted in Eq. (2).

1.1.8. Spatial Scale and Diffusion

Since volume fraction and tortuosity describe extracellular geometry by means of single, homogeneous parameters and Eqs. (1) and (2) employ smoothly changing variables, their applicability depends in some sense on averaging out local variations (*see* Fig. 1). This has been discussed rigorously by Nicholson and Phillips (1981), but suffice it to say here that the averaged parameters are applicable over distances on the order of 5 µm or more. If the behavior of substances is to be examined on a scale of 1 µm or less, then a different approach to diffusion is required (*see* Nicholson, 1995a) that takes account of local geometry in the neighborhood of the uptake site.

1.2. The Role of Uptake in Controlling Substance Movement

Volume fraction and tortuosity are morphological parameters that reflect the geometrical (extracellular space volume) and topological (connectivity) characteristics of the BEM. These two parameters, in ideal measurements, are independent of the substance or method used to measure them.

A third, substance-specific aspect of the BEM is the uptake property or the extent to which a given substance is removed from the extracellular space. This reflects the physiology of the system and may be incorporated into Eq. 1 by subtracting a term $f(C)/\alpha$. This expression indicates the fact that the uptake of the substance is most generally some function of C. This can also represent the destruction of molecules in the extracelluar space by enzymatic action. The new equation is:

$$\frac{\partial C}{\partial t} = \left(\frac{D}{\lambda^2}\right)\frac{1}{r^2}\frac{\partial}{\partial r}\left(r^2\frac{\partial C}{\partial r}\right) + \frac{Q}{\alpha}\left(\frac{\delta(r)}{r^2}\right)[H(t) - H(t-d)] - \frac{f(C)}{\alpha} \qquad (3)$$

These three factors, volume fraction, tortuosity, and uptake, determine the spatio-temporal distribution of substances in the BEM, and thus the potential for extracellular chemical signals to be transmitted in this environment. These factors are very relevant to the interpretation of in vivo electrochemical measurements and will be discussed below. These issues have also surfaced recently in attempts to arrive at accurate interstitial concentrations using microdialysis, and several detailed theoretical and experimental analyses incorporating diffusion have been made, which may be relevant in the present context (*see* Benveniste and Hüttemeier, 1990 and Kehr, 1993 for reviews).

1.2.1. Linear Uptake Kinetics

Because of the high membrane-to-extracellular volume ratio in the BEM, a certain amount of nonspecific loss of any substance occurs, and this generally is linearly related to extracellular concentration so that:

$$f(C) = kC \tag{4}$$

This is equivalent to the assumption that the internal concentration remains low during the time of measurement and the substance permeates across the membrane under the influence of the transmembrane concentration gradient. This type of behavior is seen when TMA^+ is used to measure the diffusion parameters (Nicholson, 1992). If we denote the solution to Eqs. (3) and (4) by C, then (Nicholson and Phillips, 1981):

$$C = \frac{Q\lambda^2}{8\pi D\alpha r}\{e^{r\theta}[g_1(r,t) - H(t-d)\,g_1(r,t-d)] + e^{-r\theta}[g_2(r,t) - H(t-d)\,g_2(r,t-d)]\} \tag{5}$$

where $\theta = \lambda\left(\frac{k}{\alpha D}\right)^{\frac{1}{2}}$, $g_1(r,t) = \mathrm{erfc}\left(\frac{r\lambda}{2(Dt)^{\frac{1}{2}}} + \frac{\theta}{\lambda}(Dt)^{\frac{1}{2}}\right)$, $g_2(r,t) = \mathrm{erfc}\left(\frac{r\lambda}{2(Dt)^{\frac{1}{2}}} - \frac{\theta}{\lambda}(Dt)^{\frac{1}{2}}\right)$

Note that in several earlier papers, the uptake constant was defined by k' where $k' = k/\alpha = <S>P_E/\alpha$, where $<S>$ (1/cm) is the local volume average of the membrane surface and P_E (cm/s) is the permeability of the membrane to the substance (*see* Nicholson, 1992 for a discussion of these parameters). The consequences of linear uptake are illustrated in Figs. 3 and 4. Other forms and solutions of Eq. (3) can be written down for sources in the form of brief pulses (*see* Nicholson, 1992 for details).

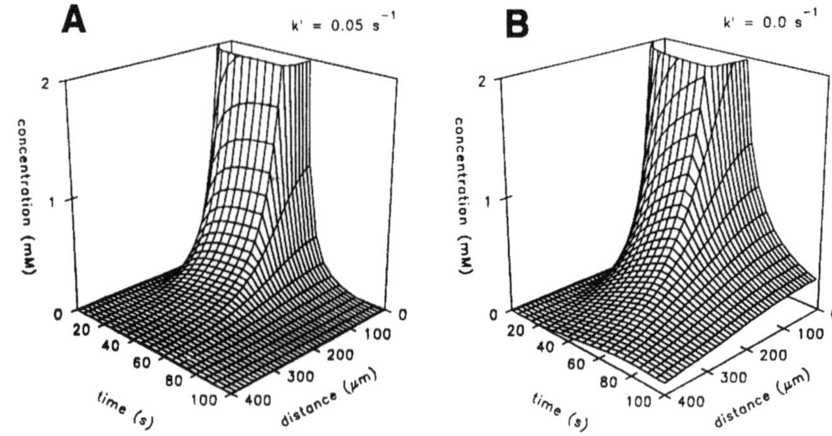

Fig. 4. Spatio-temporal distribution of TMA^+ in brain with and without linear uptake. Both sets of curves were generated using the same parameters as in Fig. 3. **(A)** Curves with $k' = 0.05/s$. **(B)** Curves without uptake ($k' = 0$). The concentration rises to high values in close proximity to the iontophoresis source, and graph is truncated at 2 m*M* so that the lower concentrations can be visualized. A typical recording distance is 100 μm from source (*see* Fig. 3). It is apparent that the diffusion surface falls more abruptly toward zero with linear uptake, but the overall characteristics are little changed from the situation with no uptake. It is important to take linear uptake into account, however, to obtain accurate curve fitting for a given distance.

1.2.2. Michaelis-Menten Uptake Kinetics

In many instances, material is actively transported across membranes by an energy-dependent uptake system that obeys nonlinear kinetics. This type of uptake is exemplified by the behavior of dopamine in the striatum, and this will be the model considered below. The model is widely applicable, however. The most generally accepted scheme for the nonlinear uptake is described by Michaelis-Menten (M-M) kinetics, which are defined in the present context by:

$$f(C) = \frac{V_{max} C}{K_m + C} \tag{6}$$

Here K_m (μ*M*) is the Michaelis constant and V_{max} (μ*M*/s) is the maximum velocity of the uptake.

The validity of M-M kinetics for the striatum is based on many studies and reviews; it will be discussed in detail in a forthcoming paper (Nicholson, 1995b). We note only the original work of Snyder and Coyle (1969) and Holz and Coyle (1974), together with the reviews of Horn (1976, 1979, 1990) and particularly the many recent studies of Wightman and colleagues using FCV (reviewed in Wightman and Zimmerman, 1990). It should be noted that the V_{max} defined here agrees with that derived from binding studies, when units are changed appropriately, but differs from that used by Wightman, which is the classical V_{max} divided by the volume fraction. This is because the classical V_{max} is measured with respect to a whole volume of tissue so the definition of concentration is modified as indicated in Section 1.1.6. We shall refer to Wightman's "V_{max}" as $V_{max}' = V_{max}/\alpha$.

Among the many assumptions required in deriving Eq. (6) to describe dopamine (DA) uptake kinetics is the requirement that the uptake proceeds only in the forward direction. This is the classical assumption that we are only dealing with the initial velocity. Measurements on tissue (*see* references cited above) suggest that this is valid for periods of at least 2 min. Incorporating Eq. (6) in Eq. (3) we obtain:

$$\frac{\partial C}{\partial t} = \left(\frac{D}{\lambda^2}\right)\frac{1}{r^2}\frac{\partial}{\partial r}\left(r^2\frac{\partial C}{\partial r}\right) + \frac{Q}{\alpha}\left(\frac{\delta(r)}{r^2}\right)[H(t) - H(t-d)] - \frac{1}{\alpha}\left(\frac{V_{max}C}{K_m + C}\right) \quad (7)$$

Equation (7) in general has no analytical solution and must be solved numerically. For the special case that $C << K_m$, then:

$$f(C) \approx (V_{max}/K_m)C \quad (8)$$

and the solution expressed in Eq. (5) may be used. Note, however, that because Eq. (7) is nonlinear this limiting case is only valid when $C << K_m$ for every point in space and every instant of time. For a point source, this requirement is almost certainly violated in the immediate neighborhood of the source.

1.2.3. Example of the Iontophoretic Method for Measuring V_{max}

Using the approach outlined in Section 1.2.2., a technique that resembles the TMA-method can be devised to look at the nonlinear uptake properties of DA. A complete numeri-

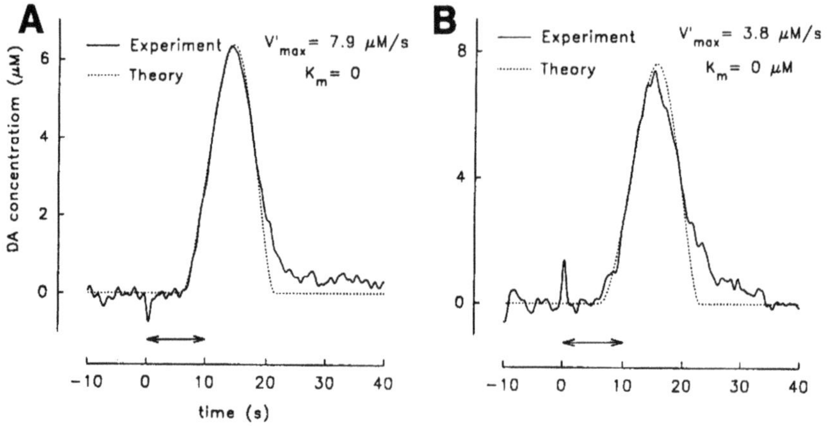

Fig. 5. Curves illustrating DA iontophoresis data from two experiments and curve fitting. Curves were obtained in slices from rat neostriatum and fitted using a numerical solution to Eq. (7) taking $\alpha = 0.21$, $\lambda = 1.54$, $D = 6.9 \times 10^{-6}$ cm²/s (DA at 32°C), $K_m = 0$ (since the concentration levels are mostly much higher than the typical value of 0.15 μ*M*) and the V'_{max} values as given on the panels. Other parameters as follows. **(A)** Distance from source 123 μm, iontophoretic current 200 nA for 10 s, transport number of iontophoretic electrode $n = 0.012$. **(B)** Distance from source 120 μm, iontophoretic current 300 nA for 10 s, transport number of iontophoretic electrode $n = 0.005$.

cal solution to Eq. (7) for an iontophoretic source has been described (Nicholson, 1995b). In the present chapter, two numerical solutions are fitted to actual data obtained with FCV and illustrated in Fig. 5. In order to use the numerical solution, it is necessary to know the free diffusion coefficient for DA (6.9×10^{-6} cm²/s for 32°C derived from Gerhardt and Adams, 1982) and the values of α (0.21) and λ (1.54), both obtained from experiments with the TMA-method in striatal slices (Rice and Nicholson, 1991). Because the concentrations DA are in the micromolar range, the numerical solution is insensitive to K_m, which is about 0.15 μ*M* (e.g., Horn, 1990), and it can be set to zero for the calculations.

It is useful to illustrate the nonlinear behavior of the solution to Eq. (7). This is shown in Fig. 6A, where a three-dimensional plot shows how DA is localized in both time and space

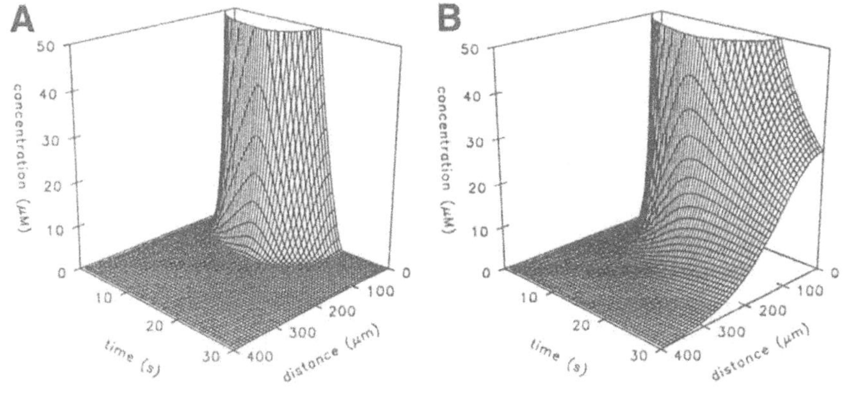

Fig. 6. Spatio-temporal distribution of DA in brain with and without Michaelis-Menten uptake. Both sets of curves were generated using the same parameters as in Fig. 5. **(A)** Curves with $V'_{max} = 7.9\ \mu M/s$ and $K_m = 0.15\ \mu M$. **(B)** Curves without uptake. With nonlinear uptake, the DA is confined to a restricted spatio-temporal domain around the release site, beyond which it is zero. Without uptake, the DA diffuses in a similar manner to TMA^+. Note the difference in time and concentration scales compared to Fig. 4.

around the source. In contrast, Fig. 6B shows that without any uptake, the diffusion source has a quite different geometry, which was shown also in Fig. 4B. It is also to be noted that the localization of DA is a nonlinear function of the source strength, so these results do not simply scale with changes in iontophoretic current or source transport number.

1.2.4. Wightman's Method of Estimating V_{max} in Brain Tissue

The iontophoretic approach may be compared with the method of Wightman and colleagues, which employs electrical stimulation to evoke "overflow" of DA from synapses in various brain regions. Although the purpose of those studies was primarily to define the release characteristics of the striatum and other brain regions, the method also provides a means of estimating V_{max}. Wightman et al. assume that DA is homogeneously raised to an equal concentration in a sufficiently large volume of tissue that the spatial or diffusional component of an equation like that depicted in Eq. (7) can be neglected. Then one simply has:

$$\frac{dC}{dt} = \frac{P(t)}{\alpha} - \frac{1}{\alpha}\left(\frac{V_{max}C}{K_m + C}\right) \quad (9)$$

where *P(t)* is the appropriate release function for the transmitter, typically based on the frequency of stimulation. Equation (9) is much easier to solve than Eq. (7), especially when a simple form is chosen for *P(t)*. The solution is particularly simple if the condition $C >> K_m$ is satisfied. The validity of this method depends on the assumption of homogeneity of DA distribution. In order for Eq. (9) to be valid and for the diffusion term to be omitted, the recording electrode must be within a region with a spatially uniform concentration of DA at each instant of the measurement. This distribution must extend sufficiently far from the electrode that "edge effects" are absent. Wightman et al. (1988) have shown significant variations in evoked DA concentration over distances of 100 μm. The required extent is itself governed by the kinetics and is a function of the concentration level. Detailed nonlinear modeling would be necessary to assess the sensitivity of the electrode to perturbations in the DA distribution. These considerations also imply that the afferent stimulation must uniformly and densely excite DA terminals in the vicinity of the recording electrode. Unfortunately, there is no way to ensure completely that this condition is fulfilled, and several papers (Ewing and Wightman, 1984; Wightman et al., 1988; Kawagoe et al., 1992) have illustrated records showing that the DA concentration continued to rise after the end of the stimulation, which is a necessary, but not sufficient condition for diffusion to play a role.

It is interesting to compare the analogous situation that arises when stimulus-evoked $[K^+]_o$ is measured with an ISM (*see* Section 3.1.1.2. for some references). It is known that extracellular $[K^+]$ increases are controlled by a variety of uptake processes. In virtually no cases have examples been shown where the rise in $[K^+]_o$ continues beyond the stimulus, and in three cases where the decay process has been analyzed from the perspective of a diffusion-mediated process (Krnjevic and

Morris, 1974; Vern et al., 1977; Cordingly and Somjen, 1978), it has been found that diffusion is not a significant determinant of the brief time-course of the curve.

Wightman and colleagues have offered two explanations for the continued rise in DA after the end of stimulation. In the first, they suggested that the carbon-fiber electrode was surrounded by a dead space, i.e., a space in which there were no cells, through which the DA had to diffuse (Wightman et al., 1988). Although initial calculations supported this conclusion, later ones did not (Kawagoe et al., 1992). It is also worth noting that in the case of diffusion measurements with ISMs, a great many records have been analyzed quantitatively without it ever being necessary to include such a dead space in the model. More recently, Wightman and colleagues have suggested that the DA overshoot is a manifestation of DA diffusion within the Nafion coat of the carbon fiber (Kawagoe et al., 1992). By using a deconvolution process, the overshoot was removed. Another issue, unique to voltammetry, is the diffusion gradient that can be created at the recording electrode because of the consumption of the electroactive species. As will be discussed later (Section 2.1.3.), this is unlikely to be a problem with small carbon-fiber electrodes and FCV.

This discussion illustrates a limitation of the stimulation method in that precise definition of the source is not possible, neither the spatial nor temporal characteristics can be defined with precision, and in any given record, the overshoot could be the result of properties of the source. A second problem is that it may not always be possible to find a suitable afferent pathway to a chosen region, so that systematic mapping of different regions is not feasible.

The iontophoretic approach outlined in Section 1.2.3. overcomes both these limitations, but the cost is that the diffusion problem must be solved, which quite taxing even for a fast PC.

1.2.5. Role of Diffusion in the Identification of a Low-Affinity, High-Capacity Uptake

Currently accepted kinetic parameters for DA uptake in the rat striatum are $V_{max} = 0.2–0.8\ \mu M/s$, $K_m = 0.15\ \mu M$. These are generally described as high-affinity, low-capacity parameters. Two early papers (Stamford et al., 1984b, 1986) using FCV identified a low-affinity ($K_m = 8\ \mu M$) system. This identification was based both on the time-course of stimulated overflow and on the reaction of this overflow to a range of uptake inhibitors.

In a later study using classical radiolabeled ligands, synaptosomes, and tissue homogenates and comparing them with tissue minces, it was claimed that kinetic data in tissue minces and slices could be misinterpreted because of the interaction of diffusion and uptake in such experiments. As a consequence of these findings, the earlier identification of a low-affinity, high-capacity system was deemed to be an artifact (Near et al., 1988; Wightman and Zimmerman, 1990), and this explanation seems to be generally accepted at present.

There is further difficulty in interpreting kinetic uptake constants in the presence of nonlinear uptake when the tissue is subject to a concentration gradient (Kelly and Wightman, 1987), and our brief analysis of iontophoresis (Sections 1.2.2.–1.2.3.) also illustrates the problem. However, as noted above, the fundamental assumption of the stimulus-evoked overflow experiments is that the tissue is not subject to a concentration gradient, so that kinetic parameters derived under those conditions should be accurate. More complex uptake mechanisms are known (*see* Justice et al., 1988), but techniques based on FCV probably cannot resolve them.

1.2.6. Diffusion vs Uptake as the Dominant Parameter in Mass Transport

It is clear from the above outline that in instances where uptake is small, the migration of a substance in the BEM can be defined by the diffusion characteristics, which consist of the free diffusion coefficient, the tortuosity, and the volume fraction. In this case, molecules of the substance can potentially reach any location if one waits long enough, although the concentration may be very small.

Alternatively, the migration may be dominated by the uptake characteristics. In the latter case, there will always be a certain distance beyond which no molecules are to be found. An estimate of this "boundary" located at b (cm) from the source point can be obtained for the extreme case of nonlinear uptake with $K_m = 0$ and a point-source that emits continuously so that a steady-state condition is attained. Then it can be shown (Nicholson, 1995b) that:

$$b = \left[\frac{3Q}{4\pi V_{max}}\right]^{\frac{1}{3}} \tag{10}$$

An interesting feature of this result is that it is independent of all the diffusion parameters.

1.2.7. Relation of Concentration at Electrode Tip to "True" Concentration

Implicit in the above discussion is the goal of interpreting measurements made with voltammetric electrodes in terms of a "true" interstitial concentration. In the case where a substance is released from micropipet in a controlled manner, the issue is simply whether the extracellular distribution follows the distribution predicted from basic theory. When the distribution is produced by stimulating synapses, it is evident that the concentration in the synaptic cleft is likely to be much higher than that in the interstitial space outside the cleft, and that in turn, there may be local gradients over distances of a few micrometers in the interstitial space, depending on proximity to a synaptic cleft and distribution of uptake sites. In the final analysis, there is no simple answer to the question of a "true" concentration for the same reason that the concept of a diffusion distance is ill-defined. A concentration is only relevant in a functional context: Does a receptor at a given location see enough to activate it? In a true diffusion regime, molecules will reach every location given sufficient time. A few will arrive very promptly, and others will come late, but there is no distance beyond which molecules are not found. There is a distance beyond which they never exceed a given concentration, however. In contrast, in the presence of uptake there is a distance beyond which no molecules are found at all.

2. Some Implications of Mass Transport Properties of the Brain Extracellular Environment for Voltammetric Measurements

2.1. *Influence of Brain Diffusion Characteristics on Mass Transport to the Voltammetric Electrode*

The complex organization and dynamic properties of the BEM provide a challenging matrix in which to make voltammetric measurements. Factors that influence the migration of substances in the BEM will have their greatest effect on the quantitative reliability of these data. In the simplest case, techniques with long sampling periods may require that diffusion parameters, α and λ, be included in the calculation of in vivo concentrations from in vitro calibration factors. Quantitative determination of release or uptake kinetics, on the other hand, not only requires accurate estimates of concentration, but depending on brain region and experimental methods, may require incorporation of diffusion parameters into the kinetic analysis, as described in Section 1.2.2.

2.1.1. The Compartment Model: Free Diffusion vs Volume Fraction and Tortuosity

From the early days of in vivo voltammetry, it was recognized that the behavior of measurements obtained in brain tissue is different from that in free solution. Typically, the first voltammetric response is larger than subsequent measurements, which decrease in size until a steady baseline is obtained. Because this equilibration process was not seen in vitro, Cheng et al. (1979) suggested the in vivo behavior reflected restricted mass transport properties of brain. They introduced a compartment model to describe the behavior of their measurements in brain tissue (Cheng et al., 1979). Although this model was introduced for chronoamperometric data, the issues involved are relevant to all voltammetric techniques and are especially important for slow-scan techniques, including differential pulse voltammetry (DPV).

Chronoamperometry is the simplest voltammetric technique in which a square-wave potential step, sufficient to

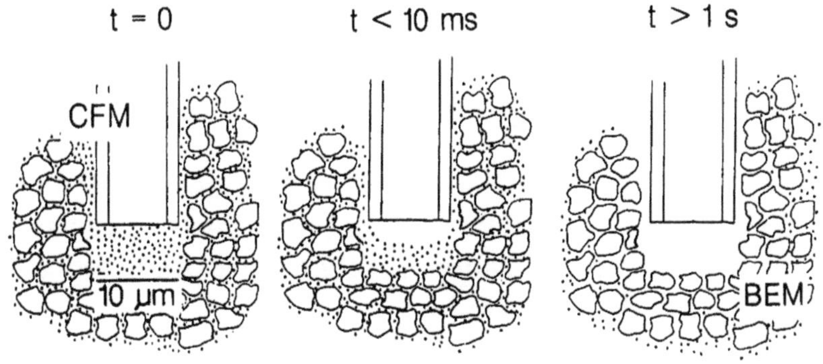

Fig. 7. Schematic representation of the proposed pool formed at the tip of a CFM implanted in brain tissue (after Cheng et al., 1979). Initially, $t = 0$; the concentration of substances in the pool is in equilibrium with that in the BEM. During a brief voltammetric measurement, the concentration of electroactive species decreases at the electrode surface during oxidation. The thickness of the resulting diffusion layer (γ) is proportional to $t^{1/2}$, so that with brief sampling times, γ is less than the radius of the pool (e.g., $t < 10$ ms). With longer sample times (e.g., $t > 1$ s), the diffusion layer extends into the surrounding BEM, so that diffusion to the CFM has mixed free solution and tissue components.

oxidize the compound of interest, is applied to the working electrode for a period of typically 10 ms to 1 s. In the compartment (or pool) model, the electrode samples from a fluid-filled compartment surrounding the electrode tip (Fig. 7). The initial concentration (*C*) of electroactive species in the pool is assumed to be in equilibrium with surrounding extracellular fluid. With unmodified carbon electrodes, the predominant electroactive substance detected in brain tissue is ascorbate (Schenk et al., 1982, 1983), which is irreversibly oxidized during the measurement. Each chronoamperometric pulse depletes the ascorbate content of the pool, so that successive measurements sample a lower concentration. Replenishment of the pool depends on diffusion of ascorbate from the surrounding extracellular fluid. Stable baseline currents are obtained when the rate of depletion is balanced by the rate of mass transport to the pool. Consequently, baseline concentration depends on sampling interval: Higher repetition rates cause lower baselines, because the dynamic steady state favors depletion. Conversely, longer sampling intervals and shorter sampling times favor equilibration with the "true" extracellular concentration. The predictive value of this compartment model was subsequently verified using digital simulation (Lindsay et al., 1980), and later made more rigorous by incorporating parameters that considered the effects of variation in pool size and tissue transport rate on the predicted electrode behavior (Cheng, 1982).

2.1.2. Electrode Calibration: When Are In Vitro Calibrations Based on Diffusion in Free Solution Valid?

A central assumption of the compartment model is that the thickness of the diffusion layer that forms at the electrode surface during a voltammetric measurement does not extend beyond the boundary of the fluid pool at the electrode tip (Fig. 7, $t < 10$ ms). With this assumption, the normal solution diffusion coefficient of a substance is appropriate for converting the measured currents to concentration, based on in vitro calibration of the electrode. Indeed, all estimates of extracellular concentration in tissue, as far as we know, have been based on this premise, although it is unlikely to be correct for all cases.

The thickness of the diffusion layer γ(cm), or the region of depletion during a voltammetric measurement, is time-dependent and can be estimated by the simple expression: $\gamma = (Dt_s)^{1/2}$, where t_s (s) is the duration of the measurement (e.g., Amatore et al., 1986). Using this expression with $D_{Asc} = 6.6 \times 10^{-6}$ cm^2/s (at 37°C, derived from Rice et al., 1985), γ will be 2.6 µm at 10 ms, 8.1 µm at 100 ms, 26 µm at 1 s, and 81 µm at 10 s. Whether γ extends beyond the boundary of the pool (Fig. 7, $t > 1$ s) depends on the dimensions of the pool, which are apparently proportional to the dimensions of the electrode used. A pool radius of 50 µm was estimated for the 200–300 µm carbon paste or carbon epoxy electrodes used in early work in Adams' laboratory (Cheng et al., 1979; Cheng, 1982), whereas a radius of only about 12 µm was calculated for the 10-µm diameter CFMs used in Wightman's laboratory (Amatore et al., 1986). The issue of a pool is not unique to voltammetric electrodes, however: Nicholson and Phillips (1981) estimated that ISMs with a tip diameter of 2–4 μm sampled a pool of similar dimensions.

It would be predicted, therefore, that a relatively large electrode surrounded by a pool of 50-μm radius would sample inside the pool during a 1-s chronoamperometric measurement (Cheng et al., 1979; Cheng, 1982), so that solution *D* values would be appropriate for calculation of concentration. By contrast, Amatore et al. (1986) used 10-ms chronoamperometric pulses ($\gamma = 2.3$ μm) to remain in the pool of a 10-μm CFM in order to determine extracellular ascorbate concentration ($[Asc]_0$). Paradoxically, although Amatore et al. estimated that the pool was 12 μm (using a model discussed further below), they found experimentally that currents obtained in vivo with 100-ms chronoamperometric steps were distorted from free solution appearance, as though they sampled beyond pool boundaries (Amatore et al., 1986). Although there are several possible explanations for this result, including variability in pool size, these data indicate that voltammetric techniques that use very short sampling times (<100 ms) are most likely to give accurate quantitation when used with in vitro calibration. Larger electrodes with larger pool dimensions may permit longer sample times, however, they also cause greater depletion, so that measured concentrations depend on sampling interval and will be lower than true extracellular levels (Cheng et al., 1979).

How are voltammetric measurements affected when the diffusion layer extends into the surrounding tissue (Fig. 7, $t > 1$ s)? In general, lower currents will be seen in vivo for a given concentration, compared to an in vitro calibration. The net result of applying in vitro calibration factors to tissue measurements, therefore, is that in vivo concentration will be underestimated. Both λ and α can contribute to decreased currents in vivo. Voltammetric current, *i*, is generally proportional to $D^{1/2}$. Because the apparent diffusion coefficient in vivo, D^* (Section 1.1.5), is smaller than solution *D*, concentration estimates based on *D*, rather than D^*, could be too low by a factor as large as λ (typically 1.5–1.6; Table 1). In addition, voltammetric equations have been derived from the assumption that the molecules available for detection are in a volume of free solution, so that *i* is proportional to *C*. This condition is not met in brain tissue, which is effectively a porous medium. Molecules reaching the electrode arrive from a volume that is roughly 20% that of free solution, so that this reduced tissue volume fraction could cause the concentration estimated to be further diminished by a factor of five (α^{-1}).

Amatore et al. (1986) re-evaluated the compartment model for the condition where the diffusion layer extended into the tissue. This situation is essentially a two-compartment model, so that diffusion to the electrode surface has both free solution and restricted diffusion components. Amatore et al. simplified this by using long chronoamperometric pulses (2.5 s), so that measured current would depend predominantly on mass transport through tissue. The model then viewed the CFM and its associated pool as a single electrode with radius R_0, which was the radius of the pool. In their treatment, the term *K* was used in place of α, and D_2 was used instead of D^*. Amatore et al. incorporated α into their concentration measurements to give an effective concentration ($C_{eff} = \alpha C$), as sometimes used to evaluate concentration (*see* Section 1.1.6.) in porous media. Rather than use a known value for *K* (e.g., 0.21), they calculated *K* to be 0.15 in rat striatum and then determined R_0 using that value, which might be one source of error in their estimates of pool dimensions. A better approach could be developed from the methods of Nicholson (1985), who considered the problem of substances diffusing from a cavity in the tissue formed during pressure ejection into brain tissue. In that case, α was introduced to describe how flux between the cavity and the tissue was affected by the discontinuity in the phase-averaged concentrations in the two compartments.

It is likely that in most cases, however, tissue measurements will reflect the more complex situation of mixed diffusion phases. Precise equations to describe measurement behavior would depend on electrode size and shape, and be specific for the techniques used. Consequently, no general formulations are attempted here.

2.1.2.1. Basal Concentration of Extracellular Ascorbate

The extent to which tissue diffusion characteristics influence concentration estimates depends on the voltammetric technique used. A clear example of deviation among methods is illustrated by literature values for $[Asc]_o$, which vary by several hundred percent. In general, measured concentration was strongly correlated with the voltammetric technique used for its determination (Table 2). The highest values (over 500 μ*M*) were found using short pulses with chronoamperometry or normal pulse voltammetry (Amatore et al., 1986; Ewing et al., 1981), whereas the lowest were determined using linear sweep or square-wave voltammetry (O'Neill et al., 1982b; Svensson et al., 1993). This is consistent with the notion that long sampling times underestimate extracellular concentration. The simplest assumption that $[Asc]_o$ is too low by a factor of $\lambda\alpha^{-1}$ would give a modified value of about 650 μ*M* for the square-wave voltammetry value of 89 μ*M* (Svensson et al., 1993), which is within the range of data reported using 10-ms chronoamperometric measurements (Amatore et al., 1986).

There are several exceptions to the correlation between sampling time and $[Asc]_o$ in Table 2. These exceptions make two important points. The first is that most voltammetric electrodes lose some sensitivity in vivo. Typically, the loss is 40–60%, although it can vary widely. We find, and other labs have noted this as well, that the greatest sensitivity loss occurs when the electrode is first exposed to tissue, with little change after that. Consequently, the use of pretissue calibration factors to calculate tissue values can also cause apparent concentration to be underestimated. This factor might contribute to the relatively low striatal $[Asc]_o$ reported with fast-scan cyclic voltammetry (Stamford et al., 1984a), although this would make CSF levels much higher than generally thought. Similarly, $[Asc]_o$ might have been be about 500 μ*M*, if post-calibration had been used to calculate the data of Schenk et

Table 2
Relationship Between Apparent $[Asc]_o$ and Voltammetric Sample Time

$[Asc]_o$, μM	Electrode tip diameter, μM	Technique	Sample time, s	Calibration	Ref.
570	10	C	0.010	Post	Amatore et al., 1986[a]
508	1–2	FCV	0.013	Pre	Stamford et al., 1984[b]
500	10	NPV	0.175	Post	Ewing et al., 1981[a]
306	8	DPV	25	Post	Gonon et al., 1981[c]
284	75–200	C	1.0	Pre	Schenk et al., 1983[d]
227	75–200	C	1.0	Pre	Schenk et al., 1982[a]
220	1–2	FCV	0.013	Pre	Stamford et al., 1984a[a]
100–200	300	LSV	50	N.S.	O'Neill et al., 1982b[e]
89	30	SWV	4	Post	Svensson et al., 1993[e]

C, chronoamperometry.
DPV, differential normal pulse voltammetry.
FCV, fast-scan cyclic voltammetry.
LSV, linear sweep voltammetry.
NPV, normal pulse voltammetry.
SWV, square-wave voltammetry.
All measurements are in rat:
[a]striatum in vivo, initial level;
[b]CSF in vivo;
[c]striatum in vivo, baseline level;
[d]striatal slice in vitro;
[e]initial or baseline concentration not specified (N.S.).

al. (1982, 1983), which were obtained from chronoamperometry with 1 s pulses. Most of the data in Table 2 were obtained by selective monitoring of $[Asc]_o$, either through the use of electrodes that discriminated between ascorbate and catecholamines (Ewing et al., 1981; Gonon et al., 1981; Amatore et al., 1986; Svensson et al., 1993) or by use of the specific enzyme ascorbate oxidase to remove ascorbate from the background response (Schenk et al., 1982, 1983; Stamford et al., 1984a). In other measurements, background measurements can include other electroactive species, like the DA metabolite DOPAC in the striatum. These other substances have been estimated to be about 55 μM (precalibration) whether in vivo or in slices (Schenk et al., 1983).

One final point is that there is no absolute answer. The $[Asc]_o$ concentration found using differential pulse voltammetry (Gonon et al., 1981) is higher than one might have predicted for a slow-scanning technique, especially with small electrodes that would do little damage and cause only a small pool to form at the electrode tip. In this study, however, the initial baseline measurements were established in awake, freely moving animals. This level dropped by about 80% when the animal was anesthetized with either chloral hydrate or halothane (Gonon et al., 1981). Consequently, the lower values are consistent with the other slow-scan data, which were also generally recorded from anesthetized animals. These data, along with the observation that there is a circadian rhythm to $[Asc]_o$ (O'Neill et al., 1982a), prompted Crespi to suggest that ascorbate is influenced by state of consciousness (Crespi et al., 1992).

2.1.2.2. Basal Concentration of Extracellular Dopamine

Few voltammetric studies have attempted to estimate extracellular DA concentration ($[DA]_o$) (Gonon and Buda, 1985; Crespi and Möbius, 1992; Witkovsky et al., 1993). The studies of DA in striatum used DPV or differential normal pulse voltammetry (DNPV; distinctions between the voltage waveforms in these techniques are discussed in Nicholson and Rice, 1988 and elsewhere in this volume) using electrochemically pretreated CFMs, which give voltammograms with separate oxidation peaks for ascorbate and catecholamines (Gonon et al., 1981). Here again, the slower technique (20 s to scan the DA oxidation peak), DPV, gave an lower estimate of $[DA]_o$ than did DNPV, with 110-ms sampling pulses (which returned to resting potential between pulses to rereduce DA; Gonon and Buda, 1985). The estimate for $[DA]_o$ in striatum was 1.5 n*M* with DPV (Crespi and Möbius, 1992) and 26 n*M* with DNPV (Gonon and Buda, 1985). One factor in these studies was that the DNPV measurements were done in pargyline pretreated animals to eliminate interference from the DA metabolite DOPAC. This causes an increase in $[DA]_o$, but only by about 50% (Crespi and Möbius, 1992), which does not explain the difference. When this drug effect is considered and the DPV result modified by $\lambda\alpha^{-1}$, however, the variance between estimates is minimized. It should be noticed that a range of DA levels (4–20 n*M*) have been estimated by microdialysis (Zetterström et al., 1983; Parsons and Justice, 1992), although the lower levels were evaluated using a more quantitative analysis (Parsons and Justice, 1992). Although it can be argued that these levels are very close in value, the concentration might be in error by an order of magnitude, which is an extremely large difference when moving into the realm of DA dynamics, discussed in Section 2.1.3.

One reason that it has been difficult to estimate $[DA]_o$ in the striatum is because active uptake and metabolism keep $[DA]_o$ low (n*M*), whereas $[DOPAC]_o$ is much higher (μ*M*). The situation is very different in the retina (Witkovsky et al., 1993). There, DA uptake is minimal, so that the ratio of DA to DOPAC is almost unity. Consequently, it is possible to determine $[DA]_o$ without pargyline or other drug treatments (Fig. 8A). Using FCV, we determined the $[DA]_o$ of the isolated retina to be about 300 n*M* (Witkovsky et al., 1993), which is at least an order of magnitude higher than in the striatum. The sensitivity of FCV measurements to background interferences in these and other measurements is discussed further below (Section 3.2.).

2.1.3. Mass-Transport Issues Involved in the Understanding of Transmitter Behavior

Voltammetric methods have been moving steadily toward faster sampling and smaller electrode size in recent years. As discussed, short sampling periods offer the best possibil-

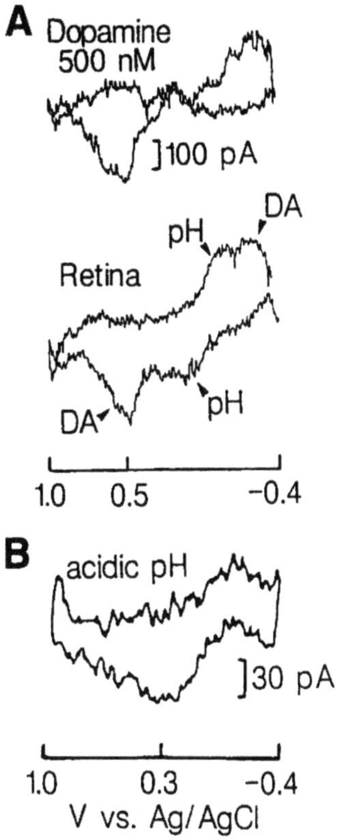

Fig. 8. The effect of pH interference on FCV voltammograms of basal $[DA]_o$ in the retina. **(A)** Subtraction voltammograms for DA in calibration solution and in the retina, and **(B)** subtraction voltammogram showing the effect of acidic pH on the background currents during iontophoresis of H^+. The apparent oxidation peak at +0.3 V vs Ag/AgCl (labeled pH in A) interfered with the DA voltammogram in the retina corresponded to the potential of the shift in background currents that accompanies acidic pH. FCV at 900 V/s was performed using an EI-400 (Ensman Instrumentation, Bloomington, IN). Electrodes were made from 8-µm diameter carbon fibers. The fibers were inserted into 2-mm Pyrex tubing, which was pulled in a vertical electrode puller. The electrode was backfilled with a drop of epoxy resin to seal the fiber into the glass. Then electrical contact between the fiber and an Ag/AgCl wire was made using 4*M* potassium acetate with 150 m*M* KCl. Complete details are described elsewhere (Rice and Nicholson, 1989). (Modified from Witkovsky et al., 1993 and Rice and Nicholson, 1989.)

ity for accurate determination of extracellular concentration. Faster sampling rates (short sampling interval) also permit measurements of dynamic changes on a more physiological time scale. In addition, electrodes made from carbon fibers, typically 5–30 μm (Feng et al., 1987; Kristensen et al., 1988; Rice and Nicholson, 1989; Millar and Williams, 1990), offer several advantages over larger, carbon-paste or graphite epoxy electrodes, not the least of which is that they do less damage to the tissue. One benefit of this is that the pool formed around the tip of a CFM is smaller, so that the time for replenishment of the compartment is minimal. Indeed, when 10-μm electrodes were used (Ewing et al., 1981), baseline currents with 92-ms chronoamperometric pulses were independent of sampling interval from 4–30 s, which indicates that the concentration of the pool was in rapid equilibrium with extracellular concentration. By contrast, larger pool size and greater consumption of larger electrodes are reflected in the strong dependence of concentration on sampling interval (Cheng et al., 1979).

Significantly, increases in $[DA]_o$ in the striatum elicited by electrical stimulation of the medial forebrain bundle (MFB) have also been reported to be independent of sampling interval when monitored with CFMs (Millar et al., 1985; Stamford, 1986). In FCV measurements, the maximum increase in $[DA]_o$ was the same whether the sampling interval was 2 s or 25 ms (Stamford, 1986). This apparently accurate monitoring of $[DA]_o$ reflects not only rapid equilibration between the small pool at the electrode tip and the BEM, but also the fact that the reverse (cathodic) sweep of the triangular FCV waveform reduces oxidized DA to minimize depletion. Measurements with larger electrodes or longer sampling times would require additional time for equilibration of the pool at the electrode tip with changes in $[DA]_o$, so that neither the time-course nor amplitude of stimulated increase can be accurately followed (Cheng, 1982).

Diffusional distortion is also seen when Nafion-coated CFMs are used for detection of DA release (May et al., 1988; Wightman et al., 1988; Kawagoe et al., 1993). In these studies, $[DA]_o$ continued to rise after stimulation ceased, which suggested a diffusional component to the response. This effect was originally attributed to distortion from the pool (Wightman et al., 1988) and was corrected by convoluting their kinetic model with diffusion (in free solution) over 10–35 μm (May et al., 1988). More recently, however, it was proposed that the delay could be accounted for by slow diffusion within the Nafion film (Engstrom et al., 1988; Kawagoe et al., 1992) rather than from a release site to the sensor, as noted in Section 1.2.4. Release curves were then corrected by deconvolution for the thickness of the film using a value of D appropriate for DA inside the film (Kawagoe et al., 1992).

With either uncoated or distortion-corrected Nafion CFMs, the assumption that the measured concentration changes mirror dynamic changes in the BEM is central to all recent kinetic studies that have attempted to model M-M uptake kinetics. When, then, are diffusion characteristics of the BEM required for evaluation of biogenic amine clearance?

For qualitative comparison of DA clearance, as in some of the pressure ejection studies by Gerhardt and colleagues (*see* Gerhardt, this volume; Van Horne et al., 1992), incorporating α and λ is not necessary. To model the nonlinear DA uptake process accurately, however, requires incorporation of diffusion characteristics, because of the large distances and the varying gradient of $[DA]_0$ between the source and electrode. A recent study from Gerhardt's laboratory on DA migration in striatum and accumbens (Cass et al., 1993) incorporates an *ad hoc* linear uptake term in the point-source pressure-injection model originally proposed by Nicholson (1985) to describe TMA diffusion. Although the model appears to represent some of the data, it has several problems:

1. The typical injection volume would have extended beyond the distance between the injection and recording site (usually <300 μm; the average volume reported in this study is a sphere of >300 μm radius in the BEM) so that a point-source model cannot be used;
2. As discussed above, uptake is not linear, which leads to the third problem;
3. The *ad hoc* extension of the equation, as stated by Cass et al. (1993), always yields negative concentrations at sufficiently long times.

In fact, Nicholson (1992) provided a complete analytical solution to the injection of a large bolus of material subject to linear uptake (in the conventional sense defined by Eq. [4], Section 1.2.1., rather than the *ad hoc* usage of Cass et al., 1993). Even that solution, however, is not adequate to deal with the experimental situation that Cass et al. are studying, which requires a full numerical solution. More seriously, the conclusion of that study, that DA regulation in the nucleus accumbens is more sensitive to uptake inhibitors than in the striatum, is called into doubt by these issues.

As noted earlier, another approach to study the kinetics of DA release and uptake has been to elicit DA release by stimulation of the MFB or other pathway (Millar et al., 1985; Michael et al., 1987; May et al., 1988; Wightman et al., 1988). Uptake calculations are based on the assumption of a gradient-free increase in $[DA]_o$ (*see* Section 1.2.5.), which if true, means that α and λ are not necessary. Adams and colleagues have focused on norepinephrine (NE) release (Renner et al., 1992; Capella et al., 1993; Ghasemzadah et al., 1993), with the same assumptions. Using Wightman's model (Wightman, 1988), they have reported the first in vivo estimates of NE release and uptake in the thalamus (Capella et al., 1993). The lower rate of uptake for NE in thalamus compared to DA in striatum led the authors to propose that diffusion might be a more significant factor in NE neurotransmission than for striatal DA (Capella et al., 1993).

These studies of release and uptake have primarily investigated regions of high catecholamine innervation. Even within the DA-rich striatum, kinetic parameters are usually determined for the site giving maximal release. In other brain areas, such as amygdala and medial prefrontal cortex (MPFC), which have lower catecholaminergic innervation, the density of uptake sites would be expected to be correspondingly decreased, so that diffusion might have a larger role in DA clearance. Indeed, in locations that give submaximal release within the striatum (Kawagoe et al., 1992), as well as in amygdala and MPFC (Garris and Wightman, 1994), $[DA]_o$ continues to rise after cessation of stimulation, even after deconvolution for the delay from the Nafion film. For such regions, it is likely that the assumption of homogeneous elevation of $[DA]_o$ does not hold, so that correct analysis of M-M kinetics should require incorporation of tissue-diffusion properties. Accurate evaluation of the amount of DA released per stimulus pulse and other kinetic factors relies on proper evaluation of these parameters as well.

3. Effects of Ionic Changes and Dynamic Characteristics of the Microenvironment on Voltammetric Measurements

The response of a voltammetric microelectrode can reflect changes in the ionic environment around the electrode, in particular in the concentrations of H^+ and Ca^{2+} and the local volume fraction of the extracellular space. All these quantities, as well as other ionic concentrations, vary with brain stimulation. In this section, we provide a very brief survey of typical ionic fluctuations in the BEM.

3.1. Chemical Composition and Ion Shifts in Extracellular Fluid

The chemical composition of the fluid that bathes the extracellular space is approximated by human cerebrospinal fluid composition (Kee and Wood, 1987, based on work by Milhorat) (concentrations in m*M*): Na: 141; K: 3.3; Ca: 1.3; Mg: 1.2; Cl: 124; HCO_3: 21: glucose: 3.4; urea 5; phosphate: 0.48. When used as an artificial bathing medium, it is usually gassed with 95% O_2/5% CO_2 to establish a pH of about 7.4. Recent measurements using improved ion-selective microelectrodes have suggested that there may be regional variation in the extracellular K^+ level in the brain with values as low as 1.9 m*M* (Moghaddam and Adams, 1987; Coles and Poulain, 1991). Possibly, if detailed studies were available of the other ions, more variation would also be evident.

3.1.1. Ion Shifts in the Minority Extracellular Ions: H^+ (pH), K^+, Ca^{2+}, Mg^{2+}

The studies discussed here are based almost exclusively on the use of ion-selective microelectrodes. The literature was surveyed by Nicholson and Rice (1988) and more recently by Syková (1992), so this section will be limited to brief comments.

3.1.1.1. $[H^+]_o$

The baseline concentration of H^+ in the extracellular space is <100 n*M* (at a typical pH of 7.4) which contrasts with that of other extracellular ions. Repetitive electrical stimulation typically produces a fast alkaline-going response followed by a slower acidification (Kraig et al., 1983; Chesler, 1990; Syková, 1992). The origin of these transients is currently under investigation, but they could represent the operation of proton- or bicarbonate-selective channels (Chesler, 1990; Chesler and Kaila, 1992; Voipio and Kaila, 1993).

3.1.1.2. $[K^+]_o$

Since the introduction of the ISMs, studies of K^+ have predominated, primarily because of early availability of a sensor for this ion and the profound influence of K^+ on the excitability of neurons. The early ISMs used the tetraphenylborate exchanger (Corning 477317), which is subject to substantial interference from sodium and other compounds, but the results are generally valid. Today ISMs for K^+ are based on valinomycin and give very reliable results (Nicholson, 1993). Whenever neurons are active, $[K^+]_o$ rises from a typical baseline level of about 3 m*M* (but *see* Section 3.1.) to a maximum of 10–12 m*M* (Vyskocil et al., 1972; Lux and Neher, 1973, Syková, 1991). Increases beyond this level are uncommon in normal conditions in the adult nervous system, unless pathophysiological conditions such as spreading depression (SD) or ischemia occur. Under the latter circumstances, $[K^+]_o$ can reach 20–80 m*M* (Kraig and Nicholson, 1978; Hansen and Zeuthen, 1981; Martins-Ferreira and do Carmo, 1987; Somjen et al., 1992). After an episode of $[K^+]_o$ increase, the baseline level frequently undershoots the original baseline for a period, owing to continued action of the Na^+-K^+ pump (Heinemann and Lux, 1975).

3.1.1.3. $[Ca^{2+}]_o$ AND $[Mg^{2+}]_o$

When neurons are active, $[Ca^{2+}]_o$ falls by about 0.2 m*M* from a normal baseline value of about 1.2 m*M* (Nicholson et al., 1976, 1977; Heinemann et al., 1977). During SD or ischemia, $[Ca^{2+}]_o$ may fall by a factor of 10 (Nicholson et al., 1977, 1978; Hansen and Zeuthen, 1981; Martins-Ferreira and do Carmo, 1987). It is thought that the lost extracellular Ca^{2+} enters cells, and that excess entry may contribute to cell damage or death, but the cascade of events during a pathophysiological episode is probably more complex than this (Rothman, 1992; Schurr and Rigor, 1992). $[Mg^{2+}]_o$ has not been measured with ion-selective microelectrodes because of the poor selectivity of these electrodes in the presence of calcium. Recently, a new neutral carrier ISM has been described (Schaller et al., 1993) and extracellular measurements may now be feasible.

3.1.2. Ion Shifts in the Majority Ions: Na^+, Cl^-

3.1.2.1. $[Na^+]_o$

It is difficult to measure changes in $[Na^+]_o$ during stimulation because of the high ambient level, but the study of Dietzel et al. (1982) demonstrated a decrease of approx 5 m*M*. This was corroborated by Coles and Orkand (1985) in a study on the retina of the bee. The majority of studies of changes in $[Na^+]_o$ have been made during SD and ischemia. In these conditions, $[Na^+]_o$ drops by about 90 m*M* (Kraig and Nicholson, 1976, 1978; Hansen and Zeuthen, 1981; Martins-Ferreira and do Carmo, 1987).

3.1.2.2. $[Cl^-]_o$

This ion has a similar concentration to Na^+ in the extracellular space, and it is equally difficult to measure changes. Dietzel et al. (1982) showed an increase of approx 7 m*M* during stimulation, but again like $[Na^+]_o$, $[Cl^-]_o$ has been primarily measured during SD and ischemia (Nicholson and Kraig, 1975; Kraig and Nicholson, 1978; Hansen and Zeuthen, 1981; Martins-Ferreira and do Carmo, 1987). Under such pathophysiological conditions, $[Cl^-]_o$ drops by about 100 m*M*.

3.1.3. Water Shifts and Changes in Volume Fraction During Stimulation and Pathophysiology

Several studies have shown that the volume of the extracellular space can change, usually decreasing, during stimulation (Dietzel et al., 1980, 1982; Svoboda and Syková 1991; Armstrong and Rice, 1993), and substantial decreases are known to occur during SD and ischemia (Phillips and Nicholson, 1979; Hansen and Zeuthen, 1981; Lundbæk and Hansen, 1992; Syková et al., 1994). These changes are usually attributed to the swelling of cells from the entry of water and consequent reduction in extracellular space volume. Following hypernatremia induced by systemic NaCl infusion, the extracellular space is reduced, but intracellular volume is regulated (Cserr et al., 1991).

3.2. Effect of Dynamic Changes in the BEM on Voltammetric Measurements

Dynamic changes in the composition and volume fraction of the extracellular fluid can introduce artifacts into voltammetric measurements of the release of biogenic amines or other electroactive species. These effects can limit qualitative assessment of the released substances, but can interfere with quantitation as well. The significance of such interference will depend on the electrode and voltammetric technique used, and on the neurotransmitter system and region under investigation. At best, the effects will be negligible; at worst, they can totally mask a signal. The possibility of interference exists for all measurements, however, and must be taken into consideration for proper interpretation of voltammetric data.

3.2.1. Background Ion Shifts

Stimulation of brain tissue induces shifts in the composition of extracellular fluid (Section 3.1.). The effects will vary depending on where recordings are made, the nature of the stimulus, e.g., electrical or chemical, and whether local or pathway stimulation is used. The strength of stimulus is also important, because the size of ion shifts is usually dependent on stimulus intensity. This discussion will focus primarily on fast-scan cyclic voltammetry, because it is the technique with which we are most familiar and because it is particularly susceptible to background changes.

Inorganic ions in the extracellular fluid are not electroactive, but they can interfere with voltammetric measurements if they interact with the surface of the electrode. Such interaction can alter charging current contributions to a voltammetric signal, as well as modulate currents arising from electroactive functionalities on the electrode surface (McCreery, this vol.). Ideally, pulse techniques, including chronoamperometry and pulse voltammetry (including DPV, NPV, DNPV) avoid charging (or non-Faradaic) current contributions by sampling after the capacitative current has decayed (Adams and Schenk, 1984; Nicholson and Rice, 1988). These techniques are relatively immune to ion shifts, which is one of their main advantages. It should be noted however, that the shorter the sampling time, the greater the charging current component. Rapid sampling techniques, like high-speed chronoamperometry (Gerhardt, this volume) offer perhaps the best compromise between speed and selectivity for in vivo monitoring. At the other extreme is the technique of FCV. Fast-scan voltammograms, recorded with scan rates of 300–900 V/s, have very large charging current components (Armstrong-James et al., 1981; Howell et al., 1986; Rice and Nicholson, 1989), so that these measurements are especially sensitive to effects from ion shifts.

3.2.2. pH

We first noticed a sensitivity of CFMs to pH when monitoring the diffusion of iontophoretically injected DA using FCV (Rice and Nicholson, 1989). Typically in FCV, a background scan, taken before an iontophoretic pulse or other stimulus, is subtracted from a voltammogram (or averaged voltammogram) recorded during the event to give a subtraction voltammogram that is characteristic of DA or other electroactive substance (*see* Stamford et al., this vol; Garris and Wightman, this vol; Rice and Nicholson, 1989). We found that initial subtraction voltammograms recorded as the iontophoresis pipet "warmed up" could not be attributed to DA, but rather to H^+ (Fig. 8B), which was emitted before the electrode tip filled with DA. The pH sensitivity of the measurement was indicated by apparent oxidation and reduction peaks that appeared on the subtraction voltammogram at about 0.3 and –0.15 V vs Ag/AgCl during the brief acidification that accompanied H^+ iontophoresis in agarose. The peak on the anodic scan was distinct from that of DA, which oxidizes at about +0.5 V on these electrodes.

Interference from pH changes between baseline and measurement periods can also accompany assessment of basal and stimulated DA release. For example, the pH of most in vitro preparations is more acidic (often 0.1–0.2 pH units) than the surrounding physiological saline because of tissue respiration (Chesler, 1990). Consequently, our measurements of basal $[DA]_o$ in the retina (Witkovsky et al., 1993) were complicated by additional pH peaks on the DA voltammogram (Fig. 8A). Interference from pH did not prevent identification of the DA component to the voltammetric response. However, if $[DA]_o$ were lower or tissue acidity greater, confidence in the DA portion of the signal would be limited. The variable degree of interference can be seen with electrical stimulation of rat striatal slices (Fig. 9; Rice and Chesler, unpublished data). Relatively mild stimulation (70 μA, 60-μs pulses at 5 Hz for 10 s delivered with a bipolar stimulating electrode) induced an increase in $[DA]_o$ and in $[K^+]_o$ with only a slight alkaline–acid shift in pH (Fig. 9A). The voltammogram recorded at the peak of the response could be clearly identified as DA. By contrast, longer stimulation (5 Hz, 40 s) caused a further increase in $[K^+]_o$ and a larger pH response, which entirely masked the DA response (Fig. 9B).

3.2.3. $[Ca^{2+}]_o$ and $[K^+]_o$

In a more detailed study, Jones et al. (1994) recently investigated contributions of ion transients to stimulated catecholamine release signals in slices of rat amygdala. Dopamine levels in the amygdala are at least 10 times lower than in striatum, and released levels are proportionally smaller (Jones et al., 1994). Consequently, alkaline–acid pH shifts confounded catecholamine (DA and NE) identification from voltammograms, whereas release signals in striatum were clearly identified as DA. This study also revealed that Ca^{2+} is another significant interferent. The fall in $[Ca^{2+}]_o$ that accompanies stimulation (Section 3.1.1.3.) was also sensed by CFMs. A decrease in $[Ca^{2+}]_o$ caused background shifts in the subtrac-

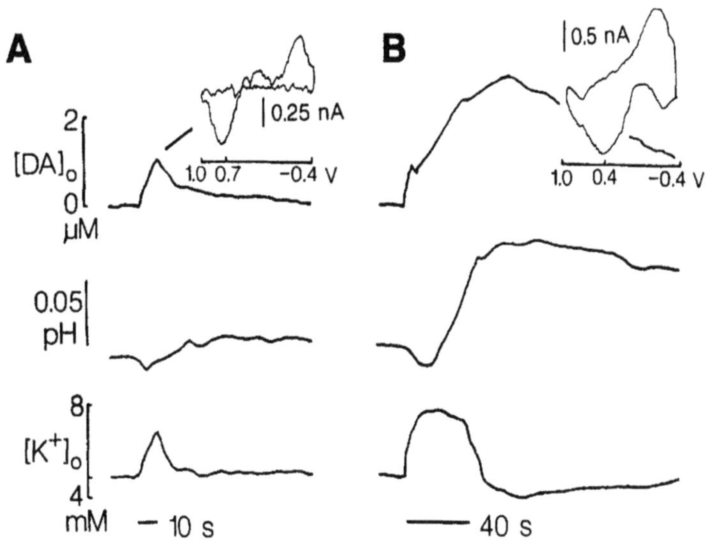

Fig. 9. Shifts in CFM current (nominally DA), extracellular K^+, and pH during electrical stimulation of the striatal slice. **(A)** Stimulus parameters were 5 Hz, 70 µA, 60-µs pulses for 10 s. The DA voltammogram above the $[DA]_o$ record is the subtraction voltammogram at the peak of the response. **(B)** Stimulation was as in (A), except duration was 40 s. The responses are larger for each of the species monitored. Although an initial increase in $[DA]_o$ was seen, this was rapidly obscured by a large extracellular acidification. The nature of the prolonged rise at the CFM was confirmed by the subtraction voltammogram and the similarity of the time-course of the response at the pH electrode. Note that the pH shift had both an alkaline and acid component. (Rice and Chesler, unpublished.)

tion voltammogram that were similar to those seen when the pH was made more alkaline (Jones et al., 1994). Changes in $[K^+]_o$ did not have any effect on electrode background currents.

Ion interferences may also be larger in brain tissue in vitro than in the same region in vivo (Jones et al., 1994). This may reflect the closer position of the stimulus to recording site in brain slices, since local stimulation, rather than stimulation of an input pathway like MFB, must usually be used with slice preparations. Ion effects might also be greater in excised tissue because the normal mechanisms of ion homeostasis, including cerebral microcirculation, are compromised in a brain slice.

3.2.4. Nafion

Nafion, a polyanionic Teflon™ derivative, has been successfully used with many electrodes and techniques (Gerhardt et al., 1984; Nagy et al., 1985). Nafion can be applied to the electrode by dip-coating (Gerhardt et al., 1984; Kristensen et al., 1988) or electroplating by applying a slight positive potential (+.0.5–1.0) between the electrode and a silver wire ring containing a drop of Nafion solution (Brazell et al., 1987; Rice and Nicholson, 1989). This polymer coating repels anions and extracts cations, so that contributions from anionic ascorbate and DOPAC to a voltammetric response are minimal, whereas those from the positively charged amines, like DA, are slightly enhanced.

We have two cautionary notes about possible interference with Nafion-coated electrodes. The first is that the Nafion film has varying affinity for positively charged species (Nagy et al., 1985). Hydrophobic cations are extracted preferentially into a Nafion film, at the expense of hydrophilic molecules of the same charge. We discovered one consequence of this when we attempted to do codiffusion studies of DA and TMA^+ in rat striatal slices. In the presence of 1 m*M* TMA^+ (included in the incubation media for calibration of the TMA^+-ISM, Nafion-coated CFMs were completely insensitive to DA. TMA^+ apparently occupied all the sites in the film, so that DA was no longer incorporated. We found no evidence of interference from TMA^+ at noncoated electrodes. In further studies of DA diffusion from a pressure ejection source, we used an anion, α-naphthalene sulfonate (ANS^-), as a control ion that was coejected with DA (Nicholson and Rice, 1991).

A second concern is that calibration of Nafion electrodes should be done in media closely resembling that in which they are to be used. We used to calibrate, as many investigators do, in phosphate-buffered saline (PBS; 50 m*M* sodium phosphate, pH 7.4, with 100 m*M* NaCl). The advantage of using PBS for testing purposes is that the pH is well buffered and constant, whereas the pH of bicarbonate-buffered solutions can drift, if not in constant equilibrium with CO_2

(e.g., 95% O2/5% CO_2). We found, however, that with Nafion-coated CFMs, the sensitivity to DA was consistently lower in physiological saline than in PBS. One possible reason is that the millimolar levels of Ca^{2+} and Mg^{2+} in physiological saline act like TMA^+ to exclude DA from the Nafion film. Because we work primarily in vitro, we can readily calibrate CFMs in the brain slice chamber at the experimental temperature, at any stage of the experiment by adding DA (typically 0.2–5 μ*M*) to the superfusing physiological saline. Dopamine is protected from oxidation in the oxygenated media, because we routinely include the antioxidants ascorbate and thiourea (Rice and Nicholson, 1991; Rice et al., 1994) in our incubation media. This procedure gives reliable calibration data for calculation of $[DA]_o$, although it would be more difficult to implement for in vivo studies.

3.2.5. Influence of Stimulus-Evoked Changes in Volume Fraction

In addition to ion shifts, decreases in extracellular volume fraction can accompany stimulation (Section 3.1.3.). The size of the decrease depends on region and stimulus strength and duration, but is usually a few percent. The concentration of substances in the extracellular space will increase as the volume decreases. In the case of stimulated DA release from striatum, where a 1-μ*M* increase in $[DA]_o$ would be a 5000–10,000% increase over baseline (for an estimated basal level of 10–20 n*M*), the contribution from extracellular volume changes would make a negligible contribution. On the other hand, in regions that have lower catecholaminergic innervation than striatum, unless electrodes are exceptionally selective for catecholamines, the potential for misinterpretation exists. For example, let us say we have a Nafion-coated electrode that has a selectivity of 100:1 for DA over ascorbate. During stimulation, a 10% volume decrease might occur, which would increase background concentrations by 10%, even without additional release. If background includes 500 μ*M* ascorbate, the resulting 50 μ*M* increase could be interpreted as a 0.5-μ*M* increase in a catecholamine. Consequently, confirmation of the response is essential for interpretation of the data. The identity of such small signals can be confirmed by their persistence after addition of ascorbate oxidase (Rice and Nicholson, 1987), by catecholamine-depleting lesions, or by the enhancement of the response in the presence of selective uptake blockers. The size of the volume decrease can be determined using ISMs to monitor the concentration of an extracellular marker ion like TMA^+ or ANS^- (Phillips and Nicholson, 1979; Armstrong and Rice, 1993). Superfusion of these extracellular markers is especially useful for slice preparations.

3.2.6. Effect of Spreading Depression, Hypoxia, and Ischemia

In each of these events, i.e., spreading depression, hypoxia, and ischemia, the ion shifts and volume changes that accompany stimulation will be amplified. The volume decrease is especially significant, since the extracellular volume fraction can fall to <50% of its normal value. It is important to note that many large depolarizing stimuli can initiate SD. Pressure ejection of K^+ (Kasser et al., 1988) and the excitatory amino acid agonist, glutamate (Moghaddam et al., 1990) can both produce SD in vivo. This can be indicated by monitoring $[K^+]_o$ or DC potential shifts some distance (at least 0.5–1 mm) from the stimulation site (Rice and Nicholson, 1987; Kasser et al., 1988; Moghaddam et al., 1990).

The large volume decrease that accompanies these pathophysiological events can affect voltammetric measurement in two main ways. First, the concentration of interfering background species will be at least doubled. For example, $[Asc]_o$ may rise to millimolar levels, when release and volume decrease combine. Less obviously, the changes in extracellular volume can alter the mass-transport characteristics of the brain. We simultaneously monitored $[Asc]_o$ and $[TMA^+]_o$ in the isolated turtle cerebellum during SD using chronoamperometry with 100-ms pulses (Rice and Nicholson, 1987). The change in $[Asc]_o$ over baseline was very similar to the change in extracellular volume fraction (Fig. 10). With large volume decreases, $[Asc]_o$ rose initially, and then decreased, with a minimum concentration at the point of maximum volume decrease. The concentration continued to rise again,

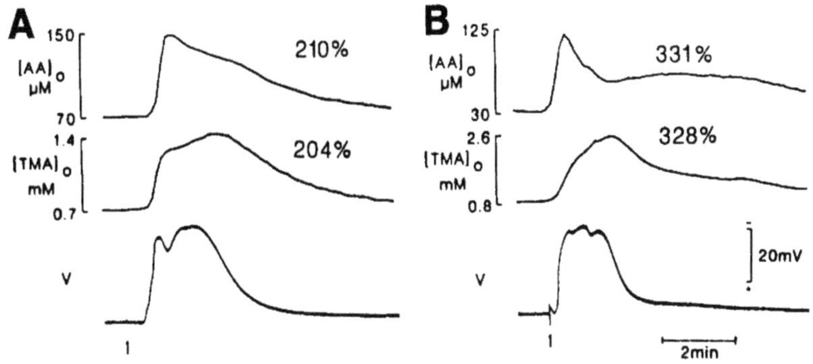

Fig. 10. Extracellular ascorbate $[Asc]_o$, volume decrease, and DC potential during spreading depression. In both **(A)** and **(B)**, the increase in $[Asc]_o$ during SD was similar to that predicted by extracellular volume decrease (a 204% increase reflects a 51% decrease in α, whereas a 328% increase indicates a 70% decrease), even though the $[Asc]_o$ signal declines before the maximum volume decrease occurs. This is especially extreme in (B), where the diffusion-limited $[Asc]_o$ record had a spike-dome shape. The data were obtained using chronoamperometry with 35-μm diameter CFMs in the isolated turtle cerebellum. The potential was stepped from 0 to +0.5 for 100 ms, and repeated at 2-s intervals. *See text* for further discussion. (Modified from Rice and Nicholson, 1987.)

as the volume relaxed back toward normal, giving a spike-dome shape to the ascorbate signal (Fig. 10B). The paradoxical concentration decrease was probably the result of the reduced rate of mass transport possible during the large volume decrease, but with a constant rate of consumption of ascorbate at the electrode with 100-ms chronoamperometric pulses. This same phenomenon is probably responsible for the apparent loss of extracellular ascorbate during terminal anoxia that was recently reported (Svensson et al., 1993). Although extracellular volume changes were not monitored, an initial rise in $[Asc]_o$ reversed at the time of the onset of a sharp increase in $[DA]_o$, which probably signaled the onset of depolarization with coincident volume decrease. In that study, $[Asc]_o$ fell below the original baseline level in two of three regions tested (Svensson et al., 1993), which would be consistent with reduced mass transport during terminal depolarization. The effects would be especially pronounced with the slow-scanning technique (square-wave voltammetry) used in that study.

4. Conclusions

In this chapter, we have described the BEM as a dynamic system, with diffusion properties and fluctuations in the composition of "supporting electrolyte" that classical electrochemists would never have imagined. In vivo voltammetric measurements are now coming to a mature stage, partly from the development of smaller and more selective electrodes and partly because of the refinement of faster sampling techniques. Both advances have served to minimize the perturbation of the BEM caused by voltammetric measurements. To realize fully the potential of voltammetry as a research tool, however, one must also understand how voltammetric measurements are perturbed by the BEM. Consequently, we have discussed here a variety of situations in which interpretation of signals can be difficult. The brain is not a beaker, yet many of us treat it as though it were. The issue of electrode calibration, especially which pool is sampled in vivo, although mundane, is critical for the proper interpretation of any quantitative data. At the very least, reliable calibration procedures verify that a response is physiologically possible. At the other extreme, as measurements become increasing quantitative in the attempt to model neurotransmission, a full understanding of the properties of the BEM will be critical.

References

Adams R. N. and Schenk J. O. (1984) Chronoamperometric measurements in the central nervous system, in *Measurement of Neurotransmitter Release In Vivo* (Marsden C. A., ed.), IBRO Handbook Series 6, Wiley Interscience, Chichester, England, pp. 193–208.

Amatore C., Kelly R. S., Kristensen E. W., Kuhr W. G., and Wightman R. M. (1986) Effects of restricted diffusion at ultramicroelectrodes in brain tissue. The pool model: theory and experiment for chronoamperometry. *J. Electroanalyt. Chem.* **213,** 31–42.

Armstrong-James M., Fox K., Kruk Z. L., and Millar J. (1981) Quantitative iontophoresis of catecholamines using multibarrel carbon fibre electrodes. *J. Neurosci. Meth.* **4,** 385–406.

Armstrong W. E. and Rice M. E. (1993) Transient extracellular volume reduction in neural lobe of rat hypophysis in response to neural stalk stimulation invitro and its relationship to extracellular potassium. *J. Neurophysiol.* **69,** 1363–1367.

Benveniste H. and Hüttemeier P. C. (1990) Microdialysis—theory and application. *Prog. Neurobiol.* **35,** 195–215.

Bondareff W. and Pysh J. J. (1968) Distribution of the extracellular space during postnatal maturation of rat cerebral cortex. *Anat. Record* **160,** 773–780.

Brazell M. P., Kasser R. J., Renner K. J., Feng J., Moghaddam B., and Adams R. N. (1987) Electrocoating carbon fiber microelectrodes with Nafion improves selectivity for electroactive transmitters. *J. Neurosci. Methods* **22,** 167–172.

Capella P., Ghasemzadeh M. B., Adams R. N., Wiedemann D. J., and Wightman R. M. (1993) Real-time monitoring of electrically stimulated-norepinephrine release in rat thalamus: 2. Modeling of release and reuptake characteristics of stimulated norepinephrine overflow. *J. Neurochem.* **60,** 449–453.

Cass W. A., Zahniser N. R., Flach K. A., and Gerhardt G. A. (1993) Clearance of exogenous dopamine in rat dorsal striatum and nucleus accumbens: role of metabolism and effects of locally applied uptake inhibitors. *J. Neurochem.* **61,** 2269–2278.

Chary S. R. and Jain R. K. (1989) Direct measurement of interstitial convection and diffusion of albumin in normal and neoplastic tissues by fluorescence photobleaching. *Proc. Natl. Acad. Sci. USA* **86,** 5385–5389.

Cheng H.-Y. (1982) Compartment model for chronoamperometric measurement *in vivo. J. Electroanalyt. Chem.* **135,** 145–151.

Cheng H.-Y., Schenk J., Huff R., and Adams R. N. (1979) In vivo electrochemistry: behavior of micro electrodes in brain tissue. *J. Electroanalyt. Chem.* **100,** 23–31.

Chesler M. (1990) The regulation and modulation of pH in the nervous system. *Prog. Neurobiol.* **34,** 401–427.

Chesler M. and Kaila K. (1992) Modulation of pH by neuronal activity. *Trends Neurosci.* **15,** 396–402.

Coles J. A. and Orkand R. K. (1985) Changes in sodium activity during light stimulation in photoreceptors, glia and extracellular space in drone retina. *J. Physiol.* **362,** 415–435.

Coles J. A. and Poulain D. A. (1991) Extracellular K^+ in the supraoptic nucleus of the rat during reflex bursting activity by oxytocin neurones. *J. Physiol.* **439,** 383–409.

Cordingley G. E. and Somjen G. G. (1978) The clearing of excess potassium from extracellular space in spinal cord and cerebral cortex. *Brain Res.* **151,** 291–306.

Cragg B. (1979) Brain extracellular space fixed for electron microscopy. *Neurosci. Lett.* **15,** 301–308.

Crespi F. and Möbius C. (1992) In vivo selective monitoring of basal levels of cerebral dopamine using voltammetry with Nafion modified (NA-CRO) carbon fibre micro-electrodes. *J. Neurosci. Methods* **42,** 149–161.

Crespi F., Möbius C., and Keane P. (1992) Evidence for a relationship between recovery from anaesthesia modified state of consciousness and striatal voltammetric levels of ascorbic acid. *Pharmacol. Res.* **26,** 55–66.

Cserr H. F. (ed.) (1986) The neuronal microenvironment. *Ann. NY Acad. Sci.* **481,** 1–391.

Cserr H. F. and Knopf P. M. (1992) Cervical lymphatics, the blood-brain barrier and the immunoreactivity of the brain: a new view. *Immunology Today* **13,** 507–512.

Cserr H. F., DePasquale M., Nicholson C., Patlak C. S., Pettigrew K. D., and Rice M. E. (1991) Extracellular volume decreases while cell volume is maintained by ion uptake in rat brain during acute hypernatremia. *J. Physiol.* **442,** 277–295.

Dietzel I., Heinemann U., Hofmeier G., and Lux H. D. (1982) Stimulus-induced changes in extracellular Na^+ and Cl^- concentration in relation to changes in the size of the extracellular space. *Exp. Brain Res.* **46,** 73–84.

Dietzel I., Heinemann U., Hofmeier G., and Lux H. D. (1980) Transient changes in the size of the extracellular space in the sensor motor cortex of cats in relation to stimulus-induced changes in potassium concentration. *Exp. Brain Res.* **40,** 432–439.

Engstrom R. C., Wightman R. M., and Kristensen E. W. (1988) Diffusional distortion in the monitoring of dynamic events. *Analyt. Chem.* **60,** 652–656.

Ewing A. G. and Wightman R. M. (1984) Monitoring the stimulated release of dopamine with in vitro voltammetry. II: clearance of released dopamine from extracellular fluid. *J. Neurochem.* **43,** 570–577.

Ewing A. G., Dayton M. A., and Wightman R. M. (1981) Pulse voltammetry with micro voltammetric electrodes. *Analyt. Chem.* **53,** 1842–1847.

Feng J.-X., Brazell M., Renner K., Kasser R., and Adams R. N. (1987) Electrochemical pretreatment of carbon fibers for in vivo electrochemistry: effect on sensitivity and response time. *Analyt. Chem.* **59,** 712–716.

Fenstermacher J. D. and Patlak C. S. (1975) The exchange of material between cerebrospinal fluid and brain, in *Fluid Environment of the Brain* (Cserr H. F., Fenstermacher J. D., and Fencl V., eds.), Academic, New York, pp. 201–214.

Fuxe K. and Agnati L. F. (eds.)(1991) *Volume Transmission in the Brain. Advances in Neuroscience,* vol. 1, Raven, New York.

Garris P. A. and Wightman R. M. (1994) Different kinetics govern dopaminergic transmission in the amygdala, prefrontal cortex, and striatum—an in vivo voltammetric study. *J. Neurosci.* **14,** 442–450.

Gerhardt G. and Adams R. N. (1982) Determination of diffusion coefficients by flow injection analysis. *Analyt. Chem.* **54,** 2618–2620.

Gerhardt G. A., Oke A. F., Nagy G., Moghaddam B., and Adams R. N. (1984) Nafion-coated electrodes with high selectivity for CNS electrochemistry. *Brain Res.* **290,** 390–395.

Ghasemzadeh M. B., Capella P., Mitchell K., and Adams R. N. (1993) Real-time monitoring of electrically stimulated-norepinephrine release in rat thalamus: I. Resolution of transmitter and metabolite signal components. *J. Neurochem.* **60,** 442–448.

Gonon F. G. and Buda M. J. (1985) Regulation of dopamine release by impulse flow and by autoreceptors as studied by *in vivo* voltammetry in the rat striatum. *Neuroscience* **14,** 765–774.

Gonon F., Buda M., Cespuglio R., Jouvet M., and Pujol J.-F. (1981) Voltammetry in the striatum of chronic freely moving rats: detection of catechols and ascorbic acid. *Brain Res.* **223,** 69–80.

Hansen A. J. and Zeuthen T. (1981) Extracellular ion concentrations during spreading depression and ischemia in the rat brain cortex. *Acta Physiol. Scand.* **113,** 437–445.

Harris E. J. and Burn G. P. (1948) The transfer of sodium and potassium ions between muscle and the surrounding medium. *Trans. Faraday Soc.* **45,** 508–528.

Heinemann U. and Lux D. D. (1975) Undershoots following stimulus-induced rises of extracellular potassium concentration in cerebral cortex of cat. *Brain Res.* **93,** 63–76.

Heinemann U., Lux H. D., and Gutnick M. J. (1977) Extracellular free calcium and potassium during paroxysmal activity in the cerebral cortex of the cat. *Exp. Brain Res.* **27,** 236–243.

Holz R. W. and Coyle J. T. (1974) The effects of various salts, temperature, and the alkaloids veratridine and batrachotoxin on the uptake of [^{3}H] dopamine into synaptosomes from rat striatum. *Mol. Pharmacol.* **10,** 746–758.

Horn A. S. (1976) Characteristics of transport in dopaminergic neurons, in *The Mechanism of Neuronal and Extraneuronal Transport of Catecholamines* (Paton D. M., ed.), Raven, New York, pp. 195–215.

Horn A. S. (1979) Characteristics of dopamine uptake, in *The Neurobiology of Dopamine* (Horn A. S., Korf J., and Westerink B. H. C., eds.), Academic, London, pp. 217–235.

Horn A. S. (1990) Dopamine uptake: a review of progress in the last decade. *Prog. Neurobiol.* **34,** 387–400.

Howell J. O., Kuhr W. G., Ensman R. E., and Wightman R. M. (1986) Background subtraction for rapid scan voltammetry. *J. Electroanalyt. Chem.* **209,** 77–90.

Jones S. R., Mickelson G. E., Collins L. B., Kawagoe K. T., and Wightman R. M. (1994) Interference by pH and Ca^{2+} ions during measurements of catecholamine release in slices of rat amygdala with fast-scan cyclic voltammetry. *J. Neurosci. Methods* **52,** 1–10.

Justice J. B., Nicolaysen L. C., and Michael A. C. (1988) Modeling the dopaminergic nerve terminal. *J. Neurosci. Methods* **22,** 239–252.

Kasser R. J., Renner K. J., Feng J.-X., Brazell M. P., and Adams R. N. (1988) Spreading depression induced by 100 mM KCl in caudate is blocked by anesthesia of the substantia nigra. *Brain Res.* **475,** 333–344.

Kawagoe K. T., Garris P. A., Wiedemann D. J., and Wightman R. M. (1992) Regulation of transient dopamine concentration gradients in the microenvironment surrounding nerve terminals in the rat striatum. *Neuroscience* **51,** 55–64.

Kawagoe K. T., Zimmerman J. B., and Wightman R. M. (1993) Principles of voltammetry and microelectrode surface states. *J. Neurosci. Methods* **48,** 225–240.

Kee D. B. and Wood J. H. (1987) Cerebrospinal fluid, in *Encylopedia of Neuroscience,* vol. 1 (Adelman G., ed.), Birkhauser, Boston, pp. 215–217.

Kehr J. (1993) A survey on quantitative microdialysis—theoretical models and practical implications. *J. Neurosci. Methods* **48,** 251–261.

Kelly R. S. and Wightman R. M. (1987) Detection of dopamine overflow and diffusion with voltammetry in slices of rat brain. *Brain Res.* **423,** 79–87.

Kraig R. P. and Nicholson C. (1976) Sodium liquid ion exchanger microelectrode used to measure large extracellular sodium transients. *Science* **194,** 725,726.

Kraig R. P. and Nicholson C. (1978) Extracellular ionic variations during spreading depression. *Neuroscience* **3,** 1045–1059.

Kraig R. P., Ferreira-Filho C. R., and Nicholson C. (1983) Alkaline and acid transients in the cerebellar microenvironment. *J. Neurophysiol.* **49,** 831–850.

Kristensen E. W., Kuhr W. G., and Wightman R. M. (1988) Temporal characterization of perfluorinated ion-exchange coated microvoltammetric electrodes for *in vivo* use. *Analyt. Chem.* **59,** 1752–1757.

Krnjevic K. and Morris M. E. (1974) Extracellular accumulation of K^+ evoked by activity of primary afferent fibers in the cuneate nucleus and dorsal horn of cats. *Can. J. Physiol. Pharmacol.* **52,** 852–871.

Kuffler S. W. and Potter D. D. (1964) Glia in the leech central nervous system: physiological properties and neuron-glia relationship. *J. Neurophysiol.* **27,** 290–320.

Lehmenkühler A., Syková E., Svoboda J., Zilles K., and Nicholson C. (1993) Extracellular space parameters in the rat neocortex and subcortical white matter during postnatal development determined by diffusion analysis. *Neuroscience* **55,** 339–351.

Levin V. A., Fenstermacher J. D., and Patlak C. S. (1970) Sucrose and inulin space measurements of cerebral cortex in four mammalian species. *Am. J. Physiol.* **219,** 1528–1533.

Lindsay W. S., Justice J. A., and Salome J. (1980) Simulation studies of *in vivo* electrochemistry. *Comp. Chem.* **4,** 19–26.

Lundbæk J. A. and Hansen A. J. (1992) Brain interstitial volume fraction and tortuosity in anoxia. Evaluation of the ion-selective micro-electrode method. *Acta Physiol. Scand.* **146,** 473–484.

Lux H. D. and Neher E. (1973) The equilibration time course of $[K^+]_o$ in cat cortex. *Exp. Brain Res.* **17,** 190–205.

Margolis R. K. and Margolis R. U. (1993) Nervous tissue proteoglycans. *Experientia* **49,** 429–446.

Martins-Ferreira H. and do Carmo R. J. (1987) Retinal spreading depression and the extracellular milieu. *Can. J. Physiol. Pharmacol.* **65,** 1092–1098.

Mathias R. T. (1983) Effect of tortuous extracellular pathways on resistance measurements. *Biophys. J.* **42,** 55–59.

May L. J., Kuhr W. G., and Wightman R. M. (1988) Differentiation of dopamine overflow and uptake processes in the extracellular fluid of the rat caudate nucleus with fast-scan in vivo voltammetry. *J. Neurochem.* **51,** 1060–1069.

McBain C. J., Traynelis S. F., and Dingledine R. (1990) Regional variation of extracellular space in the hippocampus. *Science* **249,** 674–677.

Michael A. C., Ikeda M., and Justice J. B. Jr. (1987) Mechanisms contributing to the recovery of striatal releasable dopamine following MFB stimulation. *Brain Res.* **421,** 325–335.

Millar J., Stamford J. A., Kruk Z. L., and Wightman R. M. (1985) Electrochemical, pharmacological and electrophysiological evidence of rapid dopamine release and removal in the rat caudate nucleus following electrical stimulation of the rat median forebrain bundle. *Eur. J. Pharmacol.* **109,** 341–348.

Millar J. and Williams G. V. (1990) Fast differential ramp voltammetry: a new voltammetric technique designed specifically for use in neuronal tissue. *J. Electroanalyt. Chem.* **282,** 33–49.

Moghaddam B. and Adams R. A. (1987) Regional differences in resting extracellular potassium levels of rat brain. *Brain Res.* **406,** 337–340.

Moghaddam B., Gruen R. J., Roth R. H., Bunney B. S., and Adams R. A. (1990) Effect of L-glutamate on the release of striatal dopamine: in vivo dialysis and electrochemical studies. *Brain Res.* **518,** 55–60.

Nagy G., Gerhardt G. A., Oke A. F., Rice M. E., Moore R. B. III, Szentirmay M. N., and Martin C. R. (1985) Ion exchange and transport of neurotransmitters in Nafion films on conventional and microelectrode surfaces. *J. Electroanalyt. Chem.* **188,** 85–94.

Near J. A., Bigelow J. C., and Wightman R. M. (1988) Comparison of uptake of dopamine in rat striatal chopped tissue and synaptosomes. *J. Pharmacol. Exp. Therapeut.* **245,** 921–927.

Nicholson C. (1979) Brain cell microenvironment as a communication channel, in *The Neurosciences Fourth Study Program* (Schmitt F. O. and Worden F. G., eds.), MIT Press, Cambridge, MA, pp. 457–476.

Nicholson C. (1980) Dynamics of the brain cell microenvironment. *Neurosci. Res. Prog. Bull.* **18,** 177–322.

Nicholson C. (1985) Diffusion from an injected volume of a substance in brain tissue with arbitrary volume fraction and tortuosity. *Brain Res.* **333,** 325–329.

Nicholson C. (1992) Quantitative analysis of extracellular space using the method of TMA^+ iontophoresis and the issue of TMA+ uptake. *Can. J. Physiol. Pharmacol.* **70,** S314–S322.

Nicholson C. (1993) Ion-selective microelectrodes and diffusion measurements as tools to explore the brain cell microenvironment. *J. Neurosci. Methods* **48,** 199–213.

Nicholson C. (1995a) Extracellular space as the pathway for neuron-glial cell-interaction, in *Neuroglial Cells* (Kettenmann H. and Ransom B., eds.), Oxford University Press (in press).

Nicholson C. (1995b) Interaction between diffusion and Michaelis-Menten uptake of dopamine following iontophoresis in striatum. *Biophys. J.* (in press).

Nicholson C. and Kraig R. P. (1975) Chloride and potassium changes measured during spreading depression in catfish cerebellum. *Brain Res.* **96,** 384–389.

Nicholson C. and Rice M. E. (1988) Use of ion-selective microelectrodes and voltammetric microsensors to study brain cell microenvironment, in *Neuromethods: The Neuronal Microenvironment* (Boulton A. A., Baker G. B., and Walz W., eds.), Humana, Clifton, NJ, pp. 247–361.

Nicholson C. and Rice M. E. (1991) Dopamine diffusion and uptake in slices of rat neostriatum. *Soc. Neurosci Abstr.* **17,** 88.

Nicholson C. and Tao L. (1993) Hindered diffusion of high-molecular weight compounds measured with integrated optical imaging. *Biophys. J.* **65,** 2277–2290.

Nicholson C., Miyan J. A., Potter K. T., Williamson R., and Abbott N. J. (1995) Diffusion properties of the microenvironment of cephalopod brain, in *Cephalopod Neurobiology* (Abbott N. J., Williamson R., and Maddock L., eds.), Oxford University Press, Oxford (in press).

Nicholson C. and Phillips J. M. (1981) Ion diffusion modified by tortuosity and volume fraction in the extracellular microenvironment of the rat cerebellum. *J. Physiol.* **321,** 225–257.

Nicholson C., Steinberg H., Stöckle H., and ten Bruggencate G. (1976) Calcium decrease associated with aminopyridine-induced potassium increase in cat cerebellum. *Neurosci. Lett.* **3,** 315–319.

Nicholson C., ten Bruggencate G., Steinberg R., and Stöckle H. (1977) Calcium modulation in brain extracellular microenvironment demonstrated with ion-selective micropipette. *Proc. Natl. Acad. Sci. USA* **74,** 1287–1290.

Nicholson C., ten Bruggencate G., Stöckle H., and Steinberg R. (1978) Calcium and potassium changes in extracellular microenvironment of cat cerebellar cortex. *J. Neurophysiol.* **41,** 1026–1039.

O'Neill R. D., Fillenz M., and Alberry W. J. (1982a) Circadian changes in homovanillic acid and ascorbate levels in rat striatum using microprocessor-controlled voltammetry. *Neurosci. Lett.* **34,** 189–193.

O'Neill R. D., Grunewald R. A., Fillenz M., and Albery W. J. (1982b) Linear sweep voltammetry with carbon paste electrodes in the rat striatum. *Neurosci.* **7,** 1945–1954.

Parsons L. H. and Justice J. B. Jr. (1992) Extracellular concentration and in vivo recovery of dopamine in the nucleus accumbens using microdialysis. *J. Neurochem.* **58,** 212–218.

Phillips J. M. and Nicholson C. (1979) Anion permeability in spreading depression investigated with ion-selective microelectrodes. *Brain Res.* **173,** 567–571.

Ranck J. B. and BeMent S. L. (1965) The specific impedance of the dorsal columns of cat: an anisotropic medium. *Exp. Neurol.* **11,** 451–463.

Rees S., Cragg B. G., and Everitt A. V. (1982) Comparison of extracellular space in the mature and ageing rat brain using a new technique. *J. Neurol. Sci.* **53,** 347–357.

Renner K. J., Pazos L., and Adams R. N. (1992) In vivo voltammetric evidence for the detection of norepinephrine release in the thalamus of freely moving rats. *Brain Res.* **577,** 49–56.

Rice M. E. and Nicholson C. (1989) Measurement of nanomolar dopamine diffusion using low-noise perfluorinated ionomer coated carbon fiber microelectrodes and high-speed cyclic voltammetry. *Analyt. Chem.* **61,** 1805–1810.

Rice M. E. and Nicholson C. (1991) Diffusion characteristics and extracellular volume fraction during normoxia and hypoxia in slices of rat neostriatum. *J. Neurophysiol.* **65,** 264–272.

Rice M. E. and Nicholson C. (1987) Interstitial ascorbate in turtle brain is modulated by release and extracellular volume change. *J. Neurochem.* **49,** 1096–1104.

Rice M. E., Gerhardt G. A., Nagy G., Hierl P. M., and Adams R. N. (1985) Diffusion coefficients of neurotransmitters and their metabolites in brain extracellular fluid space. *Neuroscience* **15,** 891–902.

Rice M. E., Okada Y., and Nicholson C. (1993) Anisotropic and heterogeneous diffusion in the turtle cerebellum: implications for volume transmission. *J. Neurophysiol.* **5,** 2035–2044.

Rice M. E., Pérez-Pinzón M. A., and Lee E. J. K. (1994) Ascorbic acid, but not glutathione, is taken up by brain slices and preserves cell morphology. *J. Neurophysiol.* **71,** 1591–1596.

Rothman S. A. (1992) Excitotoxins: possible mechanisms of action. *Ann. NY Acad. Sci.* **648,** 132–139.

Schaller U., Spichiger U. E., and Simon W. (1993) Novel magnesium ion-selective microelectrodes based on a neutral carrier. *Pflügers Arch.—Eur. J. Physiol.* **423,** 338–342.

Schenk J. O., Miller E., Gaddis R., and Adams R. N. (1982) Homeostatic control of ascorbate concentration in CNS extracellular fluid. *Brain Res.* **253,** 353–356.

Schenk J. O., Miller E., Rice M. E., and Adams R. N. (1983) Chronoamperometry on brain slices: quantitative evaluations of in vivo electrochemistry. *Brain Res.* **277,** 1–8.

Schmitt F. O. (1984) Molecular regulators of brain functioning: a new view. *Neuroscience* **13,** 991–1001.

Schmitt F. O. and Samson F. E. (1969) The brain cell microenvironment. *Neurosci. Res. Prog. Bull.* **7,** 277–417.

Schurr A. and Rigor B. M. (1992) The mechanism of cerebral hypoxic-ischemic damage. *Hippocampus* **2,** 221–228.

Snyder S. H. and Coyle J. T. (1969) Regional differences in H^3–norepineprine and H^3–dopamine uptake into rat brain homogenates. *J. Pharmacol. Exp. Therapeut.* **165,** 78–86.

Somjen G. G., Aitken P. G., Czèh G. L., Herreras O., Jing J., and Young J. N. (1992) Mechanisms of spreading depression: a review of recent findings and a hypothesis. *Can. J. Physiol. Pharmacol.* **70,** S248–S254.

Stamford J. A. (1986) In vivo voltammetry: some methodological considerations. *J. Neurosci. Methods* **17,** 1–29.

Stamford J. A., Kruk Z. L., and Millar J. (1984a) Regional differences in extracellular ascorbic acid levels in the rat brain determined by high speed cyclic voltammetry. *Brain Res.* **299,** 289–295.

Stamford J. A., Kruk Z. L., Millar J., and Wightman R. M. (1984b) Striatal dopamine uptake in the rat: in vivo analysis by fast cyclic voltammetry. *Neurosci. Lett.* **51,** 133–138.

Stamford J. A., Kruk Z. L., and Millar J. (1986) In vivo voltammetric characterization of low affinity striatal dopamine uptake: I Inhibition profile and relation to dopaminergic innervation density. *Brain Res.* **373,** 85–91.

Svensson L., Wu C., Hulthe P., Johannessen K., and Engel J. A. (1993) Effect of ageing on extracellular ascorbate concentration in rat brain. *Brain Res.* **609,** 36–40.

Svoboda J. and Syková E. (1991) Extracellular space volume changes in the rat spinal cord produced by nerve stimulation and peripheral injury. *Brain Res.* **560,** 216–224.

Syková E. (1991) Ionic and volume changes in neuronal microenvironment. *Physiol. Res.* **40,** 213–222.

Syková E. (1992) *Ionic and Volume Changes in the Microenvironment of Nerve and Receptor Cells. Progress in Sensory Physiology,* vol. 13, Springer-Verlag, Berlin.

Syková E., Svoboda J., Polák J., and Chvátal A. (1994) Extracellular volume fraction and diffusion characteristics during progressive ischemia and terminal anoxia in the spinal cord of the rat. *J. Cereb. Blood Flow Metab.* **14,** 301–311.

Van Harreveld A. (1972) The extracellular space in the vertebrate central nervous system, in *The Structure and Function of Nervous Tissue* (Bourne G. H., ed.), Academic, New York, pp. 447–511.

Van Horne C., Hoffer B. J., Strömberg I., and Gerhardt G. A. (1992) Clearance and diffusion of locally applied dopamine in normal and 6-hydroxydopamine-lesioned rat striatum. *J. Pharmacol. Exp. Therapeut.* **263,** 1285–1292.

Vern B. A., Schuette W. H., and Thibault L. E. (1977) $[K^+]_o$ clearance in cortex: a new analytical model. *J. Neurophysiol.* **40,** 1015–1023.

Voipio J. and Kaila K. (1993) Interstitial P_{CO2} and pH in rat hippocampal slices measured by means of a novel fast CO_2/H^+-sensitive microelectrode based on a PVC-gelled membrane. *Pflügers Arch.—Eur. J. Physiol.* **423,** 193–201.

Vyskocil F., Kriz N., and Bures J. (1972) Potassium-selective microelectrodes used for measuring the extracellular brain potassium during spreading depression and anoxic depolarization. *Brain Res.* **39,** 255–259.

Wiedemann D. J., Basse-Tomusk A., Wilson R. L., Rebec G. V., and Wightman R. M. (1990) Interference by DOPAC and ascorbate during attempts to measure drug-induced changes in neostriatal dopamine with Nafion-coated, carbon-fiber electrodes. *J. Neurosci. Methods* **35,** 9–18.

Wightman R. M., Amatore C., Engstrom R. C., Hale P. D., Kristensen E. W., Kuhr W. G., and May L. J. (1988) Real-time characterization of dopamine overflow and uptake in the rat striatum. *Neuroscience* **25,** 513–523.

Wightman R. M. and Zimmerman J. B. (1990) Control of dopamine extracellular concentration in rat striatum by impulse flow and uptake. *Brain Res. Rev.* **15,** 135–144.

Witkovsky P., Nicholson C., Rice M. E., Bohmaker K., and Meller E. (1993) Extracellular dopamine concentration in the retina of the clawed frog, *Xenopus laevis*. *Proc. Natl. Acad. Sci. USA* **90,** 5667–5671.

Zetterström T., Sharp T., Marsden C. A., and Ungerstedt U. (1983) In vivo measurement of dopamine and its metabolites by intracerebral dialysis: changes after D-amphetamine. *J. Neurochem.* **41,** 1769–1773.

Fast Cyclic Voltammetry in Brain Slices

Jonathan A. Stamford, Peter Palij, Colin Davidson, Christine M. Jorm, and Paul E. M. Phillips

1. Introduction

Brain slices have been used for several years in the study of the factors controlling neurotransmitter release in animals. Despite this, most conventional studies have drawbacks. First, it is often necessary to label the neurotransmitter pool with tritium (Farnebo and Hamberger, 1971). This assumes, probably fallaciously, that release of the radiolabel accurately reflects the neurotransmitter. For example, when endogenous and radiolabeled catecholamine releases are simultaneously measured in response to various stimuli, there are significant discrepancies between the two (Herdon and Nahorski, 1987). Second, transmitter release is frequently evoked by elevating the potassium concentration. However, some transmitter is released in a calcium-independent fashion following potassium depolarization (Okuma and Osumi, 1986). Third, many superfusion methods need to block transmitter uptake in order for efflux to be measurable (Parker and Cubeddu, 1985). These are serious limitations.

The ideal method might perhaps be one that can detect the endogenous transmitter, following calcium-dependent depolarization and without artificial enhancement of transmitter release. In this chapter, we shall try to examine the

From: *Neuromethods, Vol. 27: Voltammetric Methods in Brain Systems*
Eds: A. Boulton, G. Baker, and R. N. Adams © 1995 Humana Press Inc.

extent to which fast cyclic voltammetry (FCV) measures up to these ideals.

1.1. Fast Cyclic Voltammetry

FCV is an electrochemical technique designed to enable detection of monoamine release in real time (Stamford, 1990). The features that distinguish FCV from other voltammetric methods are the speed of the measurement and the size of the electrode used. Whereas many voltammetric methods make measurements lasting several seconds or minutes (Stamford et al., 1992), an FCV scan takes only 14–20 ms and can be repeated up to 40 times/s (Stamford et al., 1986), giving real-time detection of transmitter release (Kuhr and Wightman, 1986). This temporal resolution is matched by a comparable improvement in spatial resolution afforded by the carbon-fiber microelectrode, which has a tip approx 20–60 × 8 µm (Armstrong James and Millar, 1979) and can record in nuclei <200 µm across.

1.2. Basic Principles of FCV

In the form used in this laboratory, the input voltage to the potentiostat consists of 1½ cycles of a triangular waveform. We routinely use a –1.0 to +1.4 V (vs Ag/AgCl) waveform because of the enhanced sensitivity (Hafizi et al., 1990). The voltage scan rate used is 480 V/s with the initial scan in the cathodic direction. Usually, this waveform is applied to the potentiostat twice/s throughout the experiment. Occasionally, when higher temporal resolution of events is required (for instance, for uptake kinetics), the waveform repetition rate can be 16/s or more. The working electrode is electronically disconnected between scans and can thus be used also for electrophysiological recording *(vide infra)*. Figure 1 shows a schematic representation of the input and output waveforms.

The voltage input and current output of the working electrode are typically displayed on a digital storage oscilloscope (we use the Nicolet 310). A background current signal prior to a stimulation is stored initially, followed by one obtained in response to a stimulation. Digital subtraction of one signal

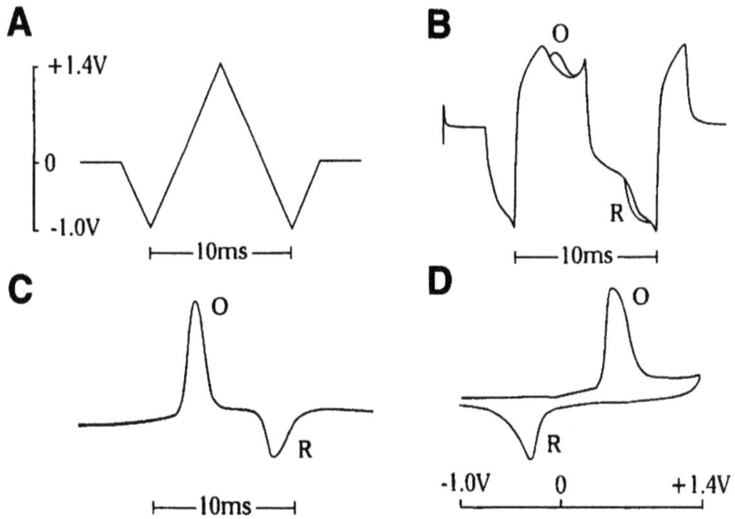

Fig. 1. Waveforms of fast cyclic voltammetry. **(A)** Input voltage waveform to the potentiostat (–1.0 to +1.4 V vs Ag/AgCl, 480 V/s). **(B)** Output current signals from a carbon-fiber microelectrode in buffer. Two signals are shown superimposed, one before and the other after addition of a catecholamine to the buffer. The areas marked (O and R) are the oxidation and reduction currents. These are exaggerated for clarity. **(C)** Subtraction of one waveform from the other yields the faradaic current, consisting of a single oxidation (O) and reduction (R) peak. **(D)** Cyclic voltammogram of catecholamine obtained by plotting the faradaic current (C) against the input voltage waveform (A).

from the other removes all features common to both signals (charging current mainly) and leaves only the faradaic (redox) current resulting from the released species. By plotting this faradaic current against the input voltage, a background current-corrected cyclic voltammogram (CV) is obtained. This voltammogram provides a "fingerprint" diagnostic of the compound and is used principally to aid identification.

To provide information on transmitter dynamics, a sample-and-hold circuit (SHC) is set to monitor current at the appropriate oxidation potential on each successive scan. The output from the SHC is displayed on a standard y-t chart recorder, storage oscilloscope, and/or computer. SHC signals are also stored on magnetic disk, and examples are shown later.

1.3. FCV Instrumentation

Because of the nature of the data, their acquisition and manipulation have been the subject of several approaches. A key consideration of the method is the extraction of tiny faradaic signals from much larger charging current signals. This necessitates high-resolution digital subtraction methods. Some laboratories use computer-based systems. Others rely on high-specification digital storage oscilloscopes. Our solution is an amalgam of both approaches. At the same time, it is always a wise precaution to have hard copies of data, and thus chart recorders are also necessary, particularly for the storage of SHC output signals. Figure 2 shows a block diagram of the apparatus used in our laboratory.

1.4. Electrodes

1.4.1. Carbon-Fiber Microelectrodes

Our carbon-fiber microelectrodes are very simple, comprising a single carbon-fiber encased in a glass insulator. Although carbon-fiber comes in a variety of sizes, we use XA-S carbon-fibers from Courtaulds, which have a diameter of 8 μm.

To make an electrode, we first fill a borosilicate capillary tube (1 mm id: Clark Electromedical) with acetone. This has two functions. First, it allows the fiber to float easily into the tube, and second, it dissolves the surface resin coating on the carbon. It is important to fill the tube with only one fiber (the more myopic among us find magnification useful). Once filled, the tube is either blotted to remove the acetone or dried under a hot light. On pulling (we use a Harvard microelectrode puller with heat and pull settings at 8–9 and 5–6, respectively), the fiber is usually completely detached from one barrel, and thus, normally only one "electrode" is obtained from each filled tube.

The electrode is then inspected under a high-power microscope (100–200× magnification is sufficient) to check that there is a good glass–carbon seal and that enough of the fiber extends back into the tube that a good electrical contact can be made later. If the glass–carbon junction is easily vis-

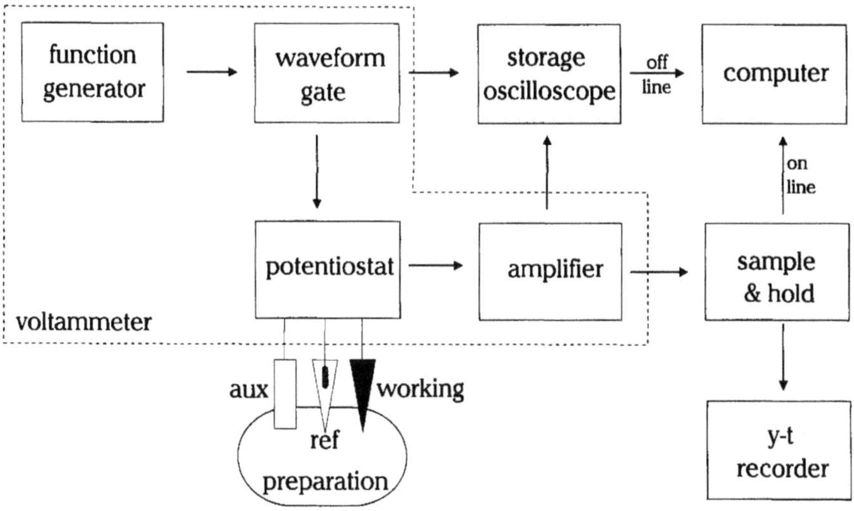

Fig. 2. Instrumentation of fast cyclic voltammetry. The function generator produces a continuous triangular output that is gated into 1½ cycle segments by a gate and then fed to the potentiostat. Current output from the working electrode is amplified, displayed on a digital storage oscilloscope, and stored on disk for off-line analysis on a computer. A sample and hold circuit monitors current at the monoamine oxidation potential, and this signal is displayed on a chart recorder and onto a computer.

ible, then there is probably a poor seal, and the electrode will be noisy. If the seal is so good that it cannot be seen, then the electrode will be quiet and stable, but it will be difficult to know where to cut the carbon fiber. The ideal result is somewhere in between. With appropriate adjustment of heat and pull settings, this can be readily achieved.

If the electrode passes muster at this stage, then the length of carbon exposed at the tip should be trimmed. We use a pair of jeweller's forceps under the guidance of a micromanipulator and a photographer's cable release to crush the fiber to an exposed tip of 20–60 μm, depending on whether the electrode is to be used for unit recording or not. Longer tips give higher electrochemical signal-to-noise ratios, but more easily pick up multiple cells in unit recording mode.

A single thread of copper (we have found multistrand telephone wire to be ideal) is then dipped in electrically con-

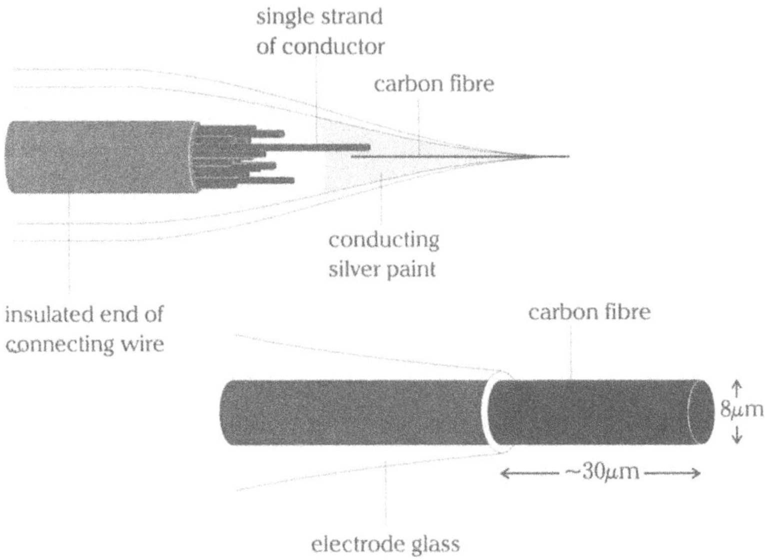

Fig. 3. A carbon-fiber microelectrode. **(Top)** Part of a carbon-fiber microelectrode. A single strand of a multistrand copper wire makes connection via conducting silver paint with the carbon-fiber. **(Bottom)** Panel shows the cut tip of the microelectrode surrounded by the glass insulation.

ductive silver paint (RS Products) and thrust as far into the electrode tip as possible to make a good contact. The silver paint is given some time (preferably overnight) to dry before connection to the voltammeter headstage and initiation of scans. The wire can be kept in place by applying cyanoacrylate glue (superglue) to the open end of the electrode. A cambion pin soldered onto the end of the wire is used to make connection with the voltammeter. These wires should be as short as possible, since they can act as electrical aerials. Figure 3 shows a schematic representation of a microelectrode and its tip geometry.

The only pretreatment to which these electrodes are subjected is a brief (ca. 5–15 min) period in which scans are applied 16 times/s with the electrode in the artificial cerebrospinal fluid (ACSF). They are then left for 10–20 min to "settle" before use. This process ("frying") reduces the background drift that is associated especially with brand-new electrodes: We do not believe that it alters the properties of the electrode, so

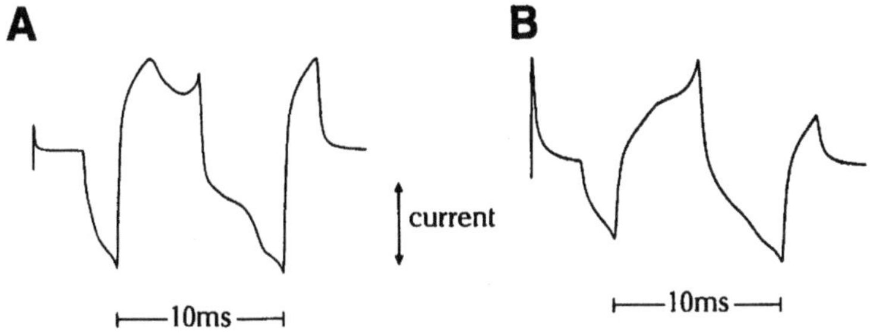

Fig. 4. Good and bad carbon-fiber microelectrode signals. Output (charging current) signals in buffer from **(A)** an electrode with high sensitivity and stability. This shows a rather squared, capacitative signal. **(B)** An insensitive, unstable electrode signal. This shows a much more resistive profile. Voltammetric parameters are as in Fig. 1.

much that it speeds equilibration. At this point, one should look closely at the shape of the charging current signal. This will change substantially during this pretreatment, especially in new electrodes. With experience, it is possible to predict with a good degree of accuracy the likely sensitivity of an electrode from the shape of its charging current signal. This is rather difficult to convey in words, so good and poor electrode signals are shown in Fig. 4 for illustration. We are currently working on ways to express these differences mathematically. If the shape is appropriate and the noise level (SHC output) is acceptable, then the electrode should be workable.

Initially, we made our carbon-fiber microelectrodes with a view to discarding them after the day's experiment. We now take a more "environmentally sensitive" approach and reuse good electrodes. Some, with appropriate care, can be used every day for weeks. Apart from handling the microelectrode carefully, there are other ways to prolong its life. For instance, leaving an electrode (tip) in distilled water between experiments particularly retards deterioration. Rather to our surprise, we have also found that the electrodes can safely be sonicated if the tip becomes encrusted with brain tissue (as it often does on repeated implantation). Together, these treatments help preserve the rapid response times of electrodes, which is of particular importance if uptake kinetics are to be measured. On a few occasions, carbon-fiber microelectrodes have even been autoclaved without apparent damage.

1.4.2. Reference Electrodes

A silver–silver chloride (Ag/AgCl) reference electrode is used. There is no particular type that we have found superior to others. Typically, we use a small chlorided silver disk (0.5-mm diameter) or cylinder enclosed in a plastic pipet tip. The tip is filled with ACSF to act as a salt bridge. The use of ACSF rather than NaCl or KCl minimizes electrode drift by preventing junction potentials between the bridge and the slice superfusate.

1.4.3. Auxiliary Electrodes

The auxiliary electrode is typically a stainless-steel wire attached to a similar binding post on the top of the slice chamber.

1.4.4. Stimulating Electrodes

The stimulating electrodes used are of a parallel-bipolar design, fabricated by sticking two epoxy-insulated tungsten microelectrodes (AM Systems, Seattle, WA) together. Experience has shown that a tip separation of 100–200 µm is most suitable. The tips are tapered to produce a fine (8–12°) point, allowing them to be inserted into the slice with minimal trauma, and are gently sanded with the finest grade abrasive paper to reduce the nominal impedance of the electrode and ensure that the required amount of current is able to pass.

1.5. What Compounds Can Be Measured by FCV in Brain Slices?

FCV is most commonly used in brain slices to measure the stimulated release and uptake of the three major monoamine neurotransmitters, dopamine (DA), noradrenaline (norepinephrine, NE) and serotonin (5-hydroxytryptamine, 5-HT), and it is to these compounds that we will restrict our discussion. Figure 5 shows representative CVs of these compounds. Several (often unidentified) compounds may contribute to the background current signal, but will not be discussed here. These include uric acid, ascorbic acid, NADH, ATP, and maybe even nitric oxide.

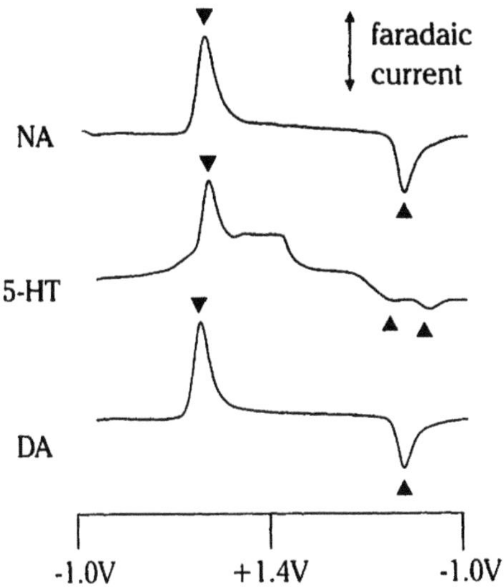

Fig. 5. Redox signals of norepinephrine (noradrenaline, NA), serotonin (5-HT), and dopamine. Background current-corrected redox signals (faradaic current only) of (top to bottom) NE, 5-HT, and DA. Representative signals showing the oxidation and reduction peaks are shown, scaled for clarity. Downward-pointing triangles show the position of oxidation peaks, whereas upward-pointing triangles mark reduction peaks.

1.5.1. Multicomponent Signals

It is not currently possible with FCV to measure selectively the concentration of one monoamine when others are also present at detectable levels. This is, in large measure, a result of the high-voltage scan rates integral to FCV, which tends to merge oxidation peaks: Chemical resolution is deliberately sacrificed for temporal resolution. Nevertheless, a number of approaches have been proposed that may ultimately enable this to be achieved. Simply, it is suggested that a compound peak may be mathematically resolved into its component peaks. This method assumes peaks to be Gaussian (Gonzales Mora et al., 1988), a valid assumption for the peaks of differential pulse voltammetry (DPV), but not of FCV (*see* Fig. 5). Modifications of this approach (Gonzales Mora et al., 1990) have yet to be truly validated.

The second approach to the problem of separation of amine mixtures involves modification of the electrode surface. The difficulty here lies in the need to exclude all but one of three very similar molecules, all cationic at physiological pH, nor do the monoamines exhibit any great difference in molecular size. Clearly, the ability to separate the components of the mixture by using a physical barrier is hindered by the similarity of the molecules. Although the separation of transmitters from their deaminated metabolites (by their exclusion) has been reported (Brazell et al., 1987), the amines cannot as yet be separated from each other by this method.

The third approach to the problem of resolution (and the one we mainly use) is to sidestep the issue and record in the several nuclei with an overwhelming predominance of one amine over the others *(vide infra)*. To some degree, this might be considered a coward's approach. We prefer to think of it as pragmatic: If chemical selectivity is a higher priority than temporal resolution for the experimenter, then another voltammetric method (such as DPV) would probably suit the experimental objectives better.

Lest this inadvertently convey too much of a laissez-faire attitude, we should state that the goal of chemical as well as temporal resolution is highly desirable and is being actively sought by this laboratory along with many others. Our best attempt so far has been only a partial resolution of DA and 5-HT by the use of reduction rather than oxidation peaks (Stamford et al., 1990). The goal of a "DA-selective" or "NE-selective" electrode is still some way off.

1.6. Signal Identification

Although this is a general issue for electrochemical techniques, it is nonetheless worth reiterating that voltammetry cannot be used alone to identify the compound detected. It is necessary in any new nucleus to characterize fully the compound detected. This should be done by answering as many as feasible of the following questions:

- Is the known sensitivity of the electrode enough to detect the likely levels released?
- Does the detected signal show electrochemical identity with its purported constituent?

- Is the detected signal anatomically specific?
- Does the signal change in a predictable manner to a battery of drugs?
- Can the signal be eliminated/enhanced by substrate-specific enzymes?
- Does another independent method give the same answer?

If one can confidently answer "yes" to each question, it is highly likely that the detected compound is as purported. Of the above tests, the most important is the pharmacology. Often, the amount of release will be too low to prevent the acquisition of a clean voltammogram, and thus the pharmacology is the primary diagnostic approach. We should stress that several drugs of differing pharmacology should be used. "Identifications" based on single drugs often lead to later retractions.

2. Voltammetry in Brain Slices

Currently, the measurement of stimulated endogenous release of monoamines is by far the most common use of FCV. This can be done both in vivo and in vitro (brain slices). The release of the neurotransmitter and its uptake from the extracellular space can be measured, after electrical stimulation, with a time resolution of milliseconds, and therefore, it is possible to monitor the kinetics of release and uptake in "realtime." Several studies have characterized the release and uptake processes of DA, NE, and 5-HT in various nuclei using brain slice preparations. Figure 6 shows a stylized example of simultaneous "real-time" measurement of amine release and uptake.

2.1. Why Brain Slices?

The measurement of stimulated neurotransmitter release by FCV was first performed in anesthetized animals, and the extension of the technique to brain slices has brought a number of major benefits. Foremost among these is the ability to control the physiological environment. Parameters, such as the ionic composition, pH, temperature, and degree of oxy-

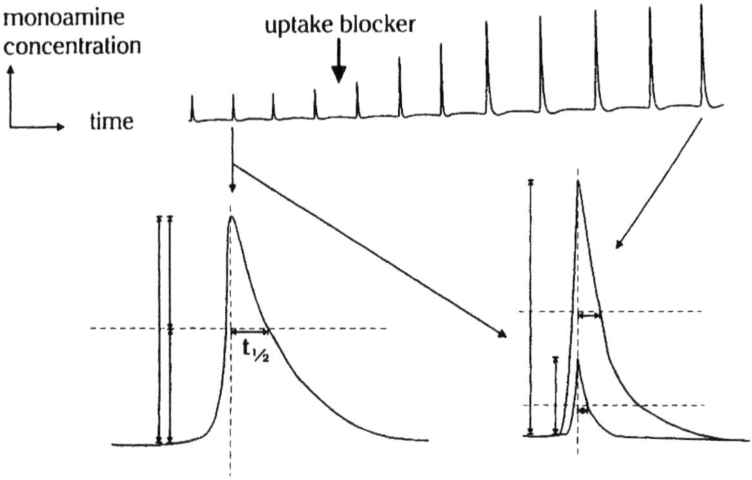

Fig. 6. Simultaneous "real-time" measurement of amine release and uptake. **(Top)** Schematic sample-and-hold output signal of stimulated monoamine release. There are four "control" stimulations before the hypothetical uptake blocker is added. The uptake blocker increases amine release and also slows its removal by uptake. The lower panels show the relation between amine release and uptake in more detail. **(Bottom left)** Expansion of the second stimulation event. Amine release is measured as the difference between the prestimulation level and the peak concentration attained. A simple measure of uptake is the "half-time," the time it takes the peak extracellular concentration to fall by half. **(Bottom right)** Illustration of the effect of the uptake blocker on the last stimulation event. The predrug stimulation is shown for comparison. The uptake blocker increases amine release and also prolongs the uptake halftime. The simultaneous recording of these two measures allows the relation between these functions to be assessed at the same anatomical site.

genation of the extracellular fluid, can be readily varied. Moreover, the concentrations of drugs in the bathing medium and, therefore, presumably also at the site of action can be easily manipulated. Consequently, it is possible to evaluate drug effects quantitatively: Agonist concentration–response curves can be constructed in the absence or presence of antagonists, and agonist EC_{50} values and antagonist pK_B and pA_2 values can be calculated. Furthermore, the direct effects of drugs can be observed without the distraction of actions mediated peripherally or at other locations within the central nervous system.

Studies in brain slices can be performed in the absence of anesthetic agents, thereby avoiding the possibility of unwanted drug interactions (this, of course, makes brain slices ideal for investigating the actions of anesthetics). The use of brain slices also enables a wide variety of drugs that cannot easily be studied in anesthetized animals to be examined. Drugs that might prejudice the survival of the whole animal, such as tetrodotoxin, can be tested, as can those whose access to the central nervous system is limited by poor penetration through the blood–brain barrier, high first-pass hepatic metabolism, or extensive plasma protein binding. Since the positioning of electrodes in brain slices is performed under visual control, electrodes can be placed accurately in very small nuclei that are difficult to access stereotaxically in the whole animal, such as the suprachiasmatic nucleus.

2.2. What Species?

For most investigations using FCV in brain slices, male animals are preferred. This is to eliminate any changes in the brain that may occur during the estrus cycle. As with other techniques used in brain slices, the vast majority of studies using FCV have used the rat (mainly Wistars and Sprague-Dawleys). Rats are probably the most widely studied laboratory species, and their neuroanatomy is well understood.

FCV has also been used in guinea pig brain slices. The guinea pig has been chosen in particular for studies on the serotonergic system, since unlike the rat, the guinea pig has the same serotonergic terminal autoreceptor subtype as the human brain, i.e., the 5-HT_{1D} receptor. Slicing is more difficult in the guinea pig and should be carried out at a slower blade advance speed, since the brain has a less firm consistency than the rat brain.

In our laboratory, we have also carried out a handful of experiments using brain slices from hamsters, and have been able to measure striatal DA release and possibly 5-HT release in other areas. Although we cannot offer any helpful methodological insights, it is clear at least that the hamster is a viable species for FCV in brain slices.

Table 1
Artificial Cerebrospinal Fluid (ACSF) Composition by Salt and by Ion

		mM
Salt		
Sodium chloride	$NaCl$	124
Potassium chloride	KCl	2
Potassium dihydrogen orthophosphate	KH_2PO_4	1.25
Magnesium sulfate heptahydrate	$MgSO_4 \cdot 7H_2O$	2
Sodium hydrogen carbonate	$NaHCO_3$	25
Calcium chloride	$CaCl_2$	2
D-Glucose	$C_6H_{12}O_6$	11
Ion		
Na^+		149
K^+		3.25
Ca^{2+}		2
Mg^{2+}		2
Cl^-		130
HCO_3^-		25
SO_4^{2-}		2
$H_2PO_4^-$		1.25
D-Glucose		11

2.3. Artificial Cerebrospinal Fluid

The composition of the ACSF used is shown in Table 1. Table 1 lists the salts used to prepare the ACSF and shows its resultant ionic composition. The main difference between the ionic composition of the ACSF and that of true CSF is in the concentrations of Ca^{2+} and Mg^{2+}: These are *prima facie* somewhat higher in ACSF. It should, however, be pointed out that the concentration of free Ca^{2+} in physiological buffers is lower than the total Ca^{2+} concentration owing to the formation of complexes with hydrogen carbonate and phosphate anions.

In order to prevent precipitation of insoluble calcium salts (carbonate, sulfate), the calcium chloride should be added only after the other salts have been dissolved. The prepared ACSF is then gassed with 95% O_2/5% CO_2 at 32°C for at least 45 min before use. Our experience has shown that, for the majority of likely applications, the addition of amino acid precursors to the ACSF does not provide any perceptible benefit; therefore, this is not routinely performed.

2.4. Preparation of Brain Slices

Animals are sacrified by cervical dislocation: No prior anesthesia is used. Hence the slices are truly drug-naive, especially important for work with anesthetics. The brain is swiftly, but carefully removed and thoroughly irrigated with ice-cold ACSF to lower the tissue temperature during dissection. Parallel coronal cuts are then made to define a basic tissue block around the target nucleus. For large bilateral nuclei (such as the caudate-putamen—CPu), a rostrocaudal incision is then usually made along the length of the midline, and one hemisphere discarded. Further trimming cuts remove adhering dura or pia mater.

We use relatively small tissue blocks (approx 7 × 7 × 7 mm) with all meninges removed. The base of the tissue block is dried by touching it lightly on a paper towel and then dropping it onto the preglued vibratome chuck. (A useful tip with some soft blocks is to touch each side with a dab of cyanoacrylate adhesive before removing the excess with a paper tissue. This provides a very thin supportive glue "shell"). Sequential trimming slices are then cut through the overlying structures until the target nucleus is visible. At this point, a 350-μm section is cut and transferred to the recording chamber. A fine paint brush is used to steady the block if necessary and to manipulate the slice when it is cut.

In an attempt to use the minimum number of animals, we often save the forebrain in a beaker with chilled ACSF while slicing the hindbrain. This needs to be kept in the refrigerator. Ideally, it is sliced within 10 min, but on occasion, delays of up to 45 min have occurred and still yielded viable slices. This enables several experimenters to share the same animal.

After the slice is placed in the bath, we allow an hour for equilibration. During this time, the electrodes are positioned in the superfusing ACSF with the voltammeter scanning to allow the recording electrode as much time as possible to equilibrate. Slices removed later from the slice saver (where they are kept at room temperature) are usually allowed a shorter time (at least 15 min) for equilibration.

The temperature of slicing is, in our view, critical. The ACSF needs to be around 0 ± 1°C, visually identified by the presence of ice floes in the ACSF. This is for both biochemical and physical reasons: Low temperatures reduce energy demands and also stiffen the tissue, making it easier to slice. Depending on the surface area of the vibratome chamber and the laboratory ambient temperature, warming of up to 8°C can occur during the time needed for slicing. This can be sufficient to cause "mushiness" or distortion of the block. The temperature for cortical slices is less critical, but the more gelatinous texture of the brainstem means that the firming effect of truly ice-cold ACSF is vital. Sometimes we have found it necessary to exchange the vibratome fluid (very rapidly) with colder ACSF midway through a sectioning sequence. Another trick for blocks that are slicing very poorly is to stop the advance of the vibratome for a few seconds before allowing it to proceed further. We use a vibration amplitude of at least 1 mm, an oscillation frequency of 50 Hz, and a blade advance rate of at least 10 mm/min.

2.5. Temperature and Superfusion Flow Rate

In preliminary experiments at 37°C, it was difficult to detect evoked DA release in striatal slices because of the dominance of uptake. However, lowering the temperature reduces the efficiency of the uptake process and thus allows a greater proportion of released transmitter to escape from the synapse. Slices are therefore generally maintained at a temperature of 32°C. This lower than physiological incubation temperature has two main benefits. First, brain slices survive longer and in a healthier state at lower temperatures. Second, measurable release can be evoked in the absence of uptake blockers.

The ACSF is pumped through the incubation chamber using standard syringe pumps. These must be capable of producing a smooth flow, since pulsation would cause vibra-

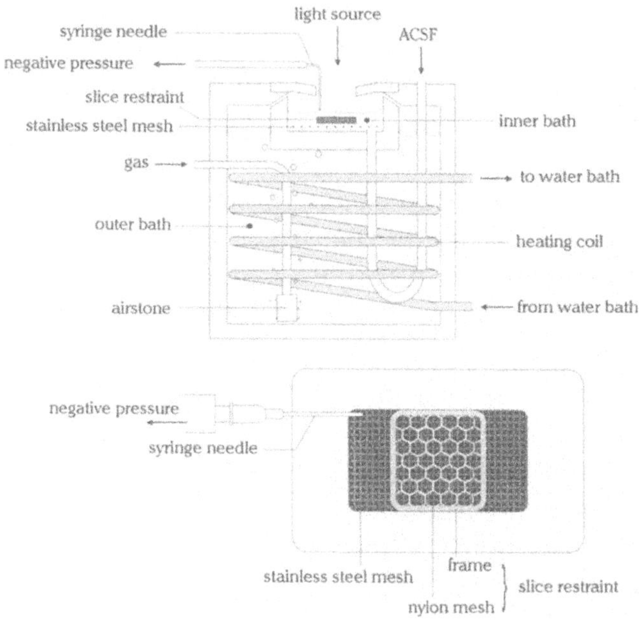

Fig. 7. The brain slice recording chamber. **(Top)** Inflow and outflow connections of the slice chamber. **(Bottom)** Top view of the inner chamber showing the stainless-steel mesh on which the slice rests and the nylon mesh that holds the slice in place. For further details, *see* text.

tion of the electrodes and fluctuations in the depth of the superfusate, thereby generating excessive noise in the electrochemical system. Flow rates in the range of 0.5–2 mL/min have been used. These play a major role in determining the equilibration time for drugs: A fast flow rate corresponds to a short equilibration time. We typically superfuse at 1 mL/min.

2.6. The Recording Chamber

For FCV in brain slices, a submersion type slice chamber is generally used, an example of which is shown in Fig. 7. The brain slice lies on the stainless-steel mesh in the inner bath and is anchored by the slice restraint above, which in turn is held in place by friction. The slice restraint consists of a taut nylon mesh stretched across the lower surface of a frame (*see* Fig. 7, lower panel).

The outer bath is responsible for the maintenance of the environment of the inner chamber, and is filled with distilled water and warmed by the heating coil. ACSF, which has been saturated with 95% O_2/5% CO_2 at 32°C, enters the inner bath via a loop of tubing passing through the outer chamber. This loop allows equilibration of temperature between the outer bath and ACSF. This and the direct warming effect of the outer bath fluid on the inner bath maintains the inner part at the desired temperature. It is nevertheless advisable to check the chamber temperature periodically to ensure constancy.

ACSF is removed from the inner bath and drained into a fluid trap by negative pressure from a vacuum pump. The ACSF is drawn from the fluid surface via a syringe needle positioned at the required inner bath fluid level. It is important that the syringe needle is bevelled to a degree that there is continuous (nonpulsatile) removal of ACSF. To achieve this, sanding down of the needle is often required.

The gas (95% O_2/5% CO_2) is bubbled through the outer bath fluid. This warms and humidifies it (and also serves to show the rate of gas flow when using chambers made of transparent material). It then passes over the surface of the ACSF in the inner bath to maintain the saturation of gas in the inner bath fluid.

Water from a thermostatically heated bath is circulated through the heating coil. Thermal equilibrium is established between this and the outer bath. This means that temperature fluctuations owing to the heater switching on and off are "smoothed out" in the inner bath. Another advantage is that two or more identical slice chambers can be heated from a single water bath.

2.7. Electrode Positioning

Both working and stimulating electrodes are mounted in micromanipulators attached to a large (30–50 kg) steel platform by magnetic bases and positioned in the slice using a binocular microscope with a maximum magnification of 20×. Depending on the slice chamber design, some systems are lit from below by a fiber-optic source, and others by a filament light from above. We have experimented with colored filters to try to improve visualization of structures without any obvious improvements. Lighting from above may be slightly superior, but can warm the bath and is thus turned off when no electrode manipulation is being performed.

The recording and stimulating electrodes are placed at a depth of approx 100 μm below the slice surface. For small nuclei, the recording electrodes are usually placed just at the edge of the area of interest to allow a maximum number of possible recording positions within the nucleus if the initial penetration proves unsatisfactory. We find that our stimulating electrodes can vary in their properties, and with some, it is necessary to place the recording electrode well away from the carbon-fiber (200–300 μm); with others, the best results are obtained with the recording electrode almost between the stimulating electrode tips.

2.8. Electrical Stimulation

Transmitter release can be elicited using either monophasic or biphasic pulses: Both evoke equivalent release, although the latter may cause less electrolytic tissue damage. Both constant-voltage and constant-current stimulations have been used to evoke release: We find that a 20-V constant-voltage pulse evokes a similar response to a 10-mA constant-current pulse (stimulated DA release evoked by single 0.1-ms pulses in CPu slices) with our stimulating electrodes. Typically, pulse widths of 0.1 or 0.2 ms have been employed together with amplitudes of 20 V or 10 mA.

For a fixed number of pulses, the frequency–response curve generally exhibits a characteristic bell shape with an optimum frequency around 50–100 Hz for DA, NE, and 5-HT. At frequencies below this, release is lower owing to the autoinhibition. At higher frequencies, the lower release is probably the result of conduction failure, since some of the applied pulses fall during a refractory period and so fail to initiate action potentials. The magnitude of evoked release increases with increasing number of pulses, although autoinhibition becomes increasingly evident with longer (>500 ms) stimuli.

Transmitter released by the first few pulses of a long train is able to activate presynaptic autoreceptors and thereby inhibit release to the later pulses. This autoinhibitory effect is most prevalent with low-frequency, long-duration trains.

The interval that must be allowed between stimulus trains in order to ensure that release is reproducible is determined by the intensity of the stimuli. In the CPu, single-pulse release may be evoked at 2-min intervals for a period of 9 h without any deterioration in the response. However, in the locus ceruleus (LC), an interval of 10 min must be allowed in order to obtain reproducible 50-pulse (0.2-ms, 10-mA pulses applied at 50 Hz) release. For intermediate stimuli, an interval of 5 min is usually sufficient.

The choice of stimulation parameters is largely determined by the nature of the experiment. When autoreceptor agonists are to be investigated, a stimulus is chosen such that the released transmitter is unable itself to activate the receptors and thus compete with the exogenous agonist: In practice, this means a "pseudo single pulse" stimulus of very short duration (<200 ms), if not a single pulse itself. However, when antagonists are to be studied, the activation of receptors by the endogenous agonist is required, and so stimuli of long duration (>1000 ms) are employed.

2.9. Slice Viability

The factors that most appear to affect slice viability are:

- Poor slicing technique, usually owing to the ACSF in the vibratome being too warm;
- Any brief period of slice drying or cessation of oxygenation;
- Excessive trauma to the tissue when placing the restraining upper mesh;
- Excessive number or depth of penetration sites in the tissue while searching for release; and
- Electrolytic damage caused by intense stimulation of the tissue.

Until recently, only the gross histological appearance of the slice, in particular, the development of swelling, gave any indication of impaired viability. We now frequently use simultaneous electrophysiology as a marker of slice viability. This is very useful when having difficulties obtaining amine release. If there are numbers of regularly firing, stable cells, the slice is healthy, if it is silent or has many dying cells, the use of a saved slice or redissection *(vide infra)* is indicated. In many respects, the electrophysiology is a more sensitive index of slice health than the voltammetry. For instance, stable drug effects on stimulated amine release may be obtained in preparations with few or unstable cells.

On some days, one is simply unable to obtain usable amine release at all. This can be because of many factors, both biological and electrochemical. For instance, poor stimulating or recording electrodes can contribute as much as bad slices or contaminants in the bath. Often, buildup of bacterial growth in the tubing (ACSF at 32°C is a fine growth medium) can indicate the latter. Under such circumstances, an acid wash of the bath and its tubing is needed. Opinions on this subject vary even within the laboratory. Some of us perform these regularly (once a week), and others only after running into experimental difficulties.

All tubing and baths are cleaned nightly with distilled water (about 500 mL) and then air is pushed through at high pressure. If drugs dissolved in unusual vehicles have been used, it is desirable also to clean the chamber with the vehicle before the distilled water. Sometimes an excessive number of floating particles are seen, which is a definite indication for an acid wash. This consists of 20–40 mL of 0.1–1*M* HCl (it does not need to be that precise) injected slowly and followed by a large amount (1–2 L) of distilled water. The pH in the bath is checked prior to placing the next slice. As a caveat on this point, we have found that not all of the connectors inside the baths have proven to be acid-resistant! This is an important specification when constructing a brain slice chamber.

2.10. The "Slice Saver"

When slicing a given nucleus, we often retain at least one slice from either side of our chosen section in a "slice saver" (*see* Fig. 8). This is little more than a perspex lidded box with the slices resting in a tea strainer, probably a rather

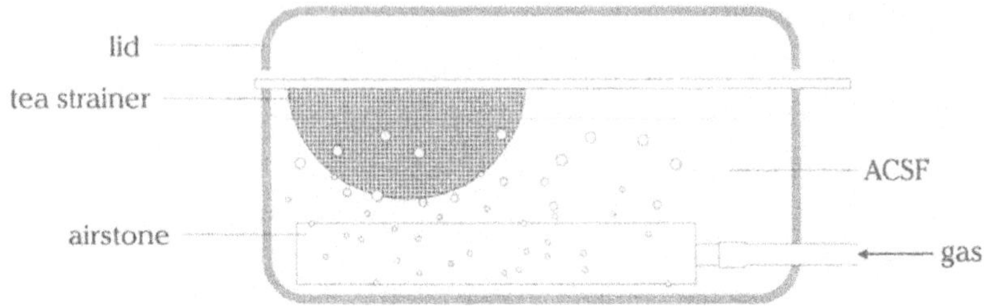

Fig. 8. A slice saver. Brain slices are placed in the tea strainer that is immersed in artificial cerebrospinal fluid. The chamber is bubbled with 95% O_2/5% CO_2 at room temperature.

British solution! The box is large and therefore contains a reasonable reservoir (500 mL) of room temperature ACSF. An oxygenator rests on the bottom and provides circulation of the fluid as well as oxygenation. The only problem we have experienced with its use occurs if the bubbling is too vigorous, since the slices can be buffeted and pushed on top of each other.

Most experiments are quite long (4–6 h), and therefore we do not always use any of the saved slices. However, it is handy for the following reasons: accidental injury to the slice in the chamber (such as letting the fluid run out) or miscalculation of the optimal section for a particular nucleus. For instance, the view under the microscope may show that the slice selected is not ideal: We are then able to retrieve an adjacent one quickly and easily, immediately or perhaps later after failing to obtain adequate amine release. Often the use of the slice saver can prevent the need for further dissections. When we did extra experiments, we have found the slices to be viable for at least 6 h in the saver, and then capable of producing consistent and adequate amine release for an experiment lasting several hours.

2.11. Electrode Calibrations

At the end of an experiment the recording electrode is withdrawn from the slice into the superfusing ACSF. When the signal has stabilized, electrode calibration is performed with concentrations of amine similar to the concentrations being measured in the slice. This enables some quantification of the amount of amine release measured. If the same recording electrode has been used for some time, it is not necessary to calibrate every day, since sensitivity changes only very slowly with time.

Our approach is to add the calibrant to the ACSF and perfuse the solution through the bath. This is not ideal because it exposes the electrodes to the solution for periods of some minutes. This can be a problem with known electrode poisons, such as 5-HT. A better approach and one we are beginning to adopt is to use a flow injection cell. In this, the electrode is exposed to a brief (1–5 s) "pulse" of calibrant. In addition to minimizing the danger of electrode poisoning, this method also has the advantage of calibrating the electrodes in a more physiologically meaningful time frame, one more consistent with the time-course of amine release and uptake events. The only real disadvantage is that it takes practice to be able to position the electrode in the flow stream without breaking it.

3. Applications of FCV in Brain Slices

In brain slices, electrical stimulation is nearly always performed locally and, thus, probably releases several compounds simultaneously. Therefore, to investigate selectively the release of a transmitter, it is necessary to choose a nucleus that contains substantially more of one of the monoamines than of the others. Under these conditions, the mixture released by stimulation has only one major monoamine component, and the signal detected by FCV is predominantly the result of that monoamine.

The matter is, however, complicated by the differing sensitivity of the recording electrodes toward the monoamines. Our microelectrodes exhibit greatest sensitivity toward 5-HT and least toward NE. It is therefore possible to monitor 5-HT clearly when its level is only marginally greater than those of DA and NE. Conversely, in order to be able to detect NE, its level must be substantially higher than those of DA and particularly 5-HT.

The measurement of stimulated DA release has been demonstrated in slices of the CPu (Palij et al., 1990), NAc, and olfactory tubercle (TuO: Trout and Kruk, 1992). In these nuclei, the levels of DA are substantially higher than those of NE and 5-HT.

Whereas there are several nuclei rich in DA, but deficient or devoid of NE and 5-HT, there are none that contain high levels of 5-HT yet low levels of DA and NE. Although *prima facie* a problem, the ability to detect stimulated 5-HT release is greatly enhanced by the much greater sensitivity of the electrodes for 5-HT. Consequently, although local electrical stimulation may only result in marginally greater release of 5-HT than of the catecholamines, the detected signal is predominantly the result of 5-HT. It has therefore been possible to demonstrate stimulated 5-HT release in slices of the suprachiasmatic (SCN), dorsal raphe (DRN) (O'Connor and Kruk, 1991), and ventral lateral geniculate (vLGN) (Davidson and Stamford, 1994) nuclei.

Although a number of nuclei contain high levels of NE, these invariably also contain lesser, but nevertheless significant amounts of DA and 5-HT. Since the sensitivity of the electrodes to NE is less than DA and 5-HT, there is a distinct likelihood that the electrochemical signal recorded following local electrical stimulation may exhibit discernible contributions from DA and particularly 5-HT. Despite these difficulties, the measurement of NE release has been demonstrated in slices of the bed nucleus of stria terminalis, ventral division (BSTV; Palij and Stamford, 1992), and locus ceruleus (LC) (Palij and Stamford, 1994).

Obviously, when stimulating a nerve terminal area, pseudo-action potentials travel down the axon, resulting in release of a given compound(s). In a cell body area, the situation is slightly more complicated. For instance, it is not known whether the detected transmitter is released from terminals innervating that particular nucleus, from dendrites in that nucleus, from the soma, or from some combination of these sites. This point should be especially noted if working in the substantia nigra or ventral tegmental area (VTA), where DA cell bodies are innervated by 5-HT terminals and, to a lesser extent, in the LC where NE cell bodies have 5-HT afferents. It is also important to remember that some nuclei may contain cell bodies and axon terminals of the same amine. For example, in the DRN, there are 5-HT cell bodies and also 5-HT terminals from other 5-HT cell body regions. This may be important if one is looking at the effects of agonists or antagonists at the 5-HT autoreceptors. The somatodendritic autoreceptor is of the 5-HT_{1A} subtype, and the terminal autoreceptor is the 5-HT_{1B} in the rat.

The following sections describe some of the areas in the rat of which we have direct experience.

3.1. Dopaminergic Nuclei

3.1.1. The Caudate Putamen

The DA fibers innervating this nucleus come from three areas. The VTA (A10 region) innervates the caudo-ventromedial CPu. The substantia nigra (A9) innervates most of the CPu. The retrorubral area (A8) innervates only very small zones in the center of the nucleus. There is also a rostrocaudal gradient of 5-HT, with more terminals in the caudal regions of the CPu, although 5-HT has never been measured by FCV in the CPu. It does not really matter which way you slice the CPu, but if the NAc is important, then slicing ventral to dorsal is better to prevent the NAc from becoming ragged. The CPu has a striated appearance caused by the fiber bundles that course through it. When implanting our stimulating electrode, we normally place the bipolar tip astride a fiber and the recording electrode outside the fiber.

In our experience, the best efflux to a single pulse stimulation is to be found in the rostral dorsolateral CPu (an almost pure A9 terminal region). There is a large difference in the regional response to longer stimulation trains, such as 20 pulses at 50 Hz. In most areas, this evokes only twice as much DA release as a single pulse. Along the medial axis, however, there are discrete areas where the 20-pulse release is often more than six times that on a single pulse. This seems to be because of clusters of A10 limbic DA terminals. Figure 9 shows a rep-

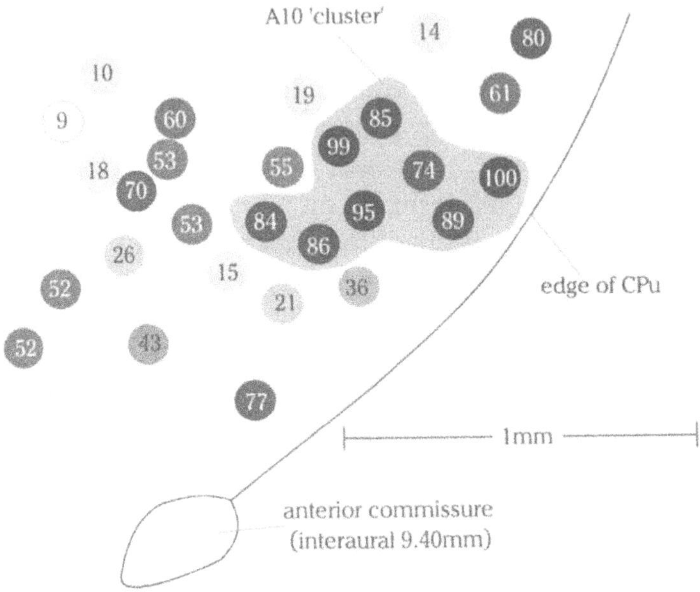

Fig. 9. Examples of typical cluster dimensions. A typical "map" of an individual high DA release cluster showing its size and location within the neostriatum. The numbers are the percentages of the maximum DA efflux obtained on train (20 pulses, 0.1 ms, 50 Hz, 10 mA) stimulation. The boundaries of the clusters are defined arbitrarily as encompassing groups of points with a ratio of 70% or more of the maximum value. The calibration bar illustrates the spatial resolution.

resentative example and illustrates the ability of the FCV method's spatial resolution to answer neuroanatomical questions.

3.1.2. The Nucleus Accumbens

This nucleus is immediately ventral to the rostral CPu and receives its DA innervation from the VTA. It has been divided into a core (immediately medial and dorsal to the anterior commissure) and a shell (further medial and dorsal). Unlike most of the CPu, the NAc seems to be made up of exclusively "high-release sites" consistent with these being related to A10 innervation. One-pulse DA release is slightly smaller in the NAc than the CPu.

Like the CPu, the NAc is a reasonably large structure, and a number of slices can be taken. Rostrally, the NAc starts at around +11.7 mm interaural (Paxinos and Watson, 1986) and extends caudally to around +9.4 mm. We normally take a slice around +11 mm and generally only record in the core, since DA release appears to be better close to the anterior commissure.

3.2. Noradrenergic Nuclei

3.2.1. The Bed Nucleus of the Stria Terminalis, pars ventralis

This nucleus is involved in cardiovascular control and in aggressive, sexual, and parental behaviors. The BSTV has the highest level of NE in the brain, together with relatively low levels of 5-HT and DA. The NE is present in terminals whose cell bodies are mainly in the A1 cell group.

The dissection is similar to that for the CPu and NAc, from the ventral to the dorsal surface. In this way, the bulk of the block lies behind the BSTV and supports the nucleus during the cutting process. At the level of the NAc, the anterior commissure appears as an almost circular structure in coronal section. Posterior to the NAc, the anterior commissure elongates laterally and medially until at the level of the BSTV it forms a structure whose shape is reminiscent of a classical "sergeant-major's moustache" (+8.7 mm interaural). Experience has shown that slices taken at this level (or marginally anterior) yield the best NE release.

We have found that the optimum location for the electrodes is at the point where the anterior commissure narrows. The working electrode should be placed within 50 μm of the anterior commissure at the point of narrowing and the stimulating electrode ventromedial to it. Although we have found that, in our hands, such positioning of the electrodes tends to yield the best release, we have also been able to detect release in other areas of the BSTV.

3.2.2. The Locus Ceruleus

This is the most prominent noradrenergic cell-body group, containing nearly half of all noradrenergic cell bodies in the brain. Although each LC consists of only about 1500 neurons, these project to all parts of the brain. The LC contains a relatively high level of NE, and comparatively lower levels of DA and 5-HT.

In terms of dissection, the block is noticeably less rigid than the forebrain, and it is therefore necessary to take some steps to assist with the cutting process *(vide supra)*. Slicing may be further facilitated by cutting from a lateral surface (from the left surface to the right or vice versa). In this way, the bulk of the tissue helps to support the LC. We have found that slices taken toward the caudal end of the LC, immediately before it begins to contract, yield the best NE release.

The LC is bound dorsally by the fourth ventricle and laterally by the mesencephalic trigeminal nucleus, which appears as a speckled area. Medially, there is no visually distinct delineation. Optimum selective NE release is obtained in the dorsal half of the LC close to the fourth ventricle. When the electrodes are positioned medial or ventral to the LC, it is sometimes possible to detect the release of other electroactive substances that do not exhibit a single sharp reduction peak, and therefore are unlikely to be the result of NE. In the ventral half of the LC, the electrochemical signals obtained during stimulation are often reminiscent of 5-HT contamination.

We have been particularly interested in the LC as a possible site of action of general anesthetics. Figure 10 shows the effect of isoflurane on stimulated NE release in the LC.

3.3. Serotonergic Nuclei

3.3.1. The Dorsal Raphe Nucleus

The DRN is by far the largest of the 5-HT cell-body regions and comprises, in the rat, around 11,500 serotonergic neurons. It projects to virtually all other brain areas, where it appears to exert a tonic modulatory influence, especially with regard to locomotion and the sleep–wake cycle.

The slice that we normally take is at +1.2 mm vs the interaural line where the DRN has a "dove" like appearance. The DRN is reasonably large, and one can probably take up to three 350-μm coronal sections, which include the DRN. We normally cut in a dorsal to ventral direction and tend to

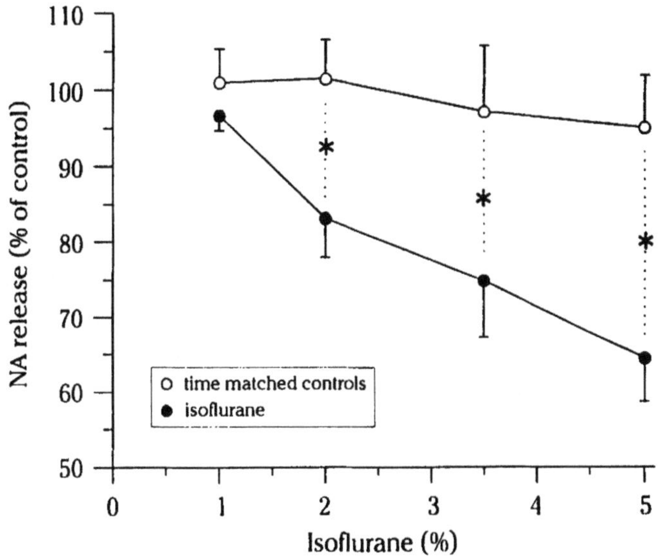

Fig. 10. Effect of isoflurane on norepinephrine (noradrenaline, NA) release in the locus ceruleus. Stimulated NE release (50 pulse trains, 50 Hz, 0.2 ms, 10 mA, every 10 min) in rat locus ceruleus slices. The slice was exposed to isoflurane for 30 min at each concentration. Isoflurane at 2–5% significantly (P <0.05) decreased stimulated NE release relative to time-matched controls.

position the block at an angle so the blade cuts through a corner of the block first. This provides maximum support for the structure (which is in one of the softer parts of the brain). We normally place the electrodes in the center of the structure, since the lateral aspects of the "wings" do contain some NE.

3.3.2. The Suprachiasmatic Nucleus

The SCN is found in the ventral hypothalamus and is a major 5-HT terminal area. It is a very small round bilateral structure found directly above the optic chiasm at the base of the brain (+7.7 mm vs interaural in the rat). Its primary function is as the putative circadian pacemaker or clock.

This is a very difficult slice to take, since the SCN is at the base of the brain and tends to be torn off with the optic chiasm unless one is extremely careful. It is also very small, and only one slice can normally be obtained. Consequently, great care is needed when removing the brain from the animal, particularly not to stretch the optic nerves. Care is also needed when trimming the block, or else the optimum slice may be missed. The best slice to take is where the SCN looks like a pair of small circles resting in an indentation of the optic chiasm. It is recommended that the slice be cut ventral to dorsal to give the SCN maximum support, although care is needed to avoid squashing the nucleus. Best results seem to be obtained when one stimulates the lateral SCN and records in the middle of the structure.

3.3.3. The Ventral Lateral Geniculate Nucleus

The vLGN is a relay station in the visual pathway receiving inputs from the retina and projecting to the visual cortex. However, it may also be associated with the circadian clock, since it has reciprocal connections with the SCN. The vLGN can be divided into a parvocellular (medial) and a magnocellular (lateral) area. Most 5-HT terminals are found in the magnocellular area. There are appreciable amounts of noradrenergic terminals in the LGN, but these are found mostly in the dorsal LGN.

The vLGN is found at the level of the thalamus and, as its name suggests, is lateral to it. If one is slicing coronally, then the LGN should become apparent shortly after the hippocampus becomes visible. The slice we take is at the level where the dorsal and ventral hippocampi have just formed a semicircle around the thalamus (+4.7 mm interaural). We slice dorsal to ventral more out of habit than method. The recording electrode is normally placed just ventral to the intergeniculate leaf (IGL) and in the lateral aspect of the vLGN. This appears to be optimal for release of 5-HT.

4. Simultaneous Electrophysiology and Electrochemistry

A useful feature of the FCV methodology is the ability to combine it with electrophysiological recording. Indeed the first reported usage of the carbon-fiber microelectrodes was not as electrochemical, but as electrophysiological microelectrodes (Armstrong James and Millar, 1979). We have recently described in detail a method for combining electrophysiology and stimulated monoamine efflux with a single carbon-fiber recording microelectrode using commercially available instrumentation (Stamford et al., 1994). Described simply, a low-noise voltage follower amplifier is used to give a record of the voltage appearing at the microelectrode tip between voltammetric scans. When the switching transients are processed out and the signal amplified and discriminated, an electrophysiological record with only tiny gaps of 50 ms twice/s (each FCV scan) can be produced. Figure 11 shows the basic circuitry.

There is no interaction between the two techniques; application of the voltage waveform to the CFM alters neither firing rate nor spike shape. We have found that with care it is possible to perform quantitative pharmacology experiments lasting several hours, examining both stimulated efflux and stimulated or spontaneous firing rate. This has enabled the comparison of the effects of drugs on cell firing and transmitter release and the correlation of these distinct indices of neuronal activity (Jorm and Stamford, 1993). Figure 12 shows the relationship between inhibition of LC cell firing and NE release by the α_2 agonist dexmedetomidine. The absence of a clear linear relation may indicate that a greater receptor reserve exists for the receptors controlling cell firing or that they are of a different subtype. The figure also illustrates the value of the combined methodology, since the electrophysiological and electrochemical phenomena occur at exactly the same site.

The combination of the two methodologies is inevitably more tricky, and it is sometimes difficult to optimize both components. In general, it is more difficult to obtain stable electrophysiological than electrochemical recordings. The electrode tip needs to be cut to to a smaller size than usual (20–30 µm) to avoid excessive multiple unit recording. It is best to place the stimulating electrode first (and allow time for the tissue to settle around it) and then "walk" the record-

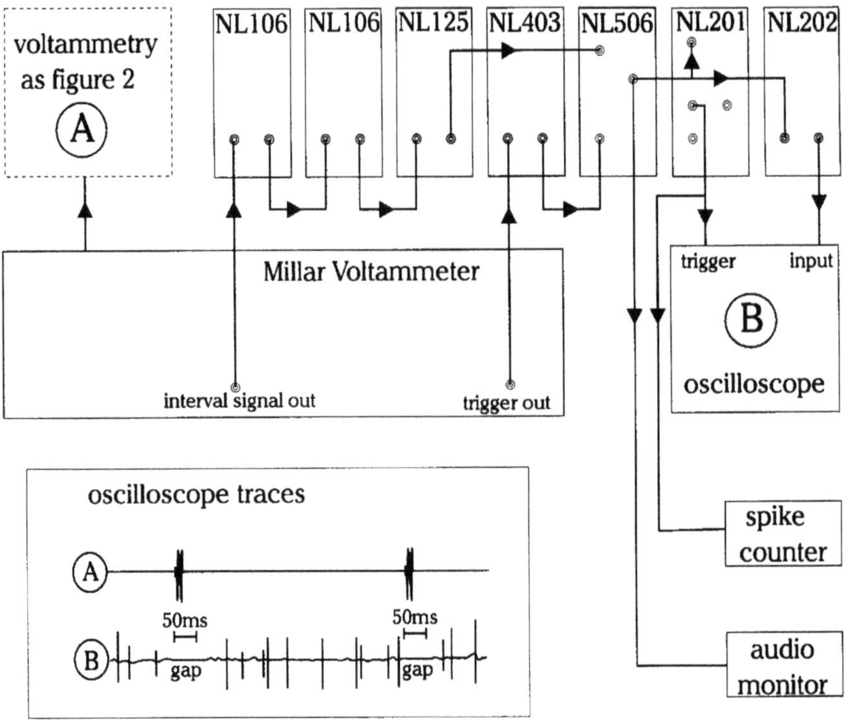

Fig. 11. Circuitry for simultaneous electrophysiological and electrochemical recording. The "interval signal" of the voltammeter is first fed to a Neurolog NL 106 module at full gain, used to bring the spike signal up to a level of a few millivolts peak-to-peak. The signal is taken to an NL 106 amplifier for further gain and offset before being fed to an NL 125 filter. This is set to give a further restriction of the signal bandwidth to 500 Hz–5 kHz. An NL 506 analog gate module (driven from an NL 403 delay-width, which, in turn, receives a pulse from the trigger output of the voltammeter) is used to blank out the switching transient. The NL 201 and NL 202 are used to differentiate individual spikes and to delay the signal to oscilloscope B, respectively. The result ("oscilloscope traces") is a spike signal **(B)** with flat "gaps" lasting 50 ms, occurring at each voltammetric scan **(A)**.

ing electrode in an arc around the optimal likely release position while looking for cells. Although it is possible to place or move the stimulating electrode later, this must be done very carefully or the tissue distortion will result in loss of the discriminated cell. Furthermore, the electrical perturbation of the milieu should be kept to the minimum compatible with the experimental aims (minimum necessary stimulation cur-

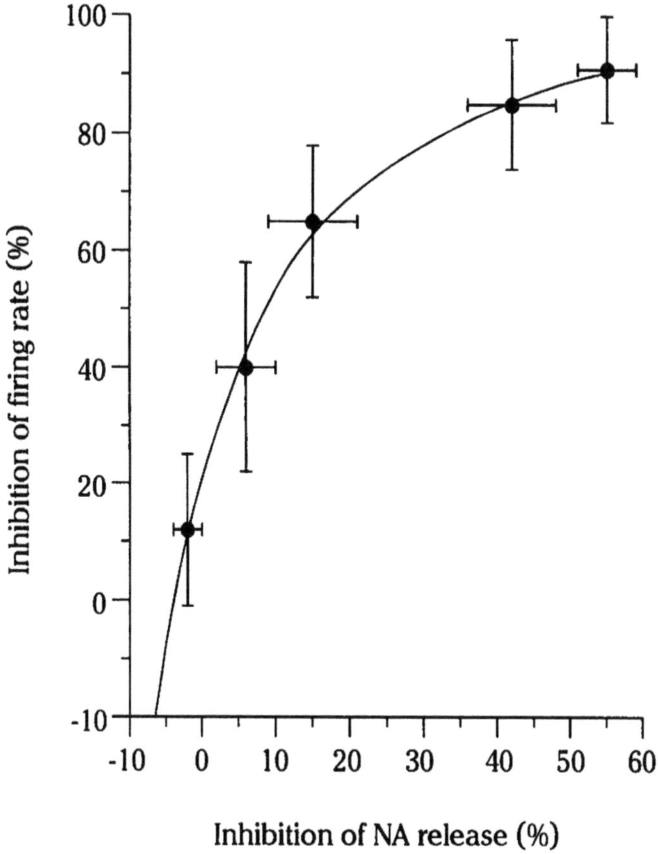

Fig. 12. Correlation of drug effects on amine release and cell firing. A representative example showing the effect of the α_2 agonist dexmedetomidine on noradrenaline release and unit activity in the rat locus ceruleus recorded at a single carbon-fiber microelectrode. Inhibition of cell firing is plotted on the ordinate, whereas the effect on noradrenaline release is shown on the abscissa. The two effects are not directly related. Dexmedetomidine shows a greater effect on cell firing than release, suggesting that either different receptor subtypes or receptor reserves are responsible for the two effects.

rent and maximum interval between stimulations in order to maintain stable unit recordings).

5. Concluding Remarks

FCV was originally used for the detection of DA in vivo, although it was soon applied to brain slices and to other transmitters. The technique circumvents many of the limitations of "classical" superfusion-based transmitter detection methods and has proven ideal for the study of monoamine release in brain slices from experimental animals. A further advantage provided by the absence of uptake blockade is that the uptake process itself may be directly studied. Monitoring extracellular transmitter levels after cessation of stimulation allows simultaneous recording of amine uptake in real-time and, thus, its correlation with (or dissociation from) transmitter release.

Our intention in this chapter was to stick very much within the range of our own experience rather than attempt to speak for others. With this in mind, we have not attempted to offer second-hand advice about methodologies or preparations that we have not ourselves actively used. We therefore hope that we do not so much dictate as illustrate how FCV may be conducted in brain slices. In many ways, our work over the last few years has reinforced the view that there are several equally valid ways of achieving the same ultimate result.

In order to concentrate on methodology, we have deliberately restricted the presentation of data in this chapter to illustrative examples. The intention here is to show the kind of data that can be obtained with FCV. We hope also that we have conveyed something of our "house style" and that, by illustration of the sort of approach we use and the data we obtain, we have highlighted some of the experimental possibilities of what, we feel, is an exciting methodology.

References

Armstrong J. M. and Millar J. (1979) Carbon fibre microelectrodes. *J. Neurosci. Methods* **1,** 279–287.

Brazell M. P., Kasser R. J., Renner K. J., Feng J., Moghaddam B., and Adams R. N. (1987) Electrocoating carbon fibre microelectrodes with nafion improves selectivity for electroactive neurotransmitters. *J. Neurosci.* **22,** 167–172.

Davidson C. and Stamford J. A. (1994) Effects of selective uptake inhibitors on monoamine efflux in rat lateral geniculate brain slices measured using fast cyclic voltammetry. *Br. J. Pharmacol.* **111,** 295P.

Farnebo L. O. and Hamberger B. (1971) Drug-induced changes in the release of ^{3}H-monoamines from field stimulated rat brain slices. *Acta. Phys. Scand.* **371,** 35–44.

Gonzales Mora J. J., Sanchez-Bruno J. A., and Mas M. (1988) Concurrent on-line analysis of striatal ascorbate, dopamine and dihydroxyphenylacetic acid concentrations by in vivo voltammetry. *Neurosci. Lett.* **86,** 61–66.

Gonzales Mora J. J., Mas M., O'Connor J. J., Trout S. J., and Kruk Z. L. (1990) Qualitative and quantitative estimation of dopamine and 5-HT in mixtures assessed by fast cyclic voltammetry (FCV) and numeric analysis. *Br. J. Pharmacol.* **100,** 399P.

Hafizi S., Kruk Z. L., and Stamford J. A. (1990) Fast cyclic voltammetry: improved sensitivity to dopamine with extended oxidation scan limits. *J. Neurosci. Methods* **33,** 41–49.

Herdon H. and Nahorski S. R. (1987) Comparison between radiolabelled and endogenous dopamine release from striatal slices: effects of electrical field stimulation and regulation by D_2 autoreceptors. *N.S. Arch. Pharmacol.* **335,** 238–242.

Jorm C. M. and Stamford J. A. (1993) Actions of the hypnotic-anesthetic dexmedetomidine on noradrenaline release and cell firing in rat locus coeruleus slices. *Br. J. Anaesth.* **71,** 447–449.

Kuhr W. G. and Wightman R. M. (1986) Real-time measurement of dopamine release in rat brain. *Brain Res.* **381,** 168–171.

O'Connor J. J. and Kruk Z. L. (1991) Fast cyclic voltammetry can be used to measure stimulated endogenous 5–hydroxytryptamine release in untreated rat brain slices. *J. Neurosci. Meth.* **38,** 25–33.

Okuma Y. and Osumi Y. (1986) KCl-induced calcium-independent release of endogenous dopamine from rat brain slices. *Brain Res.* **363,** 47–52.

Palij P. and Stamford J. A. (1992) Real-time monitoring of endogenous noradrenaline release in rat brain slices using fast cyclic voltammetry: 1. Characterization of evoked noradrenaline efflux and uptake from nerve terminals in the bed nucleus of stria terminalis, pars ventralis. *Brain Res.* **587,** 137–146.

Palij P. and Stamford J. A. (1994) Real-time monitoring of endogenous noradrenaline release in rat brain slices using fast cyclic voltammetry: 3. Selective detection of noradrenaline efflux in the locus coeruleus. *Brain Res.* **634,** 275–282.

Palij P., Bull D. R., Sheehan M. J., Millar J., Stamford J. A., Kruk Z. L., and Humphrey P. P. A. (1990) Presynaptic regulation of dopamine release in corpus striatum monitored *in vitro* by fast cyclic voltammetry. *Brain Res.* **509,** 172–174.

Parker E. M. and Cubeddu L. X. (1985) Evidence for autoreceptor modulation of endogenous dopamine release from rabbit caudate nucleus in vitro. *J. Pharmacol. Exp. Ther.* **232,** 492–500.

Paxinos G. and Watson C. (1986) *The Rat Brain in Stereotaxic Coordinates,* Academic Press, London.

Stamford J. A. (1990) Fast cyclic voltammetry: monitoring transmitter release in real-time. *J. Neurosci. Methods* **34,** 67–72.

Stamford J. A., Crespi F., and Marsden C. A. (1992) *In vivo* voltammetric methods for monitoring monoamine release and metabolism, in *Monitoring Neuronal Activity: A Practical Approach* (Stamford J. A., ed.), IRL, Oxford, pp. 113–145.

Stamford J. A., Kruk Z. L., and Millar J. (1986) Sub-second striatal dopamine release measured by *in vivo* voltammetry. *Brain Res.* **381,** 351–355.

Stamford J. A., Kruk Z. L., and Millar J. (1990) Striatal dopamine terminals release serotonin after 5HTP pretreatment: *in vivo* voltammetric data. *Brain Res.* **515,** 173–180.

Stamford J. A., Palij P., Jorm C. M., and Millar J. (1994) Simultaneous "real-time" electrochemical and electrophysiological recording in brain slices with a single carbon fibre microelectrode. *J. Neurosci. Methods* **50,** 279–290.

Trout S. J. and Kruk Z. L. (1992) Differences in evoked dopamine efflux in rat caudate putamen, nucleus accumbens and tuberculum olfactorium in the absence of uptake inhibitors: influence of autoreceptors. *Br. J. Pharmacol.* **106,** 452–458.

Rapid Chronocoulometric Measurements of Norepinephrine Overflow and Clearance in CNS Tissues

Greg A. Gerhardt

1. Introduction

The widespread innervation of norepinephrine (NE) in the mammalian central nervous system (CNS) has continually challenged the neuroscience community to explain the multifaceted roles of this important neurotransmitter. Work over the last three decades has shown that the NE appears to be involved in many CNS functions relating to arousal, fine motor control, stress, pain, learning, and memory. In addition, controversy still exists as to the direct or indirect involvement of NE in psychiatric illnesses, such as depression, schizophrenia, and mania.

The largest NE-containing nucleus in the mammalian brain is the pontine nucleus locus ceruleus (LC). This single nucleus innervates almost the entire central nervous system, including the cerebral and cerebellar cortices, the hippocampus, the hypothalamus, and the brainstem and spinal cord (Kobayashi et al., 1974; Lindvall and Bjorklund, 1974; Nygren and Olson, 1977). Although the noradrenergic afferents from the LC have been well-characterized both histochemically and physiologically (Hokfelt and Fuxe, 1969; Olson and Fuxe, 1971; Siggins et al., 1971; Bloom et al., 1972; Hoffer et al., 1973), the widespread and more diffuse innervations of NE in the CNS have limited the number of studies that have addressed

From: *Neuromethods, Vol. 27: Voltammetric Methods in Brain Systems*
Eds: A. Boulton, G. Baker, and R. N. Adams

NE release from NE-containing neurons. Clearly, very few studies have been performed on NE release processes as compared to the numerous studies that have been carried out on the dopamine (DA) systems.

Over the last decade, our laboratory has used several strategies to investigate the dynamics of NE overflow and clearance in the mammalian CNS that have taken advantage of the spatial and temporal resolution of in vivo electrochemical recording techniques. First, because certain NE neuronal populations are difficult to study directly *in situ*, we have used transplantation of central NE-containing cell populations to the anterior chamber of the rat eye to evaluate NE release and factors that influence the development and aging of such cells (Hoffer et al., 1977; Olson et al., 1979, 1983a,b; Gerhardt et al., 1984b; Granholm et al., 1987, 1992; Eriksdotter-Nilsson et al., 1989; Su et al., 1990a). These isolated brain grafts contain hyperinnervations of NE-containing fibers and large amounts of NE (Granholm et al., 1987; Eriksdotter-Nilsson et al., 1989; Su et al., 1990a), and this greatly facilitates measures of the endogenous release of NE (Gerhardt et al., 1984b, 1991; Su et al., 1990a). Second, we have taken advantage of the well-characterized LC innervation of the cerebellum in the intact rat CNS. These NE-containing fibers terminate primarily in the inner portion of the molecular layer and in the granule/Purkinje cell layers. Electrophysiological studies have shown that activation of this ceruleo-cerebellar pathway reduces the spontaneous firing rate of Purkinje cells (Siggins et al., 1971; Hoffer et al., 1973). We have combined electrophysiological studies, NE release measures, and the known distribution of NE in the cerebellum to investigate NE overflow and clearance properties in this brain area (Gerhardt et al., 1987b; Bickford-Wimer et al., 1991). Third, we have used the single hippocampal slice preparation to investigate NE overflow and clearance in this isolated brain circuit (Dunwiddie et al., 1983). Although this system lacks certain inputs and outputs analogous to the intact brain, such advantages as the precise control of the concentrations of pharmacological agents, accessibility of the preparation, freedom from anesthesia, and the speed of the preparation have made this in vitro approach a powerful tool for mechanistic studies of NE (Su et al., 1990b). Finally, we have recently developed methods that involve the controlled ejection of NE from micropipets and the in vivo electrochemical detection of NE uptake or clearance by NE-containing neurons (Cass et al., 1992, 1993a,b; Lin et al., 1993). The purpose of this chapter is to describe methodology for the measurement of NE overflow and clearance using in vivo electrochemical methods, and to discuss results obtained using the aforementioned approaches for the study of NE release processes.

2. Methods

2.1. Electrochemical Recording Electrodes

2.1.1. Description

The electrochemical working electrodes used in these experiments are the graphite epoxy capillary (A; 100–150-μ OD), single carbon-fiber (B; 30-μ OD) or multiple (C; 90-μ OD) carbon-fiber-types (Adams and Marsden, 1982; Gerhardt et al., 1984a,b; Adams, 1990; Su et al., 1990a,b; Friedemann and Gerhardt, 1992; Lin et al., 1993). A schematic diagram illustrating the different types of electrodes used is seen in Fig. 1. None of the electrodes are electrically pretreated or chemically modified. They are all fabricated, coated with the anionic polymer Nafion to enhance selectivity for cations (*see* Section 2.2.), and characterized for sensitivity and selectivity prior to their use. A clean silver wire (A-*M* Systems, Inc., Everett, WA) serves as an auxiliary electrode and an Ag/AgCl wire, prepared daily, is used as the electrochemical reference electrode. For in vitro testing and slice chamber, or *in oculo* brain graft studies, a larger Ag/AgCl reference electrode (RE-5, Bioanalytical Systems, West Lafayette, IN or MI-402, Microelectrodes Incorporated, Londonderry, NH) is used. Because of the low levels of current detected from the small graphite recording electrodes, most of the recordings are performed using a two-electrode recording scheme involving a reference electrode and carbon recording electrode.

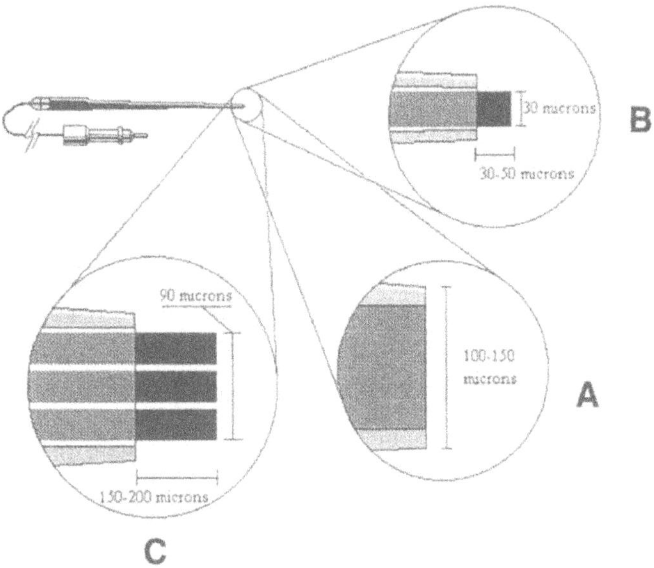

Fig. 1. Schematic diagram showing the dimensions of **(A)** graphite epoxy electrodes, **(B)** single carbon-fiber electrodes, and **(C)** multiple (three-fiber) electrodes used for in vivo electrochemical recordings.

2.1.2. Electrode Materials

In this section, some of the materials used for constructing the three different types of voltammetric electrodes are described. All of the electrodes are made using the same starting materials.

2.1.2.1. WIRE

Cut 3–4 in. lengths of 27–29 gage lacquer-coated copper wire (Radio Shack or Vector). Strip 2 mm of the coating from one end of the wire.

2.1.2.2. GLASS

Using a small triangle file, score and then break off 3–6 cm lengths of 4-mm OD Sodalime or equivalent glass. Load a piece of glass into a Narishige puller (Model PE–2) fitted with a 3.5-turn nichrome wire coil that is 6–7 mm in diameter. Adjust the heat and/or magnet settings to obtain a long uniform taper (30 μ at the tip and 1 mm at the base) that is 3–4 cm long. The current should usually read 18 A. After all the glass is pulled, use a sanding wheel on a Dremel tool to cut off the back end so that about 5 mm remain before the glass starts to taper. For fiber electrodes, bump the tip to obtain a smooth blunt opening slightly larger than the diameter of the fiber(s). If you are making graphite epoxy electrodes, the glass capillary tips should be 50–100 μ in diameter.

2.1.2.3. Adhesives

Epoxylite (#6001-M,The Epoxylite Corporation, Westerville, OH) is used to seal the fibers at the tip of the glass. This prevents the fluid from leaking inside the glass and also gives the electrode more strength.

2.1.2.4. Wood Packing Tool

This tool is used to fill pulled glass capillaries with a graphite-epoxy paste. We suggest the use of 6-in. applicator sticks (The Citmed Corp., AL) 2-mm in diameter. They can be sharpened in a pencil sharpener, and then sanded to achieve a sharp, symmetrical tip that extends 2–3 mm inside the taper of the glass capillary.

2.1.3. Construction Procedures for Carbon Electrodes

2.1.3.1. Graphite Epoxy Electrodes

The construction of graphite epoxy electrodes is a one-step process. This type of electrode is one of the oldest types of electrochemical recording probes (Adams and Marsden, 1982; Gerhardt et al., 1984a,b, 1986) and perhaps one of the simplest probes to construct. First, glass capillaries with tip diameters of 50–100 μ are prepared from 4-mm OD glass using a vertical electrode puller. The capillaries are subsequently filled with carbon-epoxy paste (PX grade GRAPHPOXY, Dylon Industries, Cleveland, OH) using a wooden stick as a tamping tool (*see* Section 2.1.2.4.). The graphite paste should be purchased every 6 mo, and a smooth "wet" consistency of the paste is needed for assembly of this type of electrode. After filling, the excess glass at the base of the capillary is removed, and a copper wire is inserted into the back of the electrode to make electrical contact with the carbon paste. Bake the electrode at 125°C for 12–16 h to cure the paste. Store the electrodes in a freezer (–20°C).

2.1.3.2. Fiber Electrodes

Construction of single- or multiple-fiber-type electrodes is a two-step process. First the fibers must be glued in place and allowed to dry. Second, the base of the electrode is packed with carbon-epoxy paste to make a good connection between the fiber and a copper wire. You should always wear gloves when handling the 30-μ diameter carbon-fibers in order to keep the carbon surfaces as clean as possible. Begin by cutting 3–4 in. lengths of fiber (30 μ carbon monofilament [as pyrolyzed], AVCO Specialty Materials, Subsidiary of Textron Inc., Lowell, MA). For multifiber (three fiber) electrodes, twist three fibers together and cut one of the ends so that all the fibers are even and tightly bundled. Insert the fiber(s) into a pulled glass capillary. The opening at the tip should have a slightly larger diameter than the diameter of the fiber(s). Hold the fiber(s) in place, and use a syringe filled with Epoxylite to fill the glass all the way to the tip. Use another syringe to draw back any excess glue from the base of the glass. Allow the Epoxylite to cure for 12–16 h at 125°C. Following curing of the cement, the electrode is filled with carbon paste (*see* Section 2.1.3.1.) to make electrical contact with the fiber(s). Introduce the paste into the back of the glass, and use the packing tool to force it part of the way up the shaft. Cut the filled electrode at the base, and insert the stripped end of a copper wire into the paste. Bake the electrode a final time at 125°C for 12–16 h. Store the electrodes in a freezer (–20°C).

2.1.4. Characterization of Electrodes

All of our electrodes are tested to determine their recording properties. Take an electrode out of the freezer, and cut it back to expose a fresh recording surface of carbon paste or fiber. Fibers can be cut back so that they either protrude about 50–300 μ from the tip of the glass, or are flush with the glass. Using a microscope, check to see if the fibers are well sealed at the tip, or if the carbon paste electrode has a flat and even surface. If the fibers are not sealed at the tip, discard this electrode. In addition, the tips of the graphite paste electrodes may need to be cut again to achieve a flat recording surface.

Connect the recording electrode and a reference electrode to your electrochemistry system. Put the electrode tip in a beaker filled with phosphate-buffered saline (PBS: 0.1*M*, pH 7.2–7.4, room temperature) to make sure that it is capable of conducting ox/red currents at an appropriate gain. If the electrode does not record any signal or the signal is "weak," throw it away. After the electrode has been recording in the beaker for 30–60 s, the measurements should drift down until they reach a steady baseline. An upward drift in the measurements or a noisy baseline can indicate problems. First, the junction at the back of the electrode (where the wire connects to the paste) may be wet, resulting in high background currents. Dry the electrode in a warm (80–85°C) oven for a few minutes, and try again. In the case of fiber electrodes, there may be a leak at the interface of the fiber and the Epoxylite inside the glass. This will cause the baseline to drift upward. Using a microscope, check to see if there is any indication that fluid is leaking into the electrode. Another potential source of noise is a bubble or particle on the tip of the electrode that interferes with diffusion at the carbon surface. Flick the electrode in the beaker to dislodge the bubble or particle. Finally, the electrode may be picking up 50 or 60 Hz (AC) noise from extrinsic sources. Turn off or move any electrical equipment that is located close to your recording system.

2.2. Coating of Carbon Electrodes with Nafion to Enhance Their Selectivity for NE Measurements

A key step in the making of all of our electrochemical probes is the coating of the surface of the electrode with the polymer, Nafion. This procedure will enhance the selectivity of the sensors for the monoamine neurotransmitters and minimize signals detected from anions, such as ascorbic acid (AA), uric acid, and monoamine metabolites (i.e., DOPAC and 5-HIAA; Gerhardt et al., 1984a,b, 1986, 1987a,b; Nagy et al., 1985). A fresh Nafion solution (Aldrich Chemical Company [Milwaukee, WI], 5% solution) should be used approximately every 2 mo. In order to get Nafion to stick to the electrode, the carbon surface must be completely dry. After checking the electrode in a beaker for electrochemical stability (*see* Section 2.1.4.), dry the electrode in an 80–85°C oven for 5 min. Remove the electrode from the oven, and swirl the electrode tip in Nafion solution for 2–3 s. Dry the electrode for 5 min. Repeat the coating and drying procedure until the electrode shows the desired selectivity and sensitivity for the monoamine neurotransmitter to be measured. We have found that 7–10 coats are sufficient for a good selectivity on carbon-fiber probes and 4–6 coats produce good selectivity on the carbon epoxy electrodes; this procedure must be optimized for a given electrode and the individual user. Dry the electrode one last time, and it is ready for calibration. We warn investigators that the Nafion can be applied too thinly as well as too thickly. If the Nafion is applied too thinly, the electrode is capable of recording very large changes in ascorbic acid. If the Nafion is too thick, the investigator will likely find that the electrode may be insensitive to all electroactive species. Thus, these are things that the individual investigator will have to explore for his or her electrodes.

2.3. Calibration of Nafion-Coated Electrodes

Place an electrode in a beaker filled with PBS, and record until a stable baseline is attained. Make a record of the baseline measurement. Add an amount of ascorbic acid that will result in a final concentration of 250 μ*M*. Stir vigorously, and record the maximal response to the ascorbic acid challenge. Calibrate the electrode using 0.1–10 μ*M* increments of NE. Be sure to stir after each addition, and record the measurement once the response reaches a stable maximum. Use the maximum measurements recorded after each addition to calculate a linear regression for the current recorded above baseline vs the micromolar concentration of neurotransmitter. The regression should have a correlation coefficient of $r^2 = 0.997$ or better. As we have previously reported, a good recording sensor must detect changes in NE in a highly linear fashion (Gerhardt et al., 1987b; Su et al., 1990a,b). Finally, the selectivity of the electrode for NE vs ascorbate should be >250:1 for its sensitivity to NE over ascorbic acid (AA). Typically, our probes used with the IVEC-5 or IVEC-10 electrochemical recording systems (*see* Section 2.4.) have detection limits of 50 n*M* (signal-to-noise ratio of 3) for 1-s measures of changes in NE. It should be noted that Nafion-coated electrodes are inherently more sensitive for dopamine than for NE by a factor of 2–4 (Gerhardt et al., 1984a; Nagy et al., 1985).

2.4. Instrumentation for Chronoamperometric/ Chronocoulometric Measurements

The electrochemical recording method we use for monitoring the overflow and clearance of monoamines is called chronoamperometry (Cheng et al., 1980; Adams and Marsden, 1982; Adams, 1990). These methods involve the application of rapid (1–25 Hz) square-wave potentials to the electrode and monitoring the current output of an electrode per unit time. We integrate the resulting current flow to enhance the signal-to-noise ratio of the recording approach, and therefore, the recording methods are more accurately called chronocoulometry.

Some of the electrochemical measurements reported in this chapter are performed using a modified version of a microcomputer-controlled chronoamperometry/chronocoulometry apparatus (Cheng et al., 1980). An oxidation potential of +0.55 V with respect to the reference is applied to the Nafion-coated carbon electrode for 1 s. A delay of 1–5 s is employed between measurements; the electrode is maintained at a resting potential of –0.2 V during this interval.

More recently, we have developed instrumentation involving high-speed (5–25 Hz) chronoamperometric/chronocoulometric electrochemical recordings that measures both the oxidation and subsequent reduction of species at the tips of the recording electrodes as a means of improving the identification of compounds that give rise to the electrochemical signals (Gerhardt et al., 1988; Gratton et al., 1988, 1989; Cass et al., 1992, 1993a,b; Friedemann and Gerhardt, 1992). The 486-based instrument uses a custom "C-based" software package. A schematic diagram of this "turnkey" recording system that is sold by Medical Systems Corporation as the

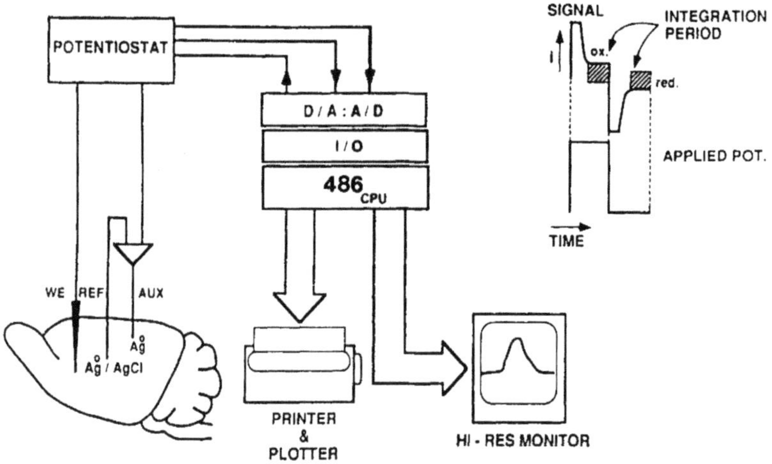

Fig. 2. Schematic diagram of the automated in vivo electrochemical recording system used for rapid chronoamperometry/chronocoulometric measurements. WE = working or carbon electrode, REF = reference electrode, AUX = auxiliary electrode.

IVEC-5 or IVEC-10 is seen in Fig. 2. The system is capable of recording real-time 5–25 Hz signals involving both oxidation and reduction current measurements.

Although Nafion-coated electrochemical electrodes minimize possible signal interference from monoamine metabolites and ascorbic acid (Gerhardt et al., 1984a,b, 1985, 1986, 1987a,b), they previously could not reliably discriminate among dopamine (DA), norepinephrine (NE), and serotonin (5-HT). However, the combination of the oxidation and reduction measurements yields a ratio that can be used for the further identification of the detected monoaminergic species. Although there is some variability in the actual values observed for individual electrodes, the largest reduction/oxidation current ratio is seen for DA (0.5–0.7), followed by NE (0.3–0.4), 5-HT (0.1–0.2), and finally AA (0.0). Examples of extracellular signals for DA, NE, 5-HT, and AA are seen in Fig. 3. Thus, the combination of oxidation and reduction current recordings can be used to identify or "finger print" further the chemical characteristics of the electrochemical signals.

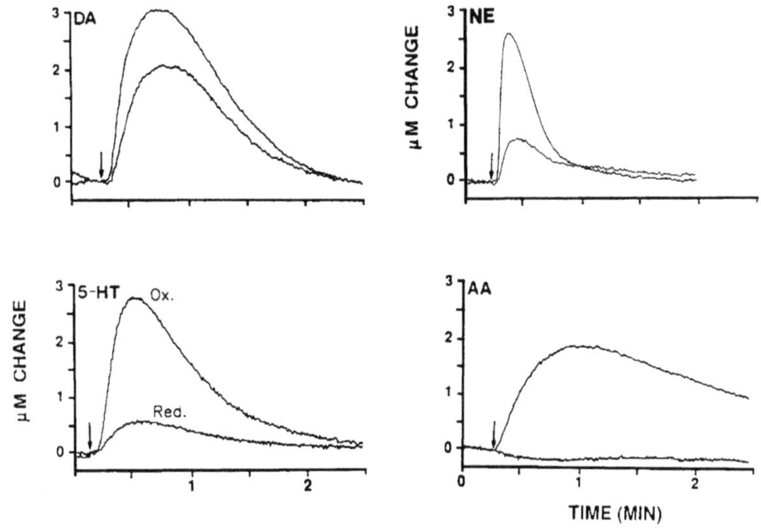

Fig. 3. Extracellular recordings from Nafion-coated multiple (3–30 μ OD fibers) carbon-fiber electrodes for pressure ejection of DA, NE, 5-HT, and large volumes of AA. The upper traces in all of the records are the oxidation signals (Ox.), and the lower traces (Red.) are the subsequent reduction current responses. All solutions are ejected at the arrows.

2.5. Pressure Ejection of Drugs Coupled to In Vivo Electrochemical Recording

2.5.1. Methods

To avoid the use of parenteral drugs to alter monoaminergic function and to study the extracellular regulation of NE directly, we developed a technique for local application of drugs coupled with in vivo electrochemical recordings (Gerhardt et al., 1991; Cass et al., 1992, 1993a,b; Friedemann and Gerhardt, 1992; van Horne et al., 1992). The voltammetric electrodes are fastened to a single- or multibarrel micropipet for these studies. A recent major improvement now allows for the volume of drug that is ejected to be measured using a dissection microscope fitted with a calibrated reticule (Cass et al., 1992, 1993a,b; Friedemann and Gerhardt, 1992). This new approach allows us to control accurately the doses of drugs delivered to the brain. However, in some experiments, we use relative measures of micropipet output, such as pressure X time parameters (Palmer et al., 1980; Gerhardt and Palmer, 1987).

2.5.2. Micropipet and Electrochemical Arrays

The in vivo electrochemical experiments involving pressure ejection of NE and drugs are performed using electrode assemblies consisting of a working electrode and a single- or double-barrel micropipet (1 mm OD, .58 mm ID glass; #6015 microfilament glass, A-M Systems, Inc., Everett, WA) with tip diameters of 5–10 μ/barrel. The electrode and the micropipet are mounted together with Sticky Wax (Kerr Brand "Sticky Wax," HEALTHCO Dental Supply, Englewood, CO) with the tips separated by 300 ± 20 μ.

Local applications of drugs *in situ* are performed by pressure ejection (Palmer et al., 1980; Gerhardt and Palmer, 1987) using a pneumatic pump (PPS-2, Medical Systems Corp.). The volumes applied can be determined using a stereomicroscope fitted with a reticule in one eyepiece to measure the movement of the meniscus in the micropipet (0.1 mm = 25 nL; Friedemann and Gerhardt, 1992).

2.6. Dissections of Fetal Locus Ceruleus and Transplantation

Young adult (2-mo-old) female Sprague-Dawley albino rats (150 g; Alab, Stockholm) are used as recipients for intraocular fetal grafts. Tissue donors are Sprague-Dawley rat fetuses of embryonic d 17. Pregnant rats are anesthetized with chloral hydrate (30 mg/kg) and sacrificed by cervical dislocation. Fetuses are removed and kept on ice until they are dissected. The fetal brain is removed, the tectal roof reflected, and small pieces of brainstem tissue are dissected bilaterally. The tissue is inserted into the anterior chamber of the eye through a razor blade slit in the cornea of anesthetized hosts and placed in the outer angle of the eye using a modified Pasteur pipet as previously described (Olson and Seiger, 1972; Olson et al., 1983a,b). At least 48 h prior to transplantation, the recipient rats are sympathetically denervated by surgical removal of the superior cervical ganglion to avoid any

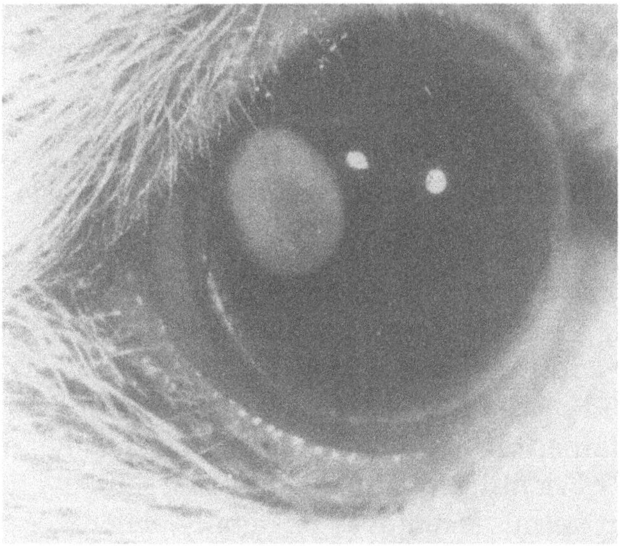

Fig. 4. Photograph of an intraocular LC transplant. The transplant is covered with blood vessels from the host iris. This transplant is 1 × 1 mm and has been growing in the anterior eye chamber for 4 mo.

interference from the host iris sympathetic ground plexus. Such procedures produce a >95% reduction of the normal NE content in host iris. Young brainstem grafts are allowed to mature for 4–6 mo (6–8-mo-old hosts) before examination, whereas transplants in the old animal group (26–30-mo-old hosts) age for 24–28 mo *in oculo*. A photograph of an LC brain graft in the anterior eye chamber of a Sprague-Dawley rat is seen in Fig. 4. Animals are kept four to a cage in a 12:12 h light:dark cycle and are given food and water ad libitum. Following surgery, the host rats are given paracetamol (6 mg/L) in their drinking water to alleviate any pain associated with the surgery.

2.7. *In Vivo Electrochemistry in* In Oculo *Transplants*

Host rats that have received *in oculo* transplants are anesthetized with urethane (1.25 g/kg, ip), tracheotomized, and placed in a stereotaxic frame. For recordings, the cornea overlying the transplant is removed and a Plexiglas™ chamber is placed over the eye (Gerhardt et al., 1984b). The eye is continually superfused with Earle's balanced salt solution maintained at 37°C (Gerhardt et al., 1984b).

2.8. Acute Animal Preparation for In Vivo Electrochemical and Electrophysiological Recordings

For in vivo electrochemical and electrophysiological recording, male Sprague-Dawley rats (275-400 g) are anesthetized with urethane (1.25 g/kg ip, in 0.9% NaCl), intubated, and placed in a stereotaxic frame. Body temperature is maintained at 37°C using an isothermal heating pad (Braintree Scientific, Braintree, MA). The skull overlying the cerebellar vermis is removed bilaterally. Reference and auxiliary electrodes (described above) are inserted into brain regions remote from the recording site and cemented into place using dental acrylic (Gerhardt et al., 1987a,b).

2.9. Electrophysiological Recording Methods

Single Nafion-coated carbon-fiber electrodes are used to record spontaneous unit activity from Purkinje cells in rat cerebellar cortex (Hoffer et al., 1973). The cells are identified by anatomical localization and their characteristic discharge of simple and complex spikes. Action potentials are amplified, separated from background activity with a window discriminator, and integrated over 1-s intervals to indicate discharge rate.

2.10. Electrical Stimulation of the Locus Ceruleus

A stimulating Electrode is placed into the nucleus LC in urethane anesthetized rats (*see* Section 2.8.) with coordinates from bregma of A.P. –12.0, M.L. 1.2–1.5, and D.V. –4–5 mm from cortical surface (atlas of Paxinos and Watson, 1986). The electrode is angled 20° from vertical toward the rat's tail. Electrical activation of the LC is accomplished using 200-µs pulses, delivered in a 10-Hz, 10-s train. Typical stimulation voltages range from 10–40 V.

2.11. Recording System for Hippocampal Slice Studies

Single hippocampal brain slices are prepared from male rats using standard techniques and incubated at 33.5°C under constant superfusion (Dunwiddie and Lynch, 1978; Dunwiddie et al., 1983). Electrophysiological recordings are made from the CA1 region, and electrochemical responses are measured in the dentate gyrus or CA1 region. Recordings of evoked responses are made from the CA1 cell-body layer with glass microelectrodes filled with 2*M* NaCl, and twisted nichrome-chrome wire stimulation electrodes are used to stimulate fibers in stratum radiatum near the border of CA1-CA2, and elicit synaptic responses. A schematic of the recording system is seen in Fig. 5. The evoked synaptic responses and electrochemical measurements are made every 0.20–5 s, and all responses are recorded by microcomputer systems for subsequent analysis. A representative data set showing combined electrochemical and electrophysiological measures is seen in Fig. 5. Drugs are added to the flow of the superfusion fluid with a calibrated Sage Model 355 syringe pump.

3. Detection of Norepinephrine in CNS Tissues

3.1. In Vivo Electrochemical Studies of Potassium-Evoked NE Overflow in Young and Aged Locus Ceruleus Brain Grafts Studied In Oculo

3.1.1. Results

The present study takes advantage of the unique experimental accessibility of transplanted LC neurons as a means of directly investigating the release of NE and potential age-induced changes in the presynaptic release processes of NE-containing neurons. A representative K^+-evoked response from a young LC (YLC) graft is seen in Fig. 6. Both the oxidation current (Ox., upper trace) and subsequent reduction current (Red., lower trace) electrochemical responses are shown. The recorded K^+-evoked overflow responses from YLC grafts averaged 6.2 ± 1.0 μM (n = 23 signals recorded

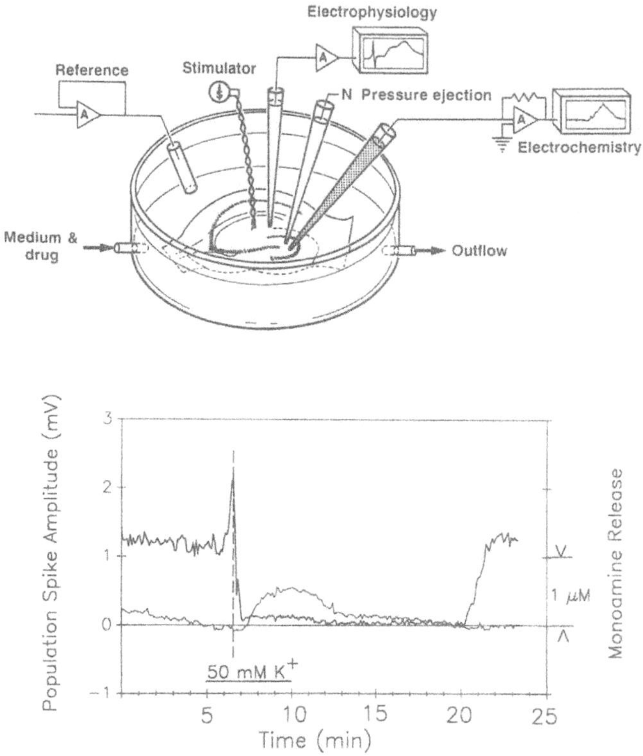

Fig. 5. **(Top)** Schematic diagram of the in vitro brain slice system used for combined electrochemical and electrophysiological measurements in single hippocampal slices. **(Bottom)** Data set illustrating combined electrochemical and electrophysiological measurements that are performed with this system.

from 10 LC grafts; Fig. 7A). In contrast, the K^+-evoked overflow magnitudes are significantly reduced in aged transplants as compared to young controls ($p < 0.001$). The average amplitudes of K^+-evoked responses are 1.6 ± 0.2 μ*M* ($n = 19$ signals recorded from 8 LC grafts) in the old LC (OLC) transplants. However, measures of NE levels in the grafts performed using high-performance liquid chromatography coupled with electrochemical detection (Hall et al., 1989) showed that NE contents of the young and aged grafts are nearly identical (Fig. 7B). The pressure ejection parameters used for the K^+-evoked overflow of NE in both YLC and OLC grafts are identical,

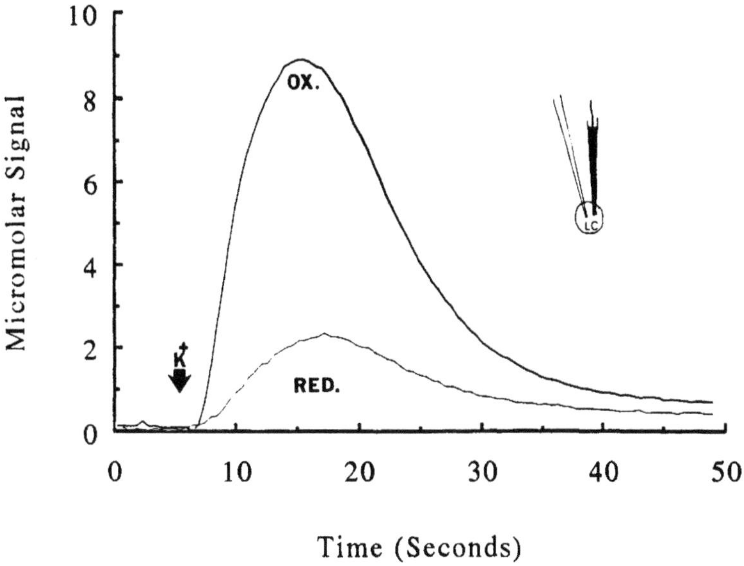

Fig. 6. K^+-evoked overflow of NE in a YLC graft. The OX. curve represents the oxidation current response, and the RED. curve is the subsequent reduction current signal. The ordinate is shown as micromolar concentration of NE based on a prior calibration curve conducted in a beaker. The inset figure shows the orientation of a pressure-ejection pipet that is used to apply K^+ locally to elicit this type of release signal. Potassium ejection is initiated at the arrow.

suggesting that the aged LC grafts are not necessarily insensitive or differentially sensitive to the K^+ solution. In addition, the K^+-evoked responses are seen to be reproducible (± 15%) in a given recording site in both age groups if a 15 min delay is employed between K^+ ejections (n = 6 YLC grafts, n = 5 OLC grafts).

Although the magnitudes of K^+-evoked overflow of NE are decreased in the OLC grafts, the electrochemical measures support that the major species that is detected in both the YLC and OLC grafts is predominantly NE. The bar graph seen in Fig. 7C shows the average reduction/oxidation current ratio recording characteristics of Nafion-coated multiple-fiber electrodes for additions of DA, NE, and AA in vitro in solutions prepared in 0.1M phosphate buffered saline at pH 7.4 for 12–22 recording electrodes. The YLC and OLC values

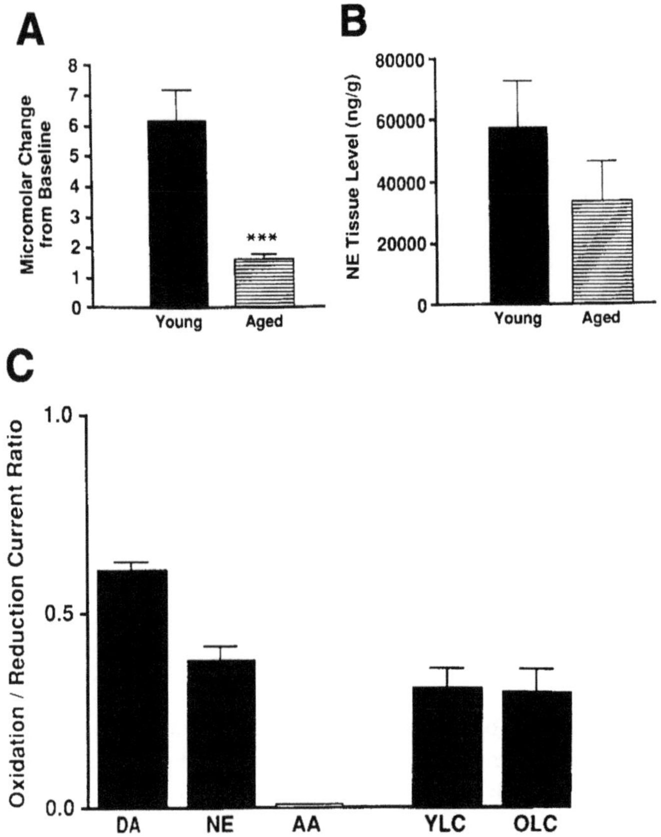

Fig. 7. **(A)** Average amplitudes (± SEM) of K^+-induced signals recorded from YLC and OLC transplants, and average whole tissue levels of NE **(B)** detected in the brainstem grafts. **(C)** Bar graph illustrating the average reverse-current ratios (reduction current/oxidation current) seen for 12–22 (± SEM) Nafion-coated carbon-fiber electrodes recorded in vitro for micromolar additions of DA, NE, and AA prepared in 0.1*M* phosphate-buffered saline (pH 7.4), and the average reverse currents recorded from K^+-evoked responses from YLC and OLC grafts. *** = $p < 0.001$ using a two-tailed Student's *t*-test.

represent the average reverse-current ratios recorded from the K^+-evoked responses seen in the YLC and OLC grafts, respectively. The average ratios are 0.31 ± 0.045 (n = 23) in the YLC group and 0.30 ± 0.063 (n = 19) in the OLC group. Thus, these current ratios indicate that NE is the predominant electroactive species detected as a consequence of the K^+ ejection in both YLC and OLC transplants.

Aside from the observed alterations in release magnitudes between the two groups, there are also changes in the temporal dynamics of the K^+-evoked responses recorded in the YLC vs OLC transplants. The average signal rise time (T_R) for YLC transplants is 21.2 ± 2.6 s (n = 23) vs 29.4 ± 2.6 s (n = 19) for signals recorded from the OLC grafts, which is statistically significant (p <0.05). On the other hand, the half decay times ($T½$) of the signals did not differ between the two groups. These averaged 53.0 ± 8.8 s (n = 23) in the YLC grafts and 55.0 ± 6.6 s (n = 19) in the OLC transplants.

3.1.2. Discussion

A major finding from the present series of experiments is that NE overflow and clearance can be reliably measured in the LC brain grafts. In addition, there are age-related alterations in K^+-evoked NE overflow from LC transplants from Sprague-Dawley rats that aged for 24–28 mo *in oculo*. The magnitudes of K^+-evoked overflow of NE-like signals are much greater in the YLC than in the OLC grafts. In addition, the rise times of the signals from OLC grafts are significantly greater than in the YLC transplants, indicating potential changes in K^+-evoked release mechanisms in these grafts. In contrast, the levels of NE are not significantly different in the YLC vs OLC tissues. However, preliminary immunohistochemical studies (Gerhardt et al., 1991) support alterations in the morphology of NE cell bodies and neurites. Thus, the electrochemical, biochemical, and histological data support a change in the NE-releasing properties of these cells that does not appear to be solely related to synthesis and storage of NE.

3.2. Electrically Evoked and Potassium-Induced Overflow of NE

3.2.1. Results

These studies are performed to take advantage of the known cerebellar distribution of NE and previous electrophysiological studies that support that NE is released after electrical stimulation of the LC. In these studies, all record-

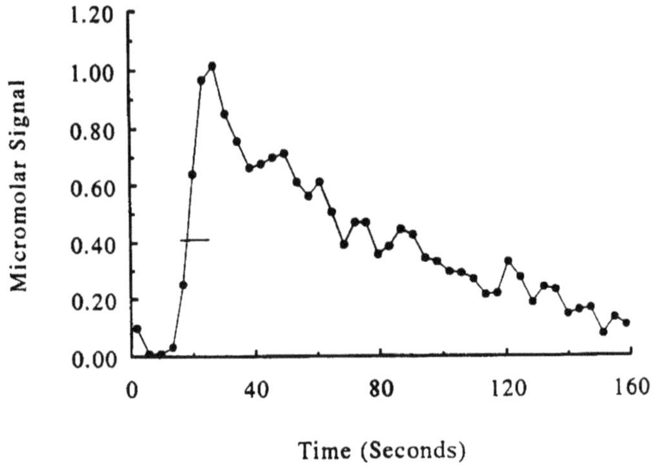

Fig. 8. In vivo electrochemical recording of an electrically induced release of electroactive species recorded from the Purkinje cell layer of the urethane anesthetized rat. The bar represents the timing of the electrical stimulation (10 Hz, 0.2 ms pulses, 10 s). The *y*-axis represents the amount of electroactive species detected in the extracellular space in nanomolar equivalents for NE.

ing and stimulating electrode placements are confirmed at the end of an experiment by histological methods (Gerhardt et al., 1987b).

Electrochemical responses are detected from the granule/Purkinje cell layer and, to a lesser extent, the molecular layer during LC stimulation. The peak response occurs slightly after the termination of the stimulus. A representative electrically evoked response recorded from the Purkinje cell layer is seen in Fig. 8. The horizontal bar represents the time period during which the LC is electrically stimulated (40 V, 10 Hz, 10 s).

The NE innervation of the cerebellum is known to be quite laminar with the greatest numbers of NE-containing fibers being located in the granule/Purkinje cell layer (Hokfelt and Fuxe, 1969; Olson and Fuxe, 1971; Bloom et al., 1972). Electrically evoked electrochemical responses are significantly greater in the granule/Purkinje cell layer than in the molecular layer (Fig. 9). The time-courses of the responses (rise time and half decay times), however, are similar in the two regions. Electrode placements in the cerebellar white

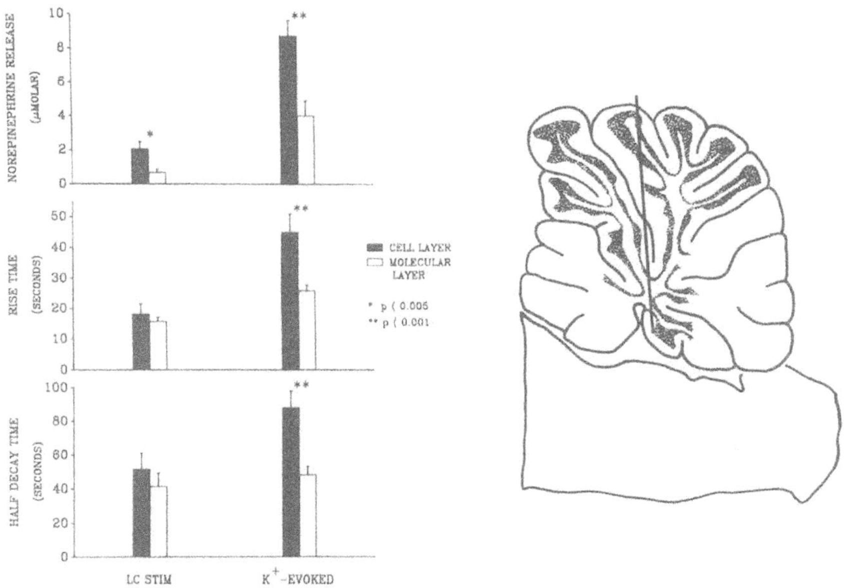

Fig. 9. **(Right)** Typical recording tract through the cerebellum showing electrode placements into the granule/Purkinje cell layer, molecular layer, and white matter. The bar graphs to the left show the average magnitudes and time dynamics of the electrochemical responses recorded from the granule/Purkinje cell layer (solid bar) and molecular layer (open bar) from both electrical stimulation of the LC and K^+-evoked release of electroactive species. Each bar represents the mean ± SEM response seen from 6–15 histologically confirmed recording electrode placements. Releases from molecular layer have a significantly lower amplitude than those recorded from the Purkinje cell layer (p <0.01 for the electrically evoked release and p <0.001 for K^+-evoked releases using a two-tailed Student's t-test).

matter did not result in detectable responses, which is consistent with the lack of NE-containing fibers in this brain area. In addition, as compared to K^+-evoked responses recorded in the different layers, release amplitudes after LC stimulation are also smaller than releases evoked by pressure ejection of K^+ (Gerhardt et al., 1987b). Moreover, the time-courses of the K^+-evoked responses are significantly different in the granule/Purkinje cell layer vs the molecular layer.

The amplitude of electrically evoked release is dependent on the stimulation voltage. Figure 10 shows results recorded from a single site within the Purkinje cell layer with

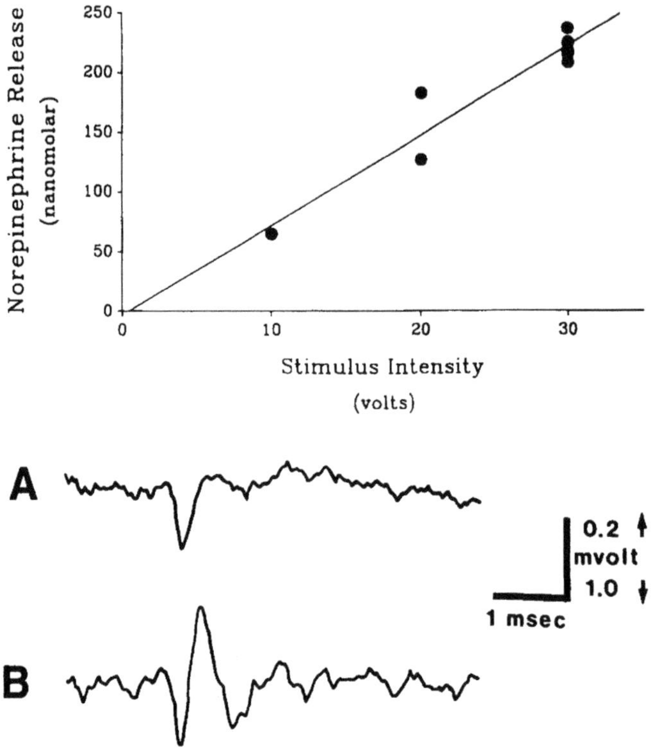

Fig. 10. **(Top)** Voltage dependence of the in vivo electrochemical responses recorded following electrical stimulation of the LC. **(Bottom)** Unfiltered (A; 0.5–10,000 Hz) and filtered (B; 500–10,000 Hz) Purkinje cell action potentials recorded using a 30-μ diameter carbon-fiber electrode. Calibration bars: horizontal = time; vertical = amplitude of signal in millivolts.

a 15-min interstimulus interval. As is seen, the amplitude of the electrically evoked response, expressed in NE equivalents, increased linearly with the stimulus intensity; correlation coefficient = 0.96. In addition, the electrically evoked responses recorded from the granule/Purkinje cell layer and the molecular layer are reproducible with at least a 10-min interval between stimulations.

The electrochemical sensors can also be used for combined electrophysiological and electrochemical studies. For these studies, Nafion-coated carbon-fiber voltammetric electrodes (30-μ in diameter) are used to record and identify cerebellar Purkinje cell action potentials. Figure 10 shows both unfiltered (A; 0.5–10,000 Hz) and filtered (B; 500–10,000 Hz) Purkinje cell action potentials that are recorded using a carbon-fiber electrode in the cerebellum of the anesthetized rat. In addition, the same probe is calibrated for electrochemical measurements. Thus, both electrophysiological responses and electrochemical signals can be recorded from the same electrode in the same discrete brain region of the rat.

The time-course of the electrochemical signals following LC activation correlates well with the inhibition of Purkinje cell discharge. Such recordings made in the Purkinje cell layer using the same carbon-fiber probe are seen in Fig. 11. Both the electrochemical response recorded following the stimulation and the rate meter recording from the same site following a 15-min time delay are shown. Identical stimulation parameters are used to activate the LC in both records. The horizontal bars above each record represent the period of stimulation. The horizontal time scales are the same in both figures.

3.2.2. Discussion

A major finding from this series of experiments is that electrical stimulation of the LC and local application of K^+ can be used to evoke the release of electroactive species from specific areas of the rat cerebellum. Electrically evoked responses observed in the granule/Purkinje cell and molecular layers are reversible, stimulus-intensity-dependent, and reproducible if a 10-min time delay is imposed between stimulation trials. As might be expected, little, if any, release is observed in white matter areas of the cerebellum. This pattern corresponds well with the known distribution of NE-containing terminals in the cell and molecular layers, and the relative paucity of such afferents in white matter (Olson and Fuxe, 1971; Bloom et al., 1972). The time dynamics of the electrically evoked responses recorded from the Purkinje cell layers correlate well with the electrophysiological rate meter recordings of the depressions in Purkinje cell firing rates resulting from the LC stimulations. Interestingly, the time-courses of NE-like electrically evoked signals are quite long

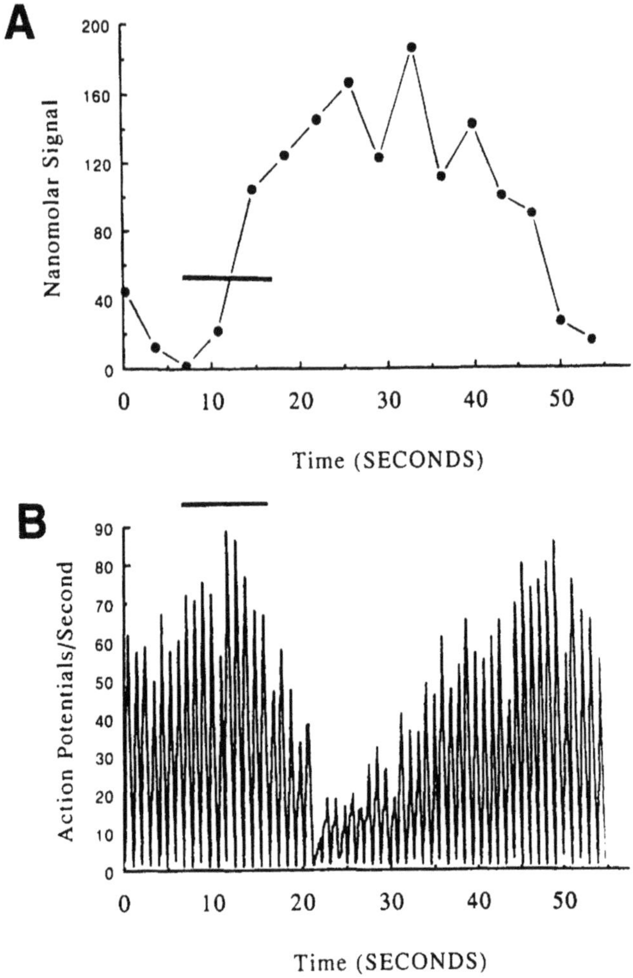

Fig. 11. Electrochemical response **(A)** and rate meter recording **(B)** of the effects of electrical stimulation of the LC recorded in the Purkinje cell layer region of the urethane-anesthetized rat cerebellum using a carbon-fiber electrode. Fifteen minutes elapsed between the two recordings. Note the rise in the electrochemical response that parallels the decline in the spontaneous discharge rate of the Purkinje cell.

as compared to the temporal properties of DA release produced by electrical stimulation of the medial forebrain bundle (Gratton et al., 1988; Wightman et al., 1988; Wightman and Zimmerman, 1990). Thus, these studies directly support the hypothesis that NE is released on electrical stimulation of the LC, and this gives rise to the depression in Purkinje cell firing rates that is observed in the cerebellum of the anesthetized rat. In addition, these studies demonstrate that both presynaptic overflow of NE and electrophysiological measures can be recorded with the same sensor.

3.3. Superfusions of Single Hippocampal Slices: Combined Electrophysiological Measures with In Vivo Electrochemical Recordings

3.3.1. Results

These studies are undertaken to take advantage of the capabilities of the single hippocampal slice preparation for studies of the release of NE combined with electrophysiological measures. Prior regional HPLC analyses from our laboratory and previous histochemical studies support that the dentate gyrus contains the highest levels of NE in the single hippocampal slice. Superfusion of single hippocampal brain slices with K^+ concentrations below 12 m*M* induce increases in the population spike amplitude with minimal increases in detectable monoamine levels recorded from the dentate gyrus. Concentrations of K^+ above 15 m*M* invariably cause initial increases in the population spike response that are immediately followed by a loss of all evoked activity and significantly larger monoamine overflow (Fig. 12). In addition, superfusion with K^+ (10, 30, and 50 m*M*) induces concentration-dependent releases of monoamines. Recovery of the electrophysiological response from superfusion with 50 m*M* K^+ often takes considerably longer than the electrochemical signal and is sometimes less complete (Su et al., 1990b).

When slices are superfused repeatedly with the same concentration of K^+ (15 or 30 m*M*), reproducible signals can be evoked. However, the synaptic response (and occasionally the monoamine release) often shows decremental responding to two or more periods of stimulation. Multiple responses to 50 m*M* K^+ can be evoked as well, provided that sufficient time, which is usually 20–30 min, elapses between stimuli. However, the recovery period is somewhat longer for 50 m*M* than for 30 m*M* K^+ stimulation.

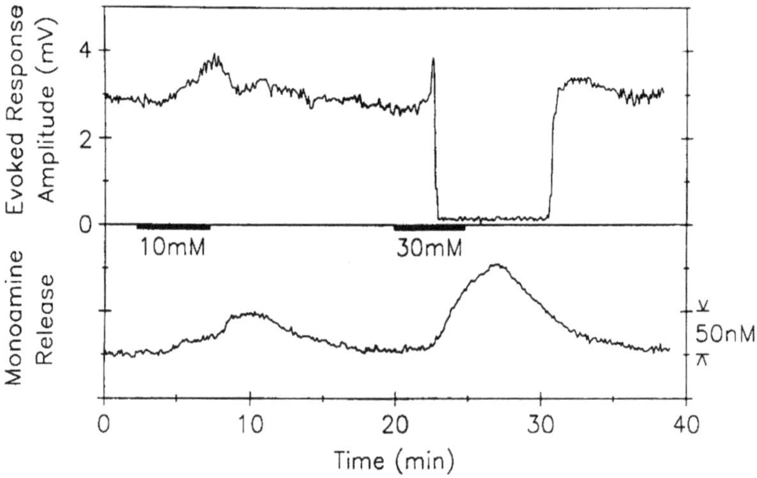

Fig. 12. Effects of K^+ superfusion (10 and 30 m*M*) on population spike responses and monoamine overflow in single hippocampal slices.

The interrelationship between the population spike amplitudes and the electrochemical response is consistent and is seen in Fig. 5. The amplitude of the population spike response increases, and then drops to 0 before there is a significant increment in the electrochemical signal. Both responses return to baseline within 10–15 min following K^+ superfusion. The preceding change in the population spike amplitude is seen in all slices studied. The concentrations of K^+ used did not alter this timing relationship.

3.3.1.1. Local Application of Potassium Studies

The long-lasting and sometimes overly refractory characteristics of the K^+ superfusions made us try a method that would allow for a more rapid and reproducible means of eliciting overflow of monoamines in the single hippocampal slice. In contrast to the K^+ superfusion studies, local application of K^+ from glass micropipets via pressure ejection produces reproducible monoamine releases (Fig. 13A). The signals recorded are produced from pipets containing 120 m*M* K^+ and 2.5 m*M* $CaCl_2$; in the case of this figure, the K^+ is ejected with 500-ms pressure pulses at 80 psi. Each release is elicited 10 min following the previous release (note the time breaks

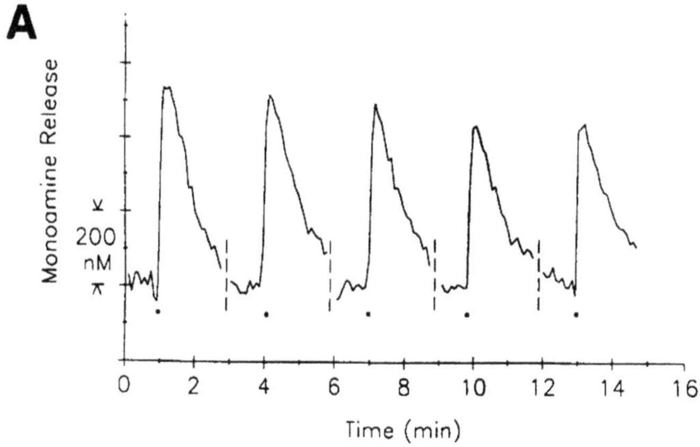

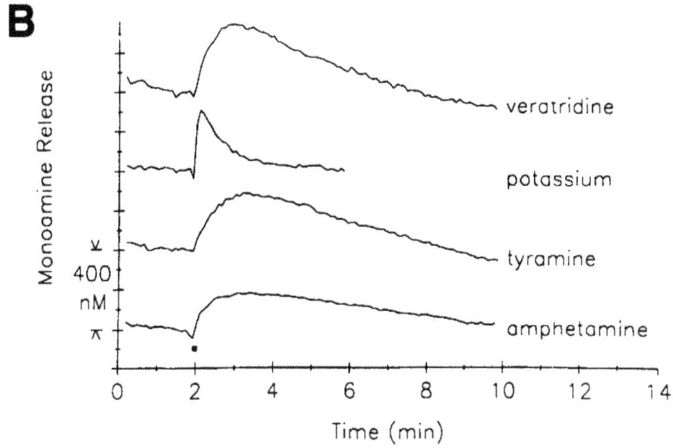

Fig. 13. **(A)** The effects of local application of K^+ via pressure ejection on monoamine release in the single hippocampal slice. Ejections of K^+ occur at the squares. Note the reproducibility of the responses seen with a 5-min interval between successive ejections of K^+. **(B)** Average release of monoamines seen following local application of veratridine, K^+, tyramine, and *N*-methylamphetamine in the dentate gyrus of single hippocampal slices. Ejections of the drugs occur at the square.

in records). Analogous to the K^+ superfusion studies, monoamine overflow produced by local application of K^+ is seen to be dose-dependent.

In contrast to the K^+ superfusion experiments, monoamine releases induced by local application of K^+ are also highly localized. Concomitant measures of CA1 pyramidal neuron population spike amplitudes show that the local applications of K^+ did not produce electrophysiological changes in the whole slice. Simultaneous electrochemical measurements of monoamine overflow made with a pair of Nafion-coated graphite GEC electrodes in the dorsal hilus and ventral hilus also show that no detectable response is recorded at the other site. Thus, the electrophysiological and electrochemical data support that the K^+ ejections produce highly localized effects in the single hippocampal slice.

3.3.1.2. Evoked Responses from Different Locally Applied Drugs

In efforts to investigate the releasing characteristics of other locally applied drugs, several other agents are locally applied analogous to local application of K^+. The average time-courses and magnitudes of monoamine releases elicited by local application of four different agents to the dentate gyrus of single hippocampal slices are illustrated in Fig. 13B. Applications are: veratridine, 200 $\mu M \times 20$ psi $\times$ 5 s; K^+, 120 m$M \times 80$ psi $\times$ 1 s; tyramine, 2 m$M \times 20$ psi $\times$ 5 s; *N*-methylamphetamine, 5 m$M \times 20$ psi $\times$ 5 s. Each release is elicited from a different slice. The data represent signals from six different slices that are averaged together. Veratridine, tyramine, and *N*-methylamphetamine elicit the evoked overflow of monoamines that have time-courses that are significantly longer than the effects of locally applied K^+ ($p < 0.01$).

3.3.2. Discussion

A major finding from the present studies is that both pre- and postsynaptic information regarding the monoaminergic neurotransmitter systems can be collected contemporaneously from single hippocampal slices. Potassium superfusion can be used to elicit changes in population spike amplitudes and electrochemical monoamine overflow responses that are recoverable, dose-dependent, and even reproducible, provided a time delay is imposed between stimulations. The electrochemical recordings support that the greatest monoamine signals can be recorded from the dentate gyrus, an area that has been previously shown to contain the largest amounts of NE and 5-HT in the hippocampal formation. Aside from K^+ superfusion, evoked overflow of monoamines could be reliably elicited in the dentate gyrus of single hippocampal slices by local application of K^+. These responses are highly localized, dose-dependent, and reproducible. Aside from K^+, other agents that have been previously shown to produce the overflow of monoamines caused detectable and recoverable responses when locally applied. Recent high-speed electrochemical recordings (data not shown) and lesion studies support that the electrochemical responses contain a significant NE contribution (Bickford-Wimer et al., 1991). However, only signals seen from amphetamine and tyramine applications appear to be more solely composed of NE. Responses recorded following K^+ or veratridine, as might be expected, contain an apparent mixture of 5-HT and NE. Thus, the data do support that both presynaptic and postsynaptic data regarding monoaminergic neurotransmitter function can be addressed using these techniques.

3.4. Clearance of Locally Applied NE from the Extracellular Space of the Rat Cerebellum

Recently, we have developed methods involving the controlled application of known amounts of DA or NE to investigate more directly the uptake of catecholamines by the high-affinity neuronal uptake carriers located on catecholamine-containing nerve fibers (Cass et al., 1992, 1993a,b; van Horne et al., 1992; Lin et al., 1993). Our studies in DA-containing brain areas support the hypothesis that the clearance of locally applied catecholamines from the extracellular space of the brain is primarily controlled by the high-affinity uptake carriers located on catecholamine-containing neurons, and not metabolism or diffusion (Cass et al., 1993b). However, preliminary results support that the clearance of NE is quite different from that seen for DA. Figure 14 shows results from the injection of small amounts (7.5 pmol) of DA into the nucleus accumbens and NE (2.5 pmol) ejections into the cerebellar cortex. As is seen, a 33% smaller ejection dose of NE

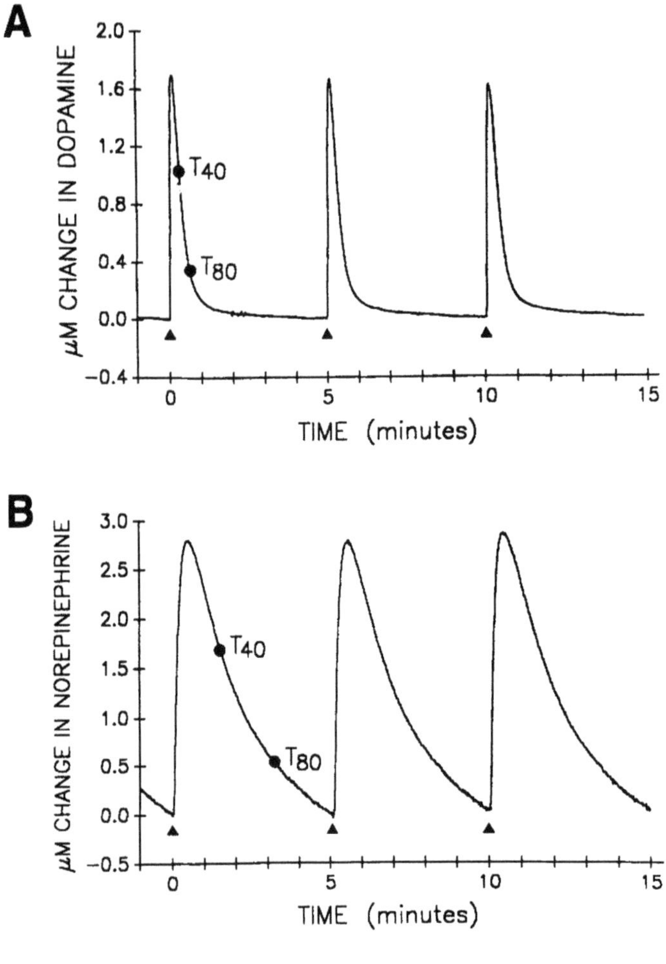

Fig. 14. Comparison between the extracellular clearance of **(A)** DA (7.5 pmol applied) in the nucleus accumbens and **(B)** NE (2.5 pmol applied) in the cerebellum. NE and DA are locally applied from micropipets positioned 300 μ from the Nafion-coated carbon-fiber electrodes.

has a significantly longer T40–80 time-course in the cerebellum. The T40–80 values average 12.8 ± 2.8 s for the 7.5-pmol ejections of DA in nucleus accumbens, whereas the 2.5-pmol ejections of NE have T-40–80 values that averaged 46 ± 2 s, $p < 0.001$. Thus, there is a dramatic difference between the regulation of DA vs NE.

Another series of studies investigates the effects of uptake inhibition on NE clearance in the cerebellum of the anesthetized rat using local application of nomifensine, a potent catecholamine uptake inhibitor (Tuomisto, 1977). Local applications of nomifensine alone are not seen to produce detectable changes in extracellular levels of NE. Such results are seen in 38 ejection attempts. However, nomifensine application 60–90 s prior to a second NE application is seen to increase the amplitude and time-course of locally applied NE in the rat cerebellum. Nomifensine produces a modest, but highly significant increase in the amplitude of extracellular NE (Lin et al., 1993). Amplitudes average 2.1 ± 0.3 μM in controls vs 2.3 ± 0.3 μM after nomifensine application, ($t = 4.0$, d.f. = 37, $p < 0.001$). Also the T50 (half decay time), and T40–80 time-course parameters of the signals are all significantly prolonged following nomifensine pretreatment. The T50 values average 104 ± 5 s vs 119 ± 5 s ($t = 6.84$, d.f = 40, $p < 0.001$) after nomifensine. The T40–80 average 46 ± 2 s in controls vs 60 ± 3 s after nomifensine application ($t = 6.55$, d.f. = 40, $p < 0.001$). Thus, nomifensine is seen to increase the amplitude and prolong the time-course of locally applied NE. Such data are consistent with the effects of catecholamine uptake inhibitors and their effects on catecholaminergic nerve endings (Cass et al., 1992, 1993a,b, van Horne et al., 1992; Gerhardt et al., 1987a,b, 1989).

4. Concluding Remarks

A major finding from these experiments is that both electrochemical and electrophysiological studies can be combined to study presynaptic events that occur during NE release events. In addition, the interpretation of NE overflow and clearance events can be greatly improved by using neurochemical, histological, and pharmacological approaches in conjunction with these recording methods. This is an important combination of technologies that should lead to a better understanding of neurotransmitter signaling processes. In addition, data regarding NE overflow and clearance in the cerebellum show that the removal of NE from the extracellular space is quite slow when compared to the rate of removal of DA in the nucleus accumbens. Such data will hopefully contribute to a better understanding of the multiple roles of NE in the CNS.

Acknowledgments

This work would not have been possible without help from Barry Hoffer, Robert Freedman, Mike Palmer, Thomas Dunwiddie, Greg Rose, Paula Bickford, Lotta Granholm, Ingrid Stromberg, Kevin Pang, Wayne Cass, Mei-Tsu Su, Anya Lin, Marilyn Friedemann, and Bill Proctor. This work is supported by USPHS Grants AG04418, AG06434, AG00441, and NS09199. Special thanks to Laura Lee Lamothe for helping prepare this manuscript and to Adam Coles for help with the artwork.

References

Adams R. N. (1990) *In vivo* electrochemical measurements in the CNS. *Prog. Neurobiol.* **35,** 297–311.

Adams R. N. and Marsden C. A. (1982) Electrochemical detection methods for monoamine measurements *in vitro* and *in vivo*, in *Handbook of Psychopharmacology*, vol. 15 (Iversen L. L., Iversen S. D., and Snyder S. H., eds.), Plenum, New York, pp. 1–74.

Bickford-Wimer P., Pang K., Rose G. M., and Gerhardt G. A. (1991) Electrically evoked release of norepinephrine in the rat cerebellum: an *in vivo* electrochemical and electrophysiological study. *Brain Res.* **558,** 305–311.

Bloom F. E., Hoffer B. J., and Siggins G. R. (1972) Norepinephrine mediated synapses: a model system for neuropharmacology. *Biol. Psychiatry* **4,** 157–177.

Cass W. A., Gerhardt G. A., Mayfield R. D., Curella P., and Zahniser N. R. (1992) Differences in dopamine clearance and diffusion in rat striatum and nucleus accumbens following systemic cocaine administration. *J. Neurochem.* **59,** 259–266.

Cass W. A., Zahniser N. R., Flach K. A., and Gerhardt G. A. (1993a) Clearance of exogenous dopamine in rat striatum and nucleus accumbens: role of metabolism and effects of locally-applied uptake inhibitors. *J. Neurochem.* **61,** 2269–2278.

Cass W. A., Gerhardt G. A., Gillespie K., Curella P., Mayfield R. D., and Zahniser N. R. (1993b) Reduced clearance of exogenous dopamine in rat nucleus accumbens, but not in dorsal striatum, following cocaine challenge in rats withdrawn from repeated cocaine administration. *J. Neurochem.* **61,** 273–283.

Cheng H.-Y., White W., and Adams R. N. (1980) Microprocessor-controlled apparatus for *in vivo* electrochemical measurement. *Anal. Chem.* **52,** 2445–2448.

Dunwiddie T. V. and Lynch G. S. (1978) Long-term potentiation and depression of synaptic responses in the rat hippocampus: localization and frequency dependency. *J. Physiol. (Lond.)* **276,** 353–367.

Dunwiddie T. V., Mueller A. L., and Basile A. S. (1983) The use of brain slices in central nervous system pharmacology. *Fed. Proc.* **42,** 2891–2898.

Eriksdotter-Nilsson M., Gerhardt G. A., Seiger A., Hoffer B. J., and Granholm A.-C. (1989) Multiple changes in noradrenergic mechanisms in the coeruleo-hippocampal pathway during aging. Structural and functional correlates from intraocular double transplants. *Neurobiol. Aging* **10,** 117–124.

Friedemann M. and Gerhardt G. A. (1992) Regional effects of aging on dopaminergic function in the Fischer-344 rat. *Neurobiol. Aging* **13,** 325–332.

Gerhardt G. A. and Palmer M. R. (1987) Characterization of the techniques of pressure ejection and microiontophoresis using *in vivo* electrochemistry. *J. Neurosci. Methods* **22,** 147–159.

Gerhardt G., Oke A., Moghaddam B., and Adams R. N. (1984a) Nafion-coated electrodes with high selectivity for CNS electrochemistry. *Brain Res.* **290,** 390–395.

Gerhardt G., Palmer M., Seiger A., Adams R. N., Olson L., and Hoffer B. J. (1984b) Adrenergic transmission in hippocampus-locus coeruleus double grafts in oculo: demonstration by in vivo electrochemical detection. *Brain Res.* **306,** 319–325.

Gerhardt G., Rose G., Stromberg I., Conboy G., Olson L., Jonsson G., and Hoffer B. (1985) Dopamine neurotoxicity of 1-methyl-4-phenyl-1,2,3,6-tetrahydropyridine (MPTP) in the mouse: an *in vivo* electrochemical study. *J. Pharmacol. Exp. Therap.* **235,** 259–265.

Gerhardt G. A., Rose G. M., and Hoffer B. J. (1986) Release of monoamines from striatum of rat and mouse evoked by local application of K^+: evaluation of a new *in vivo* electrochemical technique. *J. Neurochem.* **46,** 842–850.

Gerhardt G. A., Pang K., and Rose G. M. (1987a) *In vivo* electrochemical demonstration of the presynaptic actions of PCP in rat caudate nucleus. *J. Pharmacol. Exp. Therap.* **241,** 714–721.

Gerhardt G. A., Rose G. M., and Hoffer B. J. (1987b) *In vivo* electrochemical demonstration of potassium-evoked monoamine release from rat cerebellum. *Brain Res.* **413,** 327–335.

Gerhardt G. A., Gratton A., and Rose G. M. (1988) In vivo electrochemical studies of the effects of cocaine on dopamine nerve terminals in rat neostriatum. *Physiologia Bohemoslovaca* **37,** 249–257.

Gerhardt G. A., Friedemann M., Brodie M. S., Vickroy T. W., Gratton A. P., Hoffer B. J., and Rose G. M. (1989) The effects of cholecystokinin (CCK-8) on dopamine-containing nerve terminals in the caudate nucleus and nucleus accumbens of the anesthetized rat: an *in vivo* electrochemical study. *Brain Res.* **499,** 157–163.

Gerhardt G. A., Palmer M. R., and Granholm A.-Ch. (1991) Age-induced changes in single locus coeruleus brain transplants grown *in oculo*: an *in vivo* electrochemical study. *Neurobiol. Aging* **12,** 487–494.

Granholm A.-Ch., Gerhardt G. A., Bygdeman M., and Stromberg I. (1992) Human fetal xenografts of brainstem tissue containing locus coeruleus neurons: functional and structural studies of intraocular grafts in athymic nude rats. *Exp. Neurol.* **118,** 7–17.

Granholm A.-C., Gerhardt G. A., Eriksdotter-Nilsson M., Bickford-Wimer P. C., Palmer M. R., Seiger A., Olson L., and Hoffer B. J. (1987) Age-related changes in cerebellar noradrenergic pre- and postsynaptic mechanisms: intrinsic vs extrinsic determinants evaluated with brain grafts in oculo. *Brain Res.* **423,** 71–78.

Gratton A., Hoffer B. J., and Gerhardt G. A. (1988) Effects of electrical stimulation of brain reward sites on release of dopamine in rat: an *in vivo* electrochemical study. *Brain Res. Bull.* **21,** 319–324.

Gratton A., Hoffer B. J., and Gerhardt G. A. (1989) *In vivo* electrochemical studies of monoamine release in the medial prefrontal cortex of the rat. *Neuroscience* **29,** 57–64.

Hall M. E., Hoffer B. J., and Gerhardt G. A. (1989) Rapid and sensitive determination of catecholamines in small tissue samples by high pressure liquid chromatography coupled with dual-electrode coulometric electrochemical detection. *LC/GC* **7,** 258–265.

Hoffer B. J., Siggins G. R., Oliver A. P., and Bloom F. E. (1973) Activation of the pathway from locus coeruleus to rat cerebellar Purkinje neurons: pharmacological evidence of noradrenergic central inhibition. *J. Pharmacol. Exp. Therap.* **184,** 553–569.

Hoffer B. J., Seiger A., Taylor D., Olson L., and Freedman R. (1977) Seizures and epileptiform activity in hippocampus transplanted to the anterior chamber of the eye. *Exp. Neurol.* **54,** 233–250.

Hokfelt T. and Fuxe K. (1969) Cerebellar monoamine nerve terminals, a new type of afferent fiber to the cortex cerebella. *Exp. Brain Res.* **9,** 63–72.

Kobayashi R. M., Palkovitz M., Kopin I. J., and Jacobowitz, D. M. (1974) Biochemical mapping of noradrenergic nerves arising from the rat locus coeruleus. *Brain Res.* **77,** 269–279.

Lin A. M.-Y., Bickford P. C., Palmer M. R., and Gerhardt, G. A. (1993) Ethanol inhibits the uptake of exogenous norepinephrine from the extracellular space of the rat cerebellum. *Neurosci. Lett.* **164,** 71–75.

Lindvall O. and Bjorklund A. (1974) The organization of the ascending catecholamine neuron systems in the rat brain as revealed by the glyoxylic acid fluorescence method. *Acta Physiol. Scand. Suppl.* **412,** 1–48.

Nagy G., Gerhardt G. A., Oke A. F., Rice M. E., Adams R. N., Moore R. B. III, Szentirmay M. N., and Martin C. R. (1985) Ion exchange and transport of neurotransmitters in Nafion films on conventional and microelectrode surfaces. *J. Electroanal. Chem.* **188,** 85–94.

Nygren L.-G. and Olson L. (1977) A new major projection from locus coeruleus: the main source of noradrenergic nerve terminals in the ventral and dorsal columns of the spinal cord. *Brain Res.* **132,** 85–93.

Olson L. and Fuxe K. (1971) On projections from the locus coeruleus noradrenaline neurons: the cerebellar innervation. *Brain Res.* **28,** 165–171.

Olson L., Seiger A., Hoffer B., and Taylor D. (1979) Isolated catecholaminergic projections from substantia nigra and locus coeruleus to caudate, hippocampus and cerebral cortex formed by intraocular sequential double brain grafts. *Exp. Brain Res.* **35,** 47–67.

Olson L., Bjorklund H., Freedman R., Goldowitz D., Hoffer B., Palmer M. R., Seiger A., Taylor D., and Woodward D. (1983a) Intrinsic and extrinsic determinants of brain development as evidenced by grafting of fetal brain tissue, in *Developing and Regenerating Vertebrate Nervous Systems* (Coates, P. W., Markwald R. R., and Kenny A. D., eds.), Liss, New York, pp. 121–150.

Olson L., Stromberg I., and Seiger A. (1983b) Intraocular transplantation in rodents. A detailed account for the procedure and examples of its use in neurobiology with special reference to brain tissue grafting, in *Advances in Cellular Neurobiology,* vol. 4 (Fedoroff S. and Hertz L., eds.), Academic, New York, pp. 407–442.

Palmer M. R., Wuerthele S. M., and Hoffer B. J. (1980) Physical and physiological characteristics of micropressure ejection of drugs from multibarreled pipettes. *Neuropharmacology* **19,** 931–938.

Paxinos G. and Watson C. (1986) *The Rat Brain in Stereotaxic Coordinates* (2nd ed.), Academic, New York.

Siggins G. R., Hoffer B. J., Oliver A. P., and Bloom F. E. (1971) Activation of a central noradrenergic projection to cerebellum. *Nature* **233,** 481–483.

Su M.-T., Dunwiddie T. V., Mynlieff M., and Gerhardt G. A. (1990a) Electrochemical characterization of stimulated norepinephrine release in locus coeruleus-hippocampus double brain grafts grown *in oculo. Neurosci. Lett.* **110,** 186–192.

Su M.-T., Dunwiddie T. V., and Gerhardt G. A. (1990b) Combined electrochemical and electrophysiological studies of monoamine overflow in rat hippocampal slices. *Brain Res.* **518,** 149–158.

Tuomisto J. (1977) Nomifensine and its derivatives as possible tools for studying amine uptake. *Eur. J. Pharmacol.* **42,** 101–106.

van Horne C., Hoffer B. J., Stromberg I., and Gerhardt G. A. (1992) Clearance and diffusion of locally applied dopamine in normal and 6-hydroxydopamine-lesioned rat striatum. *J. Pharmacol. Exp. Therap.* **263,** 1285–1292.

Wightman R. M. and Zimmerman J. B. (1990) Control of dopamine extracellular concentration in rat striatum by impulse flow and uptake. *Brain Res. Rev.* **15,** 135–144.

Wightman R. M., Amatore C., Engstrom R. C., Hale P. D., Kristensen E. W., Kuhr W. G., and May L. J. (1988) Real-time characterization of dopamine overflow and uptake in the rat striatum. *Neuroscience* **25,** 513–523.

Monitoring Dopamine and Noradrenaline Release in Central and Peripheral Nervous Systems with Treated and Untreated Carbon-Fiber Electrodes

F. Gonon

1. Introduction

First attempts to monitor dopamine (DA) release in the living rat brain with an implanted carbon electrode coupled with electrochemical techniques were performed by Adams and coworkers in 1973 (Kissinger et al., 1973). However, the unequivocal monitoring of DA released by dopaminergic terminals in the rat striatum was accurately achieved only 10 yr after. In 1983, Wightman and coworkers were able to record with fast cyclic voltammetry the DA release that was evoked by brief and vigorous electrical stimulation of the dopaminergic pathway (Ewing et al., 1983). One year later, we recorded a voltammetric peak that was entirely owing to extracellular DA providing that rats were treated with an inhibitor of monoamine oxidase, pargyline (Gonon et al., 1984a). The fact that 10 yr of intense efforts were required to obtain the first reliable measurements points out how difficult the task was.

One major step along this road was the design by this author of the first carbon-fiber electrode (Gonon et al., 1978) and its improvement by electrochemical treatments (Gonon

From: *Neuromethods, Vol. 27: Voltammetric Methods in Brain Systems*
Eds: A. Boulton, G. Baker, and R. N. Adams © 1995 Humana Press Inc.

et al., 1980). In 1980, this treated carbon-fiber electrode was combined with differential pulse voltammetry, and we showed, first, that the extracellular DA level was in the nanomolar range and, second, that its electrochemical measurement was hindered by the presence in the extracellular fluid of two other easily oxidizable compounds found in much higher concentration: namely, ascorbic acid (AA) and DOPAC (a direct DA metabolite) (Gonon et al., 1980). This led us to propose a voltammetric method for monitoring simultaneously in the living rodent brain two independent signals corresponding to AA and to DOPAC. This approach has been used with some modifications by Cespuglio and coworkers for the in vivo monitoring of the main serotonin (5-HT) metabolite, 5-HIAA. These voltammetric techniques and their fields of application have been reviewed recently elsewhere (Suaud-Chagny et al., 1993) and are, thus, outside the scope of the present chapter.

During the last 10 yr several groups developed new carbon-fiber electrodes and electrochemical techniques for monitoring DA and noradrenaline (norepinephrine, NE) release. Three main streams can be distinguished. The first one combines fast cyclic voltammetry with carbon-fiber electrodes, which are often, but not always, coated with Nafion. This approach was almost exclusively used to monitor DA and NE release, which was evoked by electrical stimulation of the catecholaminergic pathway. It is extensively described in this book (chapters by Wightman and Stamford et al.). The second approach was developed by Adams and Gerhardt: Carbon-fiber electrodes were coated with Nafion and were used in combination with high-speed chronoamperometry. Chapter 4 is devoted to this second approach. The last one was mainly developed by my group: Carbon-fiber electrodes were, in most cases, electrochemically treated, and coupled with various voltammetric and amperometric techniques. This third approach was used for three purposes that will be discussed successively in this chapter: the in vitro monitoring of the NE released by sympathetic nerve terminals, the in vivo recording of the basal extracellular catecholamine

Fig. 1. Photograph of the tip of a carbon-fiber electrode after its implantation in the rat brain. The active part of the electrode is the carbon fiber (8 µm in diameter and 250 µm in length) protruding out of a glass micropipet. Notice that the tip of the fiber is coated with tissue fragments, whereas the main cylindric surface remains uncovered.

level, and finally, the in vivo monitoring of the DA and NE release evoked by brief stimulations.

2. Electrochemical Methods

2.1. Carbon-Fiber Electrodes

Although the carbon fiber is the most popular compound used for preparing electrodes, electrodes designed by different groups differ according to the number of fibers, their diameter, and their active length. In our group, we now use electrodes whose active part is only one carbon-fiber (SOFICAR, Paris, France) 8 µm in diameter and 50–250 µm in length. They were produced as previously described (Gonon et al., 1984b). Our choice was guided by one main consideration: The electrode shape must be designed to minimize the coating of the active surface by tissue fragments, which occur during the brain penetration. Figure 1 shows an electrode that has been implanted in the rat brain. One can notice that the end of the carbon-fiber is coated with tissue fragments, whereas the main part of the cylindric surface is not. Therefore, diffusion of oxidizable compounds to this cylindric surface is not hindered.

2.2. Electrochemical Treatment of Carbon-Fiber Electrodes

When untreated electrodes are combined with differential pulse voltammetry, their selectivity is poor: AA and catechol oxidations give only one broad peak (Gonon et al., 1981). Electrochemical treatments have been designed to improve this selectivity. They have already been described in detail elsewhere (Gonon et al., 1981, 1984a). Briefly, a carbon-fiber, an Ag/AgCl reference electrode, and an auxiliary stainless-steel electrode are connected to a potentiostat (Biopulse, Radiometer Analytical, Villeurbanne, France) and immersed in a saline solution. The main treatment involves the application for 20 s to the carbon-fiber electrode, vs the reference electrode, of an alternating potential of a triangular wave form (frequency 70 Hz). This potential varies between 0 V and an upper limit that is chosen between 2.0 and 2.8 V, depending on the purpose of the study. This main treatment is usually, but not always, followed by subsidiary treatments at lower potentials.

These electrochemical treatments greatly improve the catalytic properties of the carbon surface. Therefore, the oxidation reactions become much more reversible. For example, before treatment carbon-fiber electrodes combined with voltammetry give a broad peak at +0.6 V for AA oxidation, whereas after treatment, this peak becomes well defined and appears at –0.1 V (Gonon et al., 1981). This treatment allowed us to resolve catechols from AA in vitro and in the living rat brain (Gonon et al., 1980, 1981, 1984a). Unfortunately, this treatment complicates the in vivo monitoring of catecholamines. In fact, a major catechol metabolite, DOPAC, oxidizes at a potential which is close to that of the parent amine (+55 mV for DOPAC and +85 mV for DA) (Gonon et al., 1984a). Moreover, the DOPAC extracellular concentration exceeds that of the parent amine by two to three orders of magnitude (Gonon and Buda, 1985; Gonon, 1988).

In addition, these treatments enhance tremendously the sensitivity for cations, including DA, NE, and serotonin. In fact, cations are reversibly adsorbed at the treated carbon surface. Thus, the detection limit for monoamines falls below 10 n*M* in vitro and in vivo (Gonon et al., 1984a, Gonon and Buda, 1985; Mermet and Gonon, 1986). Unfortunately, since this reversible adsorption is relatively slow, this high sensitivity is paid for by a slowing down of the electrode response (time to reach 50% of the full response: 1 min) (Gonon and Buda, 1985).

2.3. Differential Normal Pulse Voltammetry

Voltammetry involves exploring a range of potentials and measuring an oxidation current. Therefore, this technique provides analytical information concerning the oxidation potential of oxidizable compounds. The most classical voltammetric technique in analytical chemistry is differential pulse voltammetry. This technique is slow (acquisition duration for one voltammogram is between 10 and 60 s) but exhibits high selectivity since oxidation reactions appear as current peaks. We used this technique for monitoring in the rat brain the extracellular levels of AA and of DOPAC (Gonon et al., 1980; Buda et al., 1983; Suaud-Chagny et al., 1986). However, this technique was not sensitive enough to monitor the basal DA extracellular level in the striatum of pargyline-treated rats (Gonon et al., 1984a). Therefore, we designed a new technique, differential normal pulse voltammetry (Fig. 2) (Gonon et al., 1984a). The only modification over differential pulse voltammetry is that between pulse measurements, the working electrode is held at the resting potential (usually –0.24 V). When this new technique is coupled with electrochemically treated electrodes, it is about five times more sensitive than differential pulse voltammetry for monitoring extracellular catecholamines (Gonon et al., 1984a).

2.4. Differential Pulse Amperometry

The main drawback of differential normal pulse voltammetry is its poor time resolution (one measurement every 1 min). When the purpose of the study is to record the variations with time of only one oxidizable compound, this technique gives too much information and is too slow. Therefore, we developed a faster technique that is derived from it: differential pulse amperometry (Fig. 3). This technique is as selective and sensitive as differential normal pulse voltammetry and has a time resolution of 1 s (Marcenac and Gonon, 1985).

2.5. Continuous Amperometry

This technique is used widely to detect oxidizable compounds at the output of liquid chromatographic columns. It con-

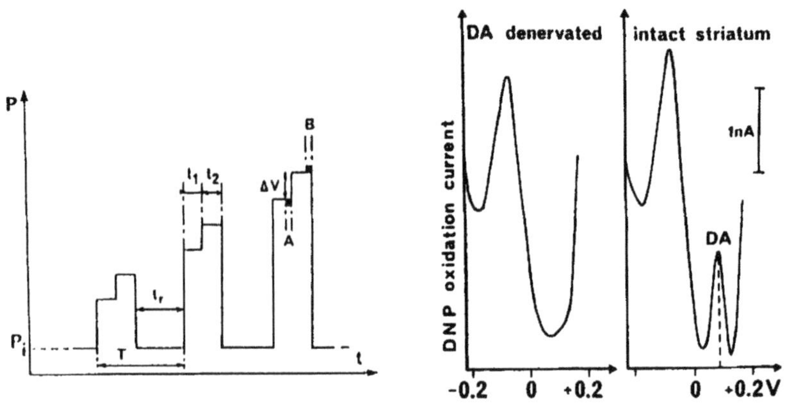

Fig. 2. Principle of differential normal pulse voltammetry. The left part of the figure shows the evolution vs time of the potential applied to the electrode during the acquisition of one voltammogram. The current is sampled just before **(A)** and at the end of the measuring pulse **(B)**. The difference between both currents is stored for each pulse, thus achieving a local differentiation of the oxidation waves. The measuring pulse (amplitude, $\Delta V = 40$ mV; duration, $t_2 = 40$ ms) is added to a "normal" pulse (duration, $t_1 + t_2 = 140$ ms) whose amplitude linearly increases step by step (4 mV every $T = 0.4$ s) in order to sweep the entire potential range from an initial potential ($P_i = -240$ mV) to a final potential (+200 mV). Therefore, oxidation reactions appear as peaks on the voltammograms. The voltammograms shown in the right part of the figure were recorded from the striatum of an anesthetized rat treated with pargyline. The same electrode was implanted in the left and then in the right striatum. The dopaminergic innervation of the left striatum had been selectively destroyed by a 6-OHDA injection in the left substantia nigra 8 d before recording (from Gonon and Buda, 1985).

sists of applying to the working electrode a constant potential and detecting the current resulting from oxidation of every compound that is oxidized at this fixed potential. This technique is poorly selective, but exhibits very high sensitivity. Its second main advantage has been pointed out recently by Wightman's group: Its time resolution is so good that it can measure the real kinetics of adrenaline secretion resulting from exocytosis of a single chromaffin granule (Wightman et al., 1991).

2.6. Instrumentation

Carbon-fiber electrodes are produced by Radiometer Analytical (Villeurbanne, France) and by World Precision

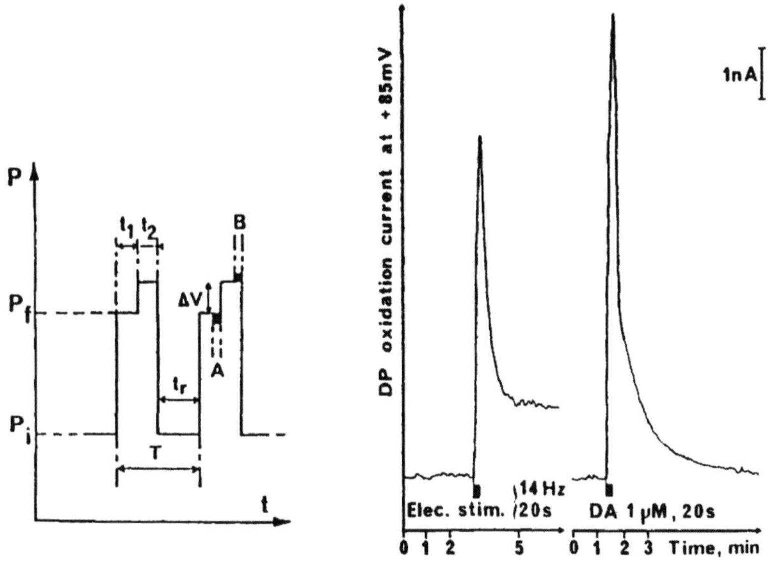

Fig. 3. Principle of differential pulse amperometry. The left part shows the evolution vs time of the potential applied to the electrode. The current is sampled just before **(A)** and at the end of the measuring pulse **(B)**. The difference between both currents is stored. The measuring pulse (amplitude, ΔV = 40 mV; duration, t_2 = 40 ms) is added to a "normal" pulse (duration, $t_1 + t_2$ = 140 ms) whose amplitude is constant from the resting potential (P_i = –240 mV) to the oxidation potential of the compound being studied (P_f = + 85 mV for DA). The right part of the figure shows two typical responses recorded with the same electrode. The left part was recorded in the olfactory tubercle and was evoked by an electrical stimulation of the dopaminergic pathway. The right part was recorded in vitro after the in vivo implantation (from Gonon, 1988).

Instruments (Sarasota, FL). The "Biopulse" from Radiometer is equipped with a complete treatment unit that permits the electrochemical treatment of carbon-fiber electrodes without the need for further instrumentation. Differential normal pulse voltammetry and differential pulse amperometry are also available with the "Biopulse" either in a manual mode or automatically repeated.

Continuous amperometry can be performed by various HPLC detectors, but usually, these instruments have a strong internal filter and, therefore, do not permit a good time resolution. A patch-clamp amplifier has been used by Neher's group

(Chow et al., 1992). Recently, Radiometer Analytical developed, with this author's collaboration, a new amperometric detector called AMU 110, which has a very fast response time (1 ms) and an excellent sensitivity (detection limit below 0.1 pA).

3. Experimental Models

3.1. Sympathetic Nerve Terminals Innervating Isolated Organs

Sympathetic nerve terminals release NE when they are stimulated. This phenomenon has been studied for a long time by classical biochemical techniques in vivo and in isolated preparations. The advantage of isolated organs over in vivo preparations is that the medium bathing the nerve terminals is artificially controlled. When a carbon-fiber electrode is in contact with the external surface of an isolated rat tail artery, the only oxidizable compound that can be detected is NE released by sympathetic nerve terminals (Mermet et al., 1990a; Gonon et al., 1993a). Therefore, this experimental model is especially suitable for investigating NE release with electrochemical techniques, including those, such as continuous amperometry, that exhibit a poor selectivity (Gonon et al., 1993b). Experimental procedures have been described previously in detail (Mermet et al., 1990a; Gonon et al., 1993a,b).

3.2. The Living Rat Brain

Obviously, the cerebral tissue of living rodents is a much more complex medium than the artificial fluid bathing isolated organs. Nevertheless, catecholamine release can also be monitored in vivo with carbon-fiber electrodes. Experimental procedures have been extensively described elsewhere (Gonon et al., 1984a,b; Gonon and Buda, 1985; Gonon, 1988; Suaud-Chagny et al., 1990, 1991). However, some critical points are stressed below.

First, in vitro calibrations of electrodes must be performed after their in vivo implantations, since this implantation alters their sensitivity. This is easy when electrodes are implanted in anesthetized animals, but more difficult with implantations in freely moving ones. Several micromanipulators that can be implanted on the rat skull have been developed (Louilot et al., 1987; Rebec et al., 1993). They allow the carbon-fiber electrode to be changed rapidly without the need of anesthesia. Therefore, in vitro calibration can be performed after in vivo recordings.

Second, one can avoid, as far as possible, the covering of the electrode surface by tissue fragments, which occurs during penetration in the living brain. Thus, after opening the skull and dura mater, the pia mater must be punctured by means of a used carbon-fiber electrode. Then, the new electrode is implanted through this hole.

Third, despite great care in the experimental procedure, the absolute amplitude of the catecholamine release appears rather variable from one experiment to the other. This variability might be partly the result of local anatomical heterogeneity in the density and activity of catecholaminergic fibers (May and Wightman, 1989; Suaud-Chagny et al., 1992a). Therefore, it can be useful to check the electrode localization after recordings. This localization can be revealed by making a small electrolytic lesion around the carbon fiber followed by a classical histological control. A suitable lesion is performed by applying an anodic potential (+5 V) to the carbon-fiber electrode for 3 s.

3.3. Methods for Evoking Catecholamine Release

The most commonly used technique for evoking catecholamine release is electrical stimulation of the catecholaminergic pathway. We used a bipolar stimulating electrode (SNEX-200, Rhodes Medical Instruments, USA), and the electrical stimulation consisted of square-wave pulses (250 μA, 0.5 ms). Higher current does not evoke larger release and might damage the tissue. These current pulses were applied by a stimulator (DS2, Digitimer, USA) that is electrically isolated in order to avoid interferences with the electrochemical measurements. However, these stimulations evoked artifacts when the electrochemical signal was continuously recorded with amperometry (*see* Figs. 4 and 5). These artifacts were eliminated with differential pulse amperometry, providing

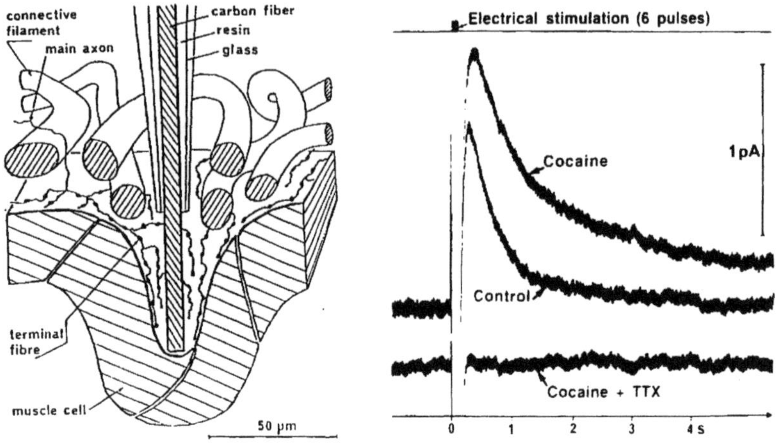

Fig. 4. Detection, by a carbon-fiber electrode coupled with amperometry, of NE released by sympathetic nerve terminals of an isolated rat tail artery. A carbon-fiber electrode whose active length was 50 µm was put in contact with the external surface of the artery as indicated in the diagram. A potential of +300 mV was continuously applied to this electrode vs an Ag/AgCl reference electrode. Electrical stimulation of the proximal end of the artery (6 pulses at 50 Hz) evoked a rapid increase in the oxidation current followed by a slower decline. Addition of cocaine (3 µ*M*) in the perfusion medium bathing the artery induced a decrease in the rate of decline, whereas further addition of tetrodotoxin (TTX, 0.5 µ*M*) completely abolished the evoked effect (Gonon et al., 1993b).

that the electrical stimulations were synchronized by the pulse measurement (Suaud-Chagny et al., 1992a).

Another method for stimulating the discharge activity of catecholaminergic neurons involves applying excitatory amino acids at the level of the neuronal cell bodies (Fig. 6). A glass pipet (external tip diameter: 40 µm) was filled with glutamate (10 m*M*) or NMDA (0.1 m*M*) dissolved in a saline solution and was implanted in the corresponding cell-body area (LC, ventral tegmental area) (Suaud-Chagny et al., 1992b; Brun et al., 1993). Controlled volume was ejected by pneumatic pressure at a rate of 10–20 nL/s. Usually, a complete stimulation of the whole nucleus was obtained with a volume of 30–100 nL.

Finally, catecholaminergic neurons can be activated by peripheral stimulations. For example, in anesthetized rats, electrical stimulation of the sciatic nerve evoked an increase

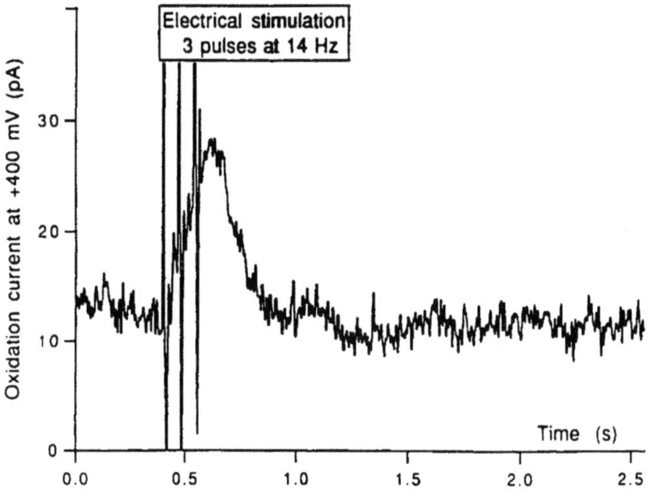

Fig. 5. In vivo monitoring of electrically evoked DA release by an untreated carbon-fiber electrode combined with amperometry. Experimental conditions were as previously described (Dugast et al., 1994). Briefly, a carbon-fiber electrode was implanted in the nucleus accumbens of an anesthetized rat and connected to an amperometric detector (AMU 110). A continuous potential of +400 mV was applied to this electrode vs an Ag/AgCl reference electrode, and the oxidation current occurring through the electrodes was continuously recorded and digitized at 1 kHz. Electrical stimulation of the dopaminergic pathway consisted of three pulses at 14 Hz that were applied every 1 min. The figure shows the average of five consecutive responses.

in the discharge activity of LC neurons that elicited a detectable release of NE in the anteroventral thalamic nucleus (Brun et al., 1993). In freely moving animals, ether stress evoked a release of NE in the paraventricular hypothalamic nucleus (Mermet and Gonon, 1988), whereas sexual activity was correlated with an increase in DA release in the nucleus accumbens (Mas et al., 1990; Louilot et al., 1991).

4. Applications

4.1. Noradrenaline (Norepinephrine) Released by Sympathetic Terminals In Vitro

It is amazing to notice that although this application of electrochemical techniques is the easiest one from the experi-

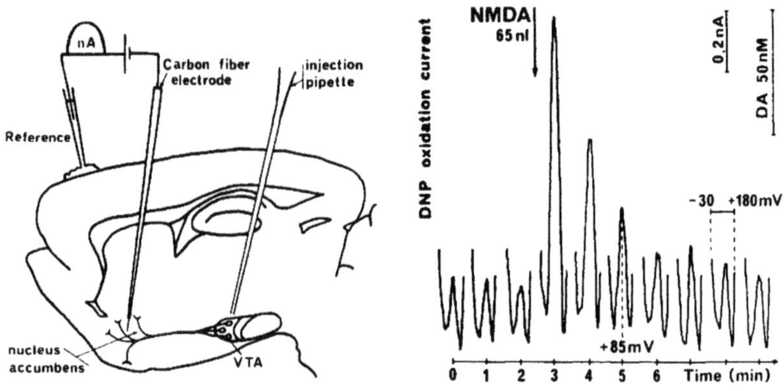

Fig. 6. Effect of a local NMDA ejection in the ventral tegmental area on the extracellular DA concentration in the nucleus accumbens. An electrochemically treated carbon-fiber electrode was implanted in the nucleus accumbens of a rat treated with pargyline and anesthetized with chloral hydrate. The oxidation peak, appearing at +85 mV and corresponding to extracellular DA, was recorded every 1 min by differential normal pulse voltammetry. NMDA was dissolved in an artificial CSF at a concentration of 0.1 m*M* and ejected through a glass pipet by air pressure (ejection duration: 10 s). After this in vivo recording the carbon-fiber electrode was calibrated in vitro for DA (from Suaud-Chagny et al., 1992b).

mental point of view, it has only recently been opened. When a carbon-fiber electrode is in close contact with sympathetic nerve terminals of the isolated rat tail artery, the spontaneous release of NE at rest is below the detection limit (10 n*M*) (Mermet et al., 1990a; Gonon et al., 1993a). Electrical stimulation at a distance from the recording site evoked an increase in the oxidation current appearing at +0.1 V, i.e., the oxidation potential for NE. This increase was closely time-correlated with the stimulation and was actually the result of NE released by sympathetic nerve terminals for the following reasons. First, it was abolished by tetrodotoxin and by guanethidine (a specific blocker of sympathetic transmission). Second, it is suppressed by omission of calcium in the perfusion medium or by the addition of cadmium, which blocked the calcium channels. Third, the amplitude and duration of this evoked oxidation current was enhanced when the reuptake of the released NE was inhibited either by cocaine or by desipramine. Fourth, the maximal increase grew with increasing duration of the stimulation until a plateau was reached. Moreover, the level of this plateau was enhanced with increasing the stimulation frequency (Mermet et al., 1990a; Gonon et al., 1993a).

Continuous amperometry allowed us to measure the kinetics of NE release and elimination with a better time resolution. We showed for the first time that these kinetics are at least 10 times faster than those of the arterial contraction caused by this release (Fig. 4) (Gonon et al., 1993b). Moreover, since we were capable of detecting the release caused by a single shock stimulation and at a level of some tens of sympathetic varicosities, we showed that this release is intermittent (Msghina et al., 1993). One main conclusion of our studies is that NE release, revealed by fast electrochemical techniques, exhibits the same properties as those of ATP, which is also released by sympathetic nerve terminals. This strongly suggests that NE and ATP are released together in the same quantal packet (Msghina et al., 1992; 1993, Msghina and Stjärne, 1993).

4.2. Basal Extracellular Concentration of Catecholamines in the Living Brain

4.2.1. Animals Treated with Pargyline

As already mentioned in the introduction, in the living brain, the electrochemical monitoring of the extracellular catecholamine concentration is complicated by the fact that other easily oxidizable compounds are present in the extracellular fluid in much higher concentration. Although the interference from AA has been resolved by using electrochemically treated carbon-fiber electrodes, the interference owing to DOPAC (a major DA metabolite) was not. Therefore, the safest solution for recording a pure signal entirely resulting from extracellular DA or NE involved a pharmacological suppression of the DOPAC contribution: The DOPAC synthesis was inhibited with pargyline, and 1 h after this treatment the extracellular DOPAC level was below the detection limit (Gonon et al., 1984a, Gonon and Buda, 1985). Under these particular experimental conditions, the oxidation current appearing at the catechol oxidation potential was actually solely the result of extracellular DA or NE, depending on the brain area under investigation, as stated by anatomical, physiological, and pharmacological evidence (Figs. 2 and 6) (Gonon et al., 1984a; Gonon and Buda, 1985; Mermet and Gonon, 1986).

Using this approach we showed that even in brain areas densely innervated by catecholaminergic terminals, the basal extracellular concentration is between 10 and 20 n*M* (Fig. 6) (Gonon et al., 1984a; Gonon and Buda, 1985; Mermet and Gonon, 1986; Gonon, 1988). This estimate is closely consistent with those obtained with microdialysis (Smith et al., 1992). Variations of the extracellular catecholamine concentration have also been observed in various experimental situations that modified the impulse flow in catecholaminergic nerves (Fig. 6). The main conclusion is that these variations are closely time-correlated with those of the impulse flow (Gonon and Buda, 1985; Gonon, 1988; Suaud-Chagny et al., 1992b). In anesthetized animals, at rest, about 70% of the basal extracellular catecholamine level is related to catecholamine released by spontaneous impulse flow (Gonon and Buda, 1985; Gonon, 1988; Suaud-Chagny et al., 1992b).

However, the main drawback of this approach is that it requires a pharmacological treatment of the animals. Pargyline treatment clearly enhances the extracellular DA concentration as revealed by microdialysis, although according to recent studies (Di Chiara et al., 1993), this effect (smaller than a doubling) seems not as large as previously reported (Imperato and Di Chiara, 1984). Therefore, this pargyline effect must always be kept in mind when this approach is used.

4.2.2. Mathematical Analysis of the Catechol Peak

Another approach for resolving DA from DOPAC has been developed by others (Gonzalez-Mora et al., 1988; Mas et al., 1990; Louilot et al., 1991). Its principle is based on the fact that, at electrochemically treated carbon-fiber electrodes, DOPAC oxidizes at +55 mV, whereas DA does so at +85 mV. Therefore, mathematical analysis of the catechol peak that was recorded with differential normal pulse voltammetry allowed them to separate the DA contribution from that of DOPAC. Pharmacological data are consistent with those obtained with microdialysis and strongly suggest that this separation procedure is valid when large variations are observed. For example, amphetamine induced the expected increase in the DA contribution and a decrease in the DOPAC one (Gonzalez-Mora et al., 1988).

This approach has also been used for monitoring the DA release in freely moving animals in various behavioral situations (Mas et al., 1990; Louilot et al., 1991). According to these studies, the observed variations of the signal attributed to extracellular DA seem slower than expected. However, it must be kept in mind that electrochemically treated carbon-fiber electrodes exhibit a very good sensitivity for DA, which is paid for by a slowing down of the response time (*see* Section 2.2.).

4.2.3. Nafion-Coated Carbon-Fiber Electrodes

Since the coating of carbon-fiber electrodes with Nafion improved their selectivity for DA vs DOPAC, this procedure has been used in an attempt to record a voltammetric peak entirely resulting from extracellular DA. However, although Crespi and Möbius (1992) reported that this goal was reached, Wiedemann et al. (1990) showed that this coating was not sufficient to reject the interferences owing to AA and DOPAC. They concluded that "the benefits of Nafion for in vivo detection of DA are limited to those situations in which DA overflow changes on a second or subsecond time scale and there is no corresponding change in the level of interferent molecules."

4.3 Catecholamine Overflow Evoked in the Living Brain by Brief Stimulations

4.3.1. Electrochemically Treated Electrodes

Electrochemically treated carbon-fiber electrodes are 100 times more sensitive to DA than to DOPAC. Nevertheless, since the extracellular DOPAC level exceeds the DA level by two to three orders of magnitude depending on the brain areas under investigation, the catechol peak recorded with these electrodes in basal conditions is still mainly the result of DOPAC. However, brief stimulations of the impulse flow in catecholaminergic nerves should evoke variations in the amplitude of the catechol oxidation current that are solely the result of the evoked release of the amine. In fact, variations of the DOPAC level are much slower, since the half-life of DOPAC is about 10 min.

According to this hypothesis, we recorded every 1 s with differential pulse amperometry the variations of the catechol oxidation current and observed that electrical stimulation of the catecholaminergic pathway for 20 s evoked variations in this current that were closely time-correlated with the stimulation. Anatomical and pharmacological arguments support the view that these evoked effects were entirely owing to the evoked release of the amine (Gonon, 1988; Suaud-Chagny et al., 1989, 1990). This approach was sensitive enough to detect the release evoked by electrical stimulations at physiological frequencies, i.e., between 3 and 20 Hz (Gonon, 1988; Suaud-Chagny et al., 1990). In these studies, the electrochemical treatment was not as strong as that used for the monitoring of the basal amine level from pargyline-treated animals (the maximal potential of the triangular treatment was 2.45 V instead of 2.7 V) (Gonon, 1988). In fact, a compromise between a response time that is as good as possible and high sensitivity must be optimized. With this lower treatment, the response time of the electrode dropped from 1 min (Gonon and Buda, 1985) to about 20 s (Fig. 3) (Gonon, 1988; Suaud-Chagny et al., 1990).

One obvious limitation of this approach in some brain areas is that it is all the more difficult to use as the DOPAC level increases. For example, in the olfactory tubercle, this level is four times lower than that of the nucleus accumbens, although the density of the dopaminergic innervation is very close in both nuclei. Therefore, the DA release evoked by electrical stimulation at low frequencies (below 5 Hz) is easier to detect in the olfactory tubercle. The monitoring of NE release is less affected by this limitation, since in brain areas densely innervated by noradrenergic terminals, the extracellelar DOPAC level is one order of magnitude lower than in dopaminergic fields exhibiting a similar density of catecholaminergic nerve terminals (Mermet and Gonon, 1986).

Despite these limitations, important observations have been made with this approach. The relationship between impulse flow and catecholamine overflow has been investigated at physiological frequencies, and an apparent nonlinearity has been described (Gonon, 1988; Suaud-Chagny et al., 1990; Brun et al., 1993). Electrical stimulations mimicking both discharge modes of dopaminergic neurons have been tested, and we showed that the bursting mode was twice as potent as the regularly spaced mode to evoke a DA overflow in the extracellular fluid (Gonon, 1988). The presynaptic autoinhibition of the catecholamine release has also been studied in detail with this approach (Mermet et al., 1990b; Suaud-Chagny et al., 1991).

Finally, this approach was also found suitable for monitoring the catecholamine release evoked by other kinds of brief stimulations. Local ejection of excitatory amino acids in cell-body areas evoked a corresponding amine release that was closely time-correlated with the evoked increase in the discharge activity of catecholaminergic neurons (Suaud-Chagny et al., 1992b; Brun et al., 1993). Electrical stimulation of the sciatic nerve evoked a detectable release of NE in the antero-ventral thalamic nucleus (Brun et al., 1993).

4.3.2. Untreated Electrodes Coupled with Differential Pulse Amperometry

As discussed above, electrochemically treated carbon-fiber electrodes are very sensitive to amines (detection limit below 10 n*M*). However, their response is linear only with submicromolar amine concentrations. Even electrical stimulations whose frequencies are still physiologic (10–20 Hz) can evoke amine overflow that exceeds 1 μ*M* after some pharmacological treatments, such as haloperidol or sulpiride (Suaud-Chagny et al., 1991). Therefore, these large effects cannot be recorded reliably with treated electrodes. Untreated electrodes coupled with differential pulse amperometry offer a more valid approach in these conditions (Suaud-Chagny et al., 1991, 1992a). Furthermore, since the potential applied to the electrode was +0.2 V, at this potential, these electrodes are totally insensitive to DOPAC and only poorly sensitive to AA (Suaud-Chagny et al., 1992a). Finally, their response time is much better than for treated electrodes. Therefore, the response time of this approach is that of differential pulse amperometry (1 s).

4.3.3. Untreated Electrodes Coupled with Continuous Amperometry

At first glance, the use of continuous amperometry in the living brain for monitoring catecholamine release seems open to criticism. In fact, as discussed above, this technique cannot resolve amines from other easily oxidizable compounds, such as AA and DOPAC, whose extracellular concentration in the living brain is much higher than that of amines. However, brief stimulations do not evoke any increase in the extracellular levels of AA (Gonon and Buda, 1985) or of DOPAC (Gonon, 1988). Therefore, it can be expected that the variations in the oxidation current, recorded at +0.4 V with amperometry and evoked by brief stimulations, might solely be the result of amine release.

Recently, we tested this assumption with regard to the electrically evoked DA release in the nucleus accumbens of anesthetized rats (Dugast et al., 1994). We observed that the evoked increases in the oxidation current were closely correlated with the stimulation and exhibited all the expected characteristics related to a DA release. First, these effects were consistent with the anatomy of the mesolimbic dopaminergic system: No effects were observed either when the carbon-fiber electrode was not located in the nucleus accumbens or when the stimulating electrode was not precisely implanted in the dopaminergic pathway. Second, these effects were abolished by a tetrodotoxin ejection (1 μM, 40 nL) in the vicinity of the carbon-fiber electrode. Third, they were strongly, but reversibly, diminished (60% decrease) when cadmium was substituted for calcium in an artificial CSF ejected close to the electrode. Fourth, consistent with previous studies, the maximal amplitude of these effects was enhanced by amphetamine, pargyline, and haloperidol. Fifth, inhibition of DA reuptake by nomifensine induced a fourfold decrease in the rate of decline of the evoked oxidation current. Finally, although the electrode was also sensitive to NE and serotonin, contribution of these amines to the observed effects seems unlikely, since specific reuptake blockers (desipramine and sertraline, respectively) were ineffective.

The main advantage of this approach is its excellent time resolution (*see* Fig. 5). It is actually so good that the DA release evoked by a single shock stimulation can be detected without any pharmacological treatment (Dugast, 1994). Therefore, the relationship between impulse flow and catecholamine release can now be studied on a pulse-by-pulse basis.

5. Conclusions

5.1. Catecholamine Release Outside the Mammalian Brain

Although electrochemical techniques were first developed for monitoring catecholamine release in the mammalian brain, the most promising advances have been performed in peripheral tissues during these last years. Wightman and his coworkers opened a new avenue when they developed the monitoring of catecholamine secretion from a single chromaffin cell (Leszczyszyn et al., 1990; Wightman et al., 1991). The electrochemical recording of NE released by sympathetic nerves also represents a major advance (Mermet et al., 1990a; Gonon et al., 1993a,b). In fact, dozens of research groups are dealing with these applications, and the advances offered by electrochemical techniques are so decisive that important findings have already resulted from their use (Chow et al., 1992, Msghina et al., 1993).

It is likely that electrochemical techniques might also be useful for monitoring catecholamines in other preparations, such as the retina, isolated catecholaminergic neurons, or giant neurons from primitive nervous systems. Thus, elementary steps of secretion and transmitter release will be studied in real time.

5.2. The Living Brain vs Brain Slices

Several groups during recent years have combined carbon-fiber electrodes with fast cyclic voltammetry for monitoring in vitro the catecholamine release evoked in brain slices. This field is covered in detail elsewhere in this volume. This in vitro approach made possible the monitoring of the release evoked by a single pulse stimulation. Therefore, some important aspects of amine release, such as its presynaptic control by autoreceptors, have been investigated on a pulse-by-pulse basis (Palij et al., 1990; Kennedy et al., 1992; Trout and Kruk, 1992). However, as underlined by Kennedy et al. (1992), some "characteristics of dopamine release in slice preparation differ from those found in vivo." These differences were previously difficult to explain since electrochemical techniques were less potent in vivo than in vitro (i.e., sensitive to the release evoked by a single pulse stimulation). Since continuous amperometry now allows us to resolve this elementary release (Dugast et al., 1994), this problem should be solved in the near future. Thus, the choice between in vivo and in vitro preparations will not be forced by limitations of electrochemical techniques.

5.3. Comparison Between In Vivo Electrochemical Techniques

Concerning the monitoring of catecholamine release in the living brain, two main goals can be distinguished: the monitoring of the basal extracellular amine level and the amine release that is evoked by brief stimulations. The first goal demands a very high sensitivity and selectivity, whereas the main requirement of the second one is time resolution. Therefore, in the present state of the art, different techniques should be used depending on the goal.

Regarding the first goal, the most appropriate technique seems to be differential normal pulse voltammetry combined with electrochemically treated carbon-fiber electrodes. However, separation of amines from DOPAC requires, in addition, either a pharmacological treatment with pargyline or a mathematical analysis of the catechol oxidation peak.

Concerning the amine release evoked by brief stimulations, two techniques have mainly been used: fast cyclic voltammetry and differential pulse amperometry coupled with electrochemically treated carbon-fiber electrodes. The first one exhibits a better time resolution, whereas the detection limit of the second one is better. However, if the capabilities of continuous amperometry are confirmed in the near future, it is clear that this last technique will tend to take its place. In this respect, this technical evolution parallels that which already occurred as regards the in vitro monitoring either of adrenaline (epinephrine) secreted by single chromaffin cells (Wightman et al., 1991) or of NE released from sympathetic nerves (Gonon et al., 1993b).

5.4. Comparison with Microdialysis

In vivo microdialysis has been proved to be a reliable technique for monitoring the basal extracellular catecholamine concentration. However, it suffers from two main limitations: its poor time resolution (usually 10 min) and the size of its probe (diameter of 0.2 mm or larger). In both respects, electrochemical techniques are far better. From an analytical point of view, however, microdialysis is still clearly safer: Catecholamines are identified without any uncertainty and can be recorded for days. Therefore, in vivo electrochemistry is more potent for monitoring the amine release evoked by brief stimulations, whereas microdialysis is still more reliable for monitoring slow variations in the basal extracellular amine level. Thus, the author agrees with Stamford's opinion that "the two methods, far from being competitive, are complementary" (Stamford, 1989).

Acknowledgments

This chapter is dedicated to the memory of my friend M. Buda, who died in an accident on August 17, 1993. I thank M. F. Suaud-Chagny for helpful discussions and M. H. Schmidt for revising the English manuscript.

References

Brun P., Suaud-Chagny M. F., Gonon F., and Buda M. (1993) In vivo noradrenaline release evoked in the anteroventral thalamic nucleus by Locus-Coeruleus activation—an electrochemical study. *Neuroscience* **52,** 961–972.

Buda M., De Simoni M. G., Gonon F., and Pujol J. F. (1983) Catecholamine metabolism in the rat locus coeruleus as studied by in vivo differential pulse voltametry: 1. Nature and origin of contributors to the oxydation current at +0,1V. *Brain Res.* **273,** 197–206.

Chow R. H., Ruden L. V., and Neher E. (1992) Delay in vesicle fusion revealed by electrochemical monitoring of single secretory events in adrenal chromaffin cells. *Nature* **356,** 60–63.

Crespi F. and Möbius C. (1992) In vivo selective monitoring of basal levels of cerebral dopamine using voltammetry with nafion modified (NACRO) carbon fibre micro-electrodes. *J. Neurosci. Methods* **42,** 149–161.

Di Chiara G., Carboni E., Morelli M., Cozzolino A., Tanda G. L., Pinna A., Russi G., and Consolo S. (1993) Stimulation of dopamine transmission in the dorsal caudate nucleus by pargyline as demonstrated by dopamine and acethylcholine microdialysis and Fos immunohistochemistry. *Neuroscience* **55,** 451–456.

Dugast C., Suaud-Chagny M. F., and Gonon F. (1994) Continuous in vivo monitoring of evoked dopamine release in the rat nucleus accumbens by amperometry *Neuroscience* **62,** 647–654.

Ewing A. G., Bigelow J. C., and Wightman R. M. (1983) Direct in vivo monitoring of dopamine released from two striatal compartments. *Science* **221,** 169–171.

Gonon F. (1988) Nonlinear relationship between impulse flow and dopamine release by rat midbrain dopaminergic neurons as studied by in vivo electrochemistry. *Neuroscience* **24,** 19–28.

Gonon F. and Buda M. (1985) Regulation of dopamine release by impulse flow and autoreceptors as studied by in vivo voltammetry in the rat striatum. *Neuroscience* **14,** 765–774.

Gonon F., Navarre F., and Buda M. (1984a) In vivo monitoring of dopamine release in the rat brain with differential normal pulse voltammetry. *Anal. Chem.* **56,** 573–575.

Gonon F., Buda M., and Pujol J. F. (1984b) Treated carbon fiber electrodes for measuring catechols and ascorbic acid, in *Measurement of Neurotransmitter Release In Vivo IBRO Handbook Series: Methods in Neurosciences* (Marsden C. A., ed.), Wiley, New York, pp. 153–171.

Gonon F., Fombarlet C., Buda M., and Pujol J. F. (1981) Electrochemical treatment of pyrolytic carbon fiber electrodes. *Anal. Chem.* **53,** 1386–1389.

Gonon F., Buda M., Cespuglio R., Jouvet M., and Pujol J. F. (1980) In vivo electrochemical detection of catechols in the neostriatum of anesthetized rats: dopamine or DOPAC? *Nature* **286,** 902–904.

Gonon F., Bao J. X., Msghina M., Suaud-Chagny M. F., and Stjärne L. (1993a) Fast and local electrochemical monitoring of noradrenaline release from sympathetic terminals in isolated rat tail artery. *J. Neurochem.* **60,** 1251–1257.

Gonon F., Msghina M., and Stjärne L. (1993b) Kinetics of noradrenaline released from sympathetic nerves. *Neuroscience* **56,** 535–538.

Gonon F., Cespuglio R., Ponchon J. L., Buda M., Jouvet M., Adams R. N., and Pujol J. F. (1978) Mesure électrochimique continue de la libération de dopamine réalisée in vivo dans le néostriatum du rat. *C. R. Acad. Sci. (III)* **286,** 1203–1206.

Gonzalez-Mora J. L., Sanchez-Bruno J. A., and Mas M. (1988) Concurrent on-line analysis of striatal ascorbate dopamine and dihydroxyphenylacetic acid concentrations by in vivo voltammetry. *Neurosci. Lett.* **86,** 61–66.

Imperato A. and Di Chiara G. (1984) Trans-striatal dialysis coupled to reverse phase high performance liquid chromatography with electrochemical detection: a new method for the study of the in vivo release of endogenous dopamine and metabolites. *J. Neurosci.* **4,** 966–977.

Kennedy R. T., Jones S. R., and Wightman R. M. (1992) Dynamic observation of dopamine autoreceptor effects in rat striatal slices. *J. Neurochem.* **59,** 449–455.

Kissinger P. T., Hart J. B., and Adams R. N. (1973) Voltammery in brain tissue—a new neurophysiological measurement. *Brain Res.* **55,** 209–213.

Leszczyszyn D., Jankowski J. A., Viveros O. H., Diliberto E. J., Near J. A., and Wightman R. M. (1990) Nicotinic-receptor mediated cathecholamine secretion from individual chromaffin cells: chemical evidence for exocytosis. *J. Biol. Chem.* **265,** 14,736–14,737.

Louilot A., Serrano A., and D'Angio M. (1987) A novel carbon fiber implantation assembly for cerebral voltammetric measurements in freely moving rats. *Physiol. Behav.* **41,** 227–231.

Louilot A., Gonzalez-Mora J. L., Guadalupe T., and Mas M. (1991) Sex-related olfactory stimuli induce a selective increase in dopamine release in the nucleus accumbens of male rats—a voltammetric study. *Brain Res.* **553,** 313–317.

Marcenac F. and Gonon F. (1985) Fast in vivo monitoring of dopamine release in the rat brain with differential pulse amperometry. *Anal. Chem.* **57,** 1778,1779.

Mas M., Gonzalez-Mora J. L., Louilot A., Sole C., and Guadalupe T. (1990) Increased dopamine release in the nucleus accumbens of copulating male rats as evidence by in vivo voltammetry. *Neurosci. Lett.* **110,** 303–308.

May L. J. and Wightman R. M. (1989) Heterogeneity of stimulated dopamine overflow within rat striatum as observed with in vivo votammetry. *Brain Res.* **487,** 311–320.

Mermet C. and Gonon F. (1986) In vivo voltammetric monitoring of noradrenaline release and catecholamine metabolism in the hypothalamic paraventricular nucleus. *Neuroscience* **19,** 829–838.

Mermet C. and Gonon F. G. (1988) Ether stress stimulates noradrenaline release in the hypothalamic paraventricular nucleus. *Neuroendocrinology* **47,** 75–82.

Mermet C., Gonon F. G., and Stjarne L. (1990a) On-line electrochemical monitoring of the local noradrenaline release evoked by electrical stimulation of the sympathetic nerves in isolated rat tail artery. *Acta Physiol. Scand.* **140,** 323–329.

Mermet C., Suaud-Chagny M. F., and Gonon F. (1990b) Electrically evoked noradrenaline release in the rat hypothalamic paraventricular nucleus studied by in vivo electrochemistry: autoregulation by alpha-2 receptors. *Neuroscience* **34,** 423–432.

Msghina M. and Stjärne L. (1993) Sympathetic transmitter release in rat tail artery and mouse vas deferens: facilitation and depression during high frequency stimulation. *Neurosci. Lett.* **155,** 37–41.

Msghina M., Gonon F., and Stjärne L. (1993) Intermittency of noradrenaline release from sympathetic nerves of rat tail artery and effects of high frequency stimulation. *Neuroscience* **57,** 887–890.

Msghina M., Mermet C., Gonon F., and Stjärne L. (1992) Electrophysiological and electrochemical analysis of the secretion of ATP and noradrenaline from the sympathetic nerves in rat tail artery: effects of α_2-adrenoceptor agonists and antagonists and noradrenaline reuptake blockers. *Naunyn-Schmiedeberg's Arch. Pharmacol.* **346,** 173–186.

Palij P., Bull D. R., Sheehan M. J., Millar J., Stamford J., Kruk Z. L., and Humphrey P. P. A. (1990) Presynaptic regulation of dopamine release in corpus striatum monitored in vitro in real time by fast cyclic voltammetry. *Brain Res.* **509,** 172–174.

Rebec G. V., Langley P. E., Pierce R. C., Wang Z. R., and Heidenreich B. A. (1993) A simple micromanipulator for multiple uses in freely moving rats—electrophysiology, voltammetry, and simultaneous intracerebral infusions. *J. Neurosci. Methods* **47,** 53–59.

Smith A. D., Olson R. J., and Justice J. B. (1992) Quantitative microdialysis of dopamine in the striatum—effect of circadian variation. *J. Neurosci. Methods* **44,** 33–41.

Stamford J. A. (1989) In vivo voltammetry—prospects for the next decade. *Trends Neurosci.* **12,** 407–412.

Suaud-Chagny M. F., Buda M., and Gonon F. G. (1989) Pharmacology of evoked dopamine release studied in the olfactory tubercle by in vivo electrochemistry. *Eur. J. Pharmacol.* **164,** 273–283.

Suaud-Chagny M. F., Mermet C., and Gonon F. (1990) Electrically evoked noradrenaline release in the rat hypothalamic paraventricular nucleus studied by in vivo electrochemistry: characterization and facilitation by increasing the stimulation frequency. *Neuroscience* **34,** 411–422.

Suaud-Chagny M. F., Ponec J., and Gonon F. (1991) Presynaptic autoinhibition of the electrically evoked dopamine release studied in the rat olfactory tubercle by in vivo electrochemistry. *Neuroscience* **45,** 641–652.

Suaud-Chagny M. F., Brun P., Buda M., and Gonon F. (1992a) Fast in vivo monitoring of electrically evoked dopamine release by differential pulse amperometry with untreated carbon fibre electrodes. *J. Neurosci. Methods* **45,** 183–190.

Suaud-Chagny M. F., Chergui K., Chouvet G., and Gonon F. (1992b) Relationship between dopamine release in the rat nucleus accumbens and the discharge activity of dopaminergic neurons during local in vivo application of amino-acids in the ventral tegmental area. *Neuroscience* **49,** 63–72.

Suaud-Chagny M. F., Cespuglio R., Rivot J. P., Buda M., and Gonon F. (1993) High sensitivity measurement of brain catechols and indoles in vivo using electrochemically treated carbon-fiber electrodes. *J. Neurosci. Methods* **48,** 241–250.

Suaud-Chagny M. F., Steinberg R., Mermet C., Biziere K., and Gonon F. (1986) In vivo voltammetric monitoring of catecholamine metabolism in the A1 and A2 regions of the rat medulla oblongata. *J. Neurochem.* **47,** 1141–1147.

Trout S. J. and Kruk Z. L. (1992) Differences in evoked dopamine efflux in rat caudate putamen, nucleus accumbens and tuberculum olfactorium in the absence of uptake inhibition: influence of autoreceptors. *Br. J. Pharmacol.* **106,** 452–458.

Wiedemann D. J., Basse Tomusk A., Wilson R. L., Rebec G. V., and Wightman R. M. (1990) Interference by DOPAC and ascorbate during attempts to measure drug-induced changes in neostriatal dopamine with Nafion-coated, carbon-fiber electrodes. *J. Neurosci. Methods* **35,** 9–18.

Wightman R. M., Jankowski J. A., Kennedy R. T., Kawagoe K. T., Schroeder T. J., Leszczyszyn D. J., Near J. A., Diliberto E. J., and Viveros O. H. (1991) Temporally resolved catecholamine spikes correspond to single vesicle release from individual chromaffin cells. *Proc. Natl. Acad. Sci. USA* **88,** 10,754–10,758.

Regional Differences in Dopamine Release, Uptake, and Diffusion Measured by Fast-Scan Cyclic Voltammetry

Paul A. Garris and R. Mark Wightman

1. Introduction

The neurotransmitter dopamine (DA) is present in high concentrations in select areas of the central nervous system (CNS), including the striatal and limbic regions. Appropriate control of its role in communication between neurons is clearly important for normal function. Possibly the most recognizable aspect of improper regulation of DA is Parkinson's disease. The symptoms of this neurodegenerative disorder, which affects the control of motor activity and results from deterioration of nigrostriatal DA neurons, are treated by various means that attempt to restore dopaminergic capacity. Dopamine is also thought to subserve cognitive, emotive, and even endocrine functions, and has been implicated in other neuropathologies, such as schizophrenia and substance addiction.

Despite a wealth of studies in DA neurobiology, little is known concerning the fundamental principles of dopaminergic transmission. To achieve a clearer understanding, a real-time view of DA molecules at the level of the individual DA synapse in the intact brain is required. In this way, the complex series of events that occur in the synaptic microenvironment, both temporal and spatial, can be observed both

From: *Neuromethods, Vol. 27: Voltammetric Methods in Brain Systems*
Eds: A. Boulton, G. Baker, and R. N. Adams © 1995 Humana Press Inc.

before and after the binding of DA to a receptor. An understanding of these processes is important not only for our fundamental knowledge of the nature of chemical communication in the CNS, but also to understand in more detail the interaction of drugs of abuse and clinical use with dopaminergic transmission.

The study of DA neurotransmission has been advanced by the ability to monitor DA with unprecedented temporal and spatial resolution using chemical microsensors (Adams, 1990). These probes take advantage of the property of DA to be oxidized electrochemically at very low potentials. This feature was first recognized by Ralph Adams, an editor of this volume, who initiated efforts to develop electrochemical techniques for the detection of neurotransmitters in the extracellular milieu of the brain in the early 1970s. Today, measurements of DA concentrations can be made with resolution in the millisecond temporal range and micron spatial range. Thus, chemical in vivo measurements of DA provide information that is comparable to a single unit recording. This chapter focuses on the specific methods we use in our laboratory to monitor dopaminergic transmission, and describes some of the insights that we have obtained with this technique.

2. Fast-Scan Cyclic Voltammetry as a Rapid Sensing Technique

2.1. Electrodes and Instrumentation

The microelectrodes we employ use the surface of a single carbon fiber ($r = 5$ µm, Thornel P-55 Amoco, Greenville, SC) as the element to sense DA. The fiber is inserted into a microfilament glass capillary (0.6 mm ID, 1.2 mm OD, A.M. Systems, Evertt, WA; Kawagoe et al., 1993) with the aid of suction. The capillary is placed in a standard microelectrode puller (Model PE-2, Narashige, Tokyo, Japan), and two micropipets are obtained, each containing a carbon fiber. To ensure a water-tight seal between the glass and carbon fiber, the tips are trimmed under a microscope with a scalpel to a diameter of approx 15 µm, slightly larger than the fiber itself. Then the tips are dipped in liquid epoxy (Epon 828 with *m*-phenylenediamine as a hardener, Miller-Stevenson, Danbury, CT), and the tips fill by capillary action. The epoxy is thermally cured, and the electrodes are stored indefinitely in this form until they are to be used.

The electrodes we use differ from others described in this volume in that the sensing element is an elliptical surface defined by a cross-section of the carbon fiber. The day before an experiment, the electrodes are beveled at an angle of 25° on a diamond-crust-embedded wheel (KT Brown Type, Sutter Instruments, Navato, CA) to give the elliptical shape. An electron micrograph of a microelectrode at this stage is shown in Fig. 1. Electrodes are then soaked in isopropanol to remove debris deposited during polishing and dipped in a solution of 2.5% Nafion. Nafion is a cation-exchange polymer that can form a film on the surface that retards access of negatively charged species to the electrode surface and deters electrode fouling during implantation in biological tissues (Gerhardt et al., 1984; Baur et al., 1988). The film is formed by evaporation of the solvent with a heat gun. Electrical connection to the carbon fiber is provided by colloidal graphite (Bio-Rad, Cambridge, MA) and a metal wire inserted in the back end of the micropipet.

The voltammetric experiments employ a waveform generator and current transducer (El 400, Ensman Instrument, Bloomington, IN) that can simultaneously acquire data from two carbon-fiber microelectrodes. The counterelectrode is a sodium-saturated calomel electrode. The small currents (nA) involved in these experiments can be passed through the reference electrode without disturbing its potential, so that a more complicated potentiostatic mode of operation is not required. The instrument includes an adjustable second-order, low-pass Bessel filter. The instrument is interfaced to a PC computer equipped with a digital input/output board (Labmaster, Scientific Solutions, Solon, OH) using locally written software.

We primarily use fast-scan cyclic voltammetry in our experiments, a procedure that is described in more detail in

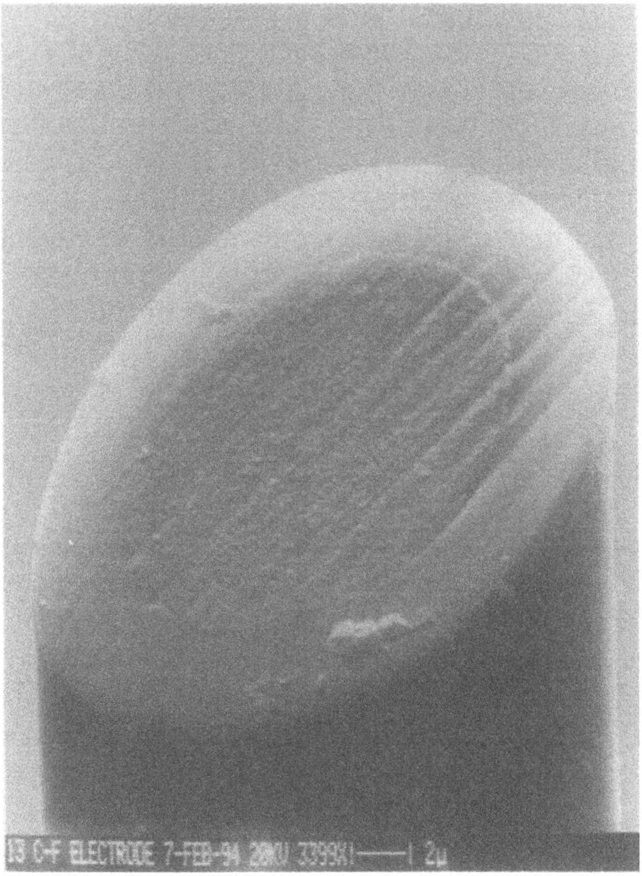

Fig. 1. Electron micrograph of a beveled carbon-fiber microelectrode.

this volume by Stamford et al. The potential is applied to the electrode in the form of a triangular wave, typically from –400 to 1000 mV for DA detection, at a scan rate of 300 V/s, and the voltammograms are repeated every 100 ms (Fig. 2). In between scans, the working electrode rests at a bias of –400 mV preventing adsorption of negatively charged proteins when implanted in the brain. The scan takes approx 10 ms, that allows 90 ms for relaxation of the diffusion layer between scans. For each voltammetric scan, 150 data points are digitized and stored in a computer file. Since each voltammogram has a large background current, in large part

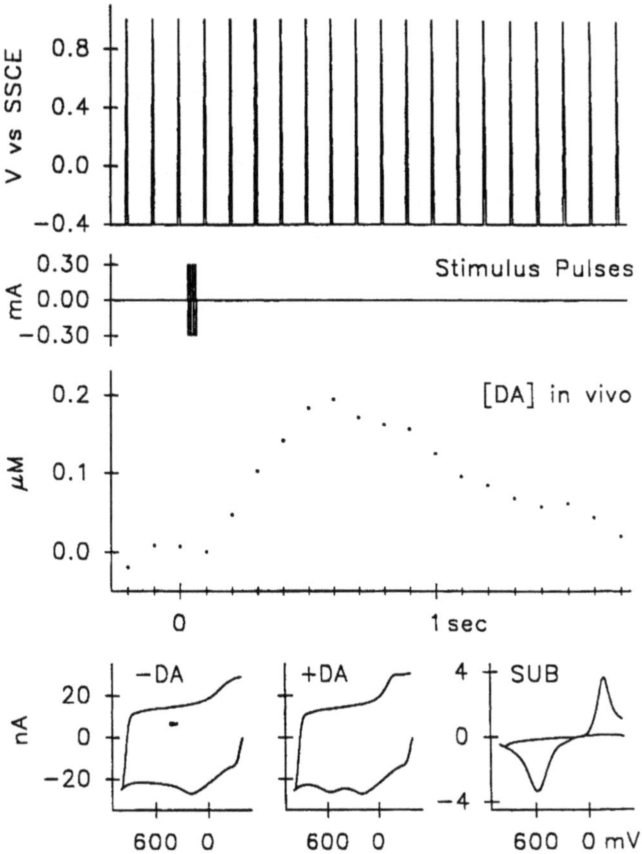

Fig. 2. Fast-scan cyclic voltammetry. First panel: Oscilloscope trace of the potential of a triangle wave (–400 to 1000 mV at a scan rate of 300 V/s) applied every 100 ms. Second panel: Oscilloscope trace of the current applied at a stimulating electrode during a 100 Hz stimulation for four stimulus pulses. Third panel: Evoked DA overflow recorded in vivo in the nucleus accumbens of the rat elicited by four stimulus pulses at 100 Hz. Current was monitored over the peak oxidation potential for DA (500–700 mV vs SSCE) in successive voltammograms and converted to concentration based on calibration of the working electrode postexperiment. Bottom panel: Averaged background voltammograms collected during baseline recording (left) and during an imposed change of DA concentration (middle; note the small change in features as a result of DA). The resulting subtracted voltammogram is shown at the right.

owing to charging the electrode double layer, the difference in concentration of solution species as a function of time is revealed by subtraction of scans. The stability of backgrounds at carbon electrodes allows this procedure to be extremely reproducible. Since the individual voltammograms are saved, subtracted voltammograms can be evaluated after the measurement for any time point to provide voltammetric identification of the detected species. The current at a particular potential from successive voltammograms provides a record of the temporal concentration changes. For DA, we use the potential range of the oxidation peak for DA, which normally occurs between 500 and 700 mV. Similar information can be obtained directly from the EI-400, which contains two sample-and-hold amplifiers for this purpose. Their output can be sent to a strip-chart recorder for real-time evaluation of the responses. In either case, conversion of the current to concentration employs postexperiment calibration.

2.2. Sensitivity and Selectivity

Based on theoretical considerations, the detection limit for a substance with electrochemical kinetics similar to DA (Deakin et al., 1986) during a single cyclic voltammetric scan at 300 V/s is approx 1 µ*M* (Michael et al., 1989). This concentration limit is in part determined by the area of the electrode, which is quite small for the microelectrodes we employ. This limit is far above physiological concentrations of extracellular DA *in situ*. For example, basal levels of DA in the densely innervated striatum are only 5 n*M* according to the best estimates using microdialysis (Parsons and Justice, 1992). Fortunately, a number of factors increase the magnitude of the signal for the oxidation of DA. A major contributor to this increase is the Nafion film on the electrode surface, which accumulates DA in the time between scans. This, coupled with adsorption to the electrode surface, results in a 16-fold amplification of the signal with respect to the theoretically calculated value (Baur et al., 1988).

The detection limits for DA can be further improved by digital filtering and ensemble averaging (Rice and Nicholson, 1989; Wiedeman et al., 1991). During data collection, the data are filtered at a frequency of 2 KHz to retain the shape of the cyclic voltammogram. This results in underfiltering of the concentration changes recorded in each voltammogram, which, when collected at 100-ms intervals, have a maximum frequency component of 5 Hz as set by the Nyquist criterion. To compensate, these data are filtered by digital means. This is done by Fourier transformation of the current from successive voltammograms, multiplication in the Fourier domain with a trapezoidal function, followed by inverse transformation. In many situations, ensemble averaging can be used, which increases the signal-to-noise ratio by the square root of the number of repetitions. Signal averaging of in vivo results obtained during electrical stimulation of axons or cell bodies is feasible, because the stimulus is computer controlled and precisely coordinated with the voltammetry. The combined approaches to signal-to-noise improvements results in detection limits for DA between 25 and 50 n*M* (Wiedemann et al., 1992).

Extracellular fluid in the brain contains a plethora of electroactive species that can interfere with the electrochemical measurement of DA. The concentration of ascorbic acid and the DA metabolite dihydroxyphenylacetic acid (DOPAC), both of which are easily oxidized, is at least 1000 times greater than DA (Wightman et al., 1987; Marsden et al., 1988). Although some selectivity against these anions is conferred by Nafion, this is inadequate to discriminate the low concentration of DA under basal conditions. More recent findings indicate that changes in the ionic composition of extracellular fluid, particularly H^+ and Ca^{2+}, also interfere with the voltammetric detection of DA (Jones et al., 1994). The concentration of these ions varies during different physiological states and alters the redox potential of chemical groups on the carbon surface (Kawagoe et al., 1993).

Recognizing these issues, the following criteria have been established for identifying the extracellular species detected with in vivo voltammetry (Wightman et al., 1987; Marsden et al., 1988): electrochemical characterization, independent chemical analysis, anatomical specificity, and physiological and pharmacological evidence. Fast-scan cyclic voltammetry is an important tool in this regard, because the chemical information provided by the voltammogram is sufficient to distinguish DA from its metabolites and from ascorbic acid and serotonin (Baur et al., 1988). Although resolution of the individual constituents of a "mixed" response may not be possible, the power of the voltammogram is that the investigator is alerted to the problem of multiple signals. With these approaches, the measurement of extracellular DA in the striatum, elicited by electrical stimulation of the medial forebrain bundle, has previously been established (Ewing et al., 1983; Kuhr et al., 1984; Millar et al., 1985; Michael et al., 1987; Stamford et al., 1988b; Suaud-Chagny et al., 1991).

The measurement of extracellular DA in areas with a sparse innervation of DA is more difficult. For example, although the cortex and amygdala receive functionally important DA innervations, DA is not the predominant monoamine present in tissue (Palkovits et al., 1979; Kilts and Anderson, 1987), and voltammograms for catecholamines recorded with fast-scan cyclic voltammetry are similar (Baur et al., 1988). However, the monoamine innervations to these regions are not uniform, and at a microanatomical level, have different origins and terminations (Ungerstedt, 1971; Bjorklund and Lindvall, 1984; Moore and Card, 1984). We have exploited this heterogeneity to measure evoked extracellular DA in the medial prefrontal cortex (Garris et al., 1993) and basolateral amygdaloid nucleus (Garris and Wightman, 1994b). Our strategy was to activate ascending DA fibers selectively using electrical stimulation and to record selectively in DA-rich regions by making use of the exquisite spatial resolution of carbon-fiber microelectrodes.

The use of anatomy to assist in the identification of a voltammetric measurement is necessary when examining DA systems exhibiting unique pharmacological characteristics. Classical blockers of DA autoreceptors and uptake transporters robustly increase evoked extracellular DA in the striatum (Wightman and Zimmerman, 1990), and have been used to verify the determination of DA. However, these drugs are ineffective in altering extracellular DA in the amygdala, for example (Jones et al., submitted; Garris and Wightman, submitted). Furthermore, mesoamygdaloid DA neurons are insen-

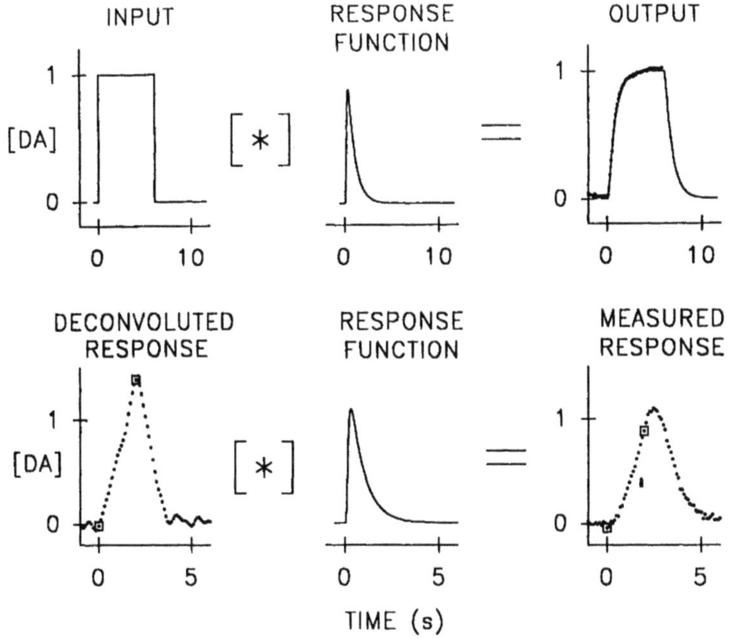

Fig. 3. Temporal distortion in measured responses. **(Top)** Convolution of an applied square pulse of DA by the diffusion response function calculated for a Nafion-coated carbon fiber microelectrode. The measured response collected in vitro is shown by the solid circles. **(Bottom)** Deconvolution of the temporal distortion present in the in vivo measured response owing to Nafion. Since the measured response in vivo is the convolution of the true response and the response function of the microelectrode, it is recovered by deconvolution.

sitive to chemical ablation using the neurotoxin 6-hydroxydopamine (Eldon et al., 1994).

2.3. Temporal Response (Deconvolution Techniques)

Curves for evoked DA overflow in vivo measured by fast-scan cyclic voltammetry at Nafion-coated carbon-fiber microelectrodes are temporally distorted (Figs. 2 and 3). This is seen as the continued rise in extracellular DA after the stimulation has ceased. Simultaneous voltammetric monitoring of evoked DA in the striatum and electrophysiological recording of DA cell bodies in the ventral mesencephalon show that this should not be the case, since DA neurons stop firing immediately after termination of electrical stimulation (Kuhr et al., 1987). Thus, the lag time appears to be a manifestation of diffusional barriers between the working electrode and the location of DA release. There are three potential regions where this diffusion could occur: the intact tissue, a damage region created by electrode implantation, or the Nafion film (Kawagoe et al., 1992).

A major contributor to the time lag has been shown to be the Nafion film (Kristensen et al., 1987). This is because the diffusion coefficient of DA in Nafion is three orders of magnitude less than in free solution (Nagy et al., 1985). The temporal response of a Nafion-coated electrode can be measured by observing the response time to a concentration step of DA in vitro. This typically reveals a response time of >500 ms at an electrode that has previously been used in vivo. Diffusion through the Nafion-coating on the electrode can be described by equations for transport through a thin film (Crank, 1975; Engstrom et al., 1988) because the Nafion-coating of the working electrode is only 200–500 nm thick (Kristensen et al., 1987). The effective thickness can be determined by fitting the measured response to the appropriate diffusion equation using the value of the diffusion coefficient (Crank, 1975).

Once the thickness is known, the delay caused by the Nafion can be evaluated by deconvolution (Bracewell, 1986). Deconvolution can be accomplished with the impulse response function of the film, which is obtained by differentiation of the equation for a step response, and evaluation with the known thickness and diffusion coefficient of DA in Nafion (Kawagoe et al., 1992). Then, the impulse response function and the in vivo data are converted to the frequency domain by Fourier transformation, and deconvolution is accomplished by division of the measured data with the impulse response function followed by the inverse Fourier transformation of the quotient. If the deconvoluted in vivo response is devoid of temporal distortion (Fig. 3), then the Nafion film is responsible for the delay. The importance of deconvolution is that instantaneous concentration dynamics at time zero, e.g., application of the stimulus pulse, can be determined.

3. Electrically Stimulated Overflow of Dopamine in the Caudate-Putamen

3.1. Stimulation Procedures

We employ electrical stimulation of ascending DA fibers to alter transiently extracellular DA levels in the telencephalon in vivo. The constant-current stimulation is computer-generated and coordinated with the electrochemistry, such that voltammograms are never recorded during a stimulus pulse (Fig. 2). To minimize stimulus artifact further, electrical stimulation is optically isolated (Neurolog NL 800, Medical Systems, Great Neck, NY). Bipolar stimulating electrodes, which are insulated to the tips, are purchased from a commercial source (Plastic Products, Roanoke, VA). The tips have a diameter of 250 μm and are separated by 1 mm. Stimulus pulses are biphasic (2 ms and 300 μA each phase) and typically applied at rates between 10 and 60 Hz. For more flexibility, an external Grass S88 stimulator (Grass Instruments, Quincy, MA) can be computer triggered to create, for example, one pulse or four pulses at 100-Hz stimulation.

3.2. Characterization of Dopamine Uptake

Electrical stimulation of the medial forebrain bundle results in a transient increase in the extracellular concentration of DA in the striatum followed by the rapid return to baseline levels as measured by fast-scan cyclic voltammetry at Nafion-coated carbon-fiber microelectrodes (Fig. 4). The extracellular clearance of evoked DA is far too fast to be accounted for by diffusion alone and is sensitive to DA uptake inhibitors. Therefore, this rate reflects the cellular uptake of DA (Ewing and Wightman, 1984).

Dopamine uptake in the striatum is described by Michaelis-Menten kinetics:

$$\{d[DA]/dt\}_{uptake} = V_{max}/(\{K_m/[DA] + 1\}) \qquad (1)$$

with a high-affinity K_m of 0.2 μM (Horn, 1990). The value of V_{max} is proportional to the number of transporter sites. The kinetic parameters for DA uptake operant at the tip of the

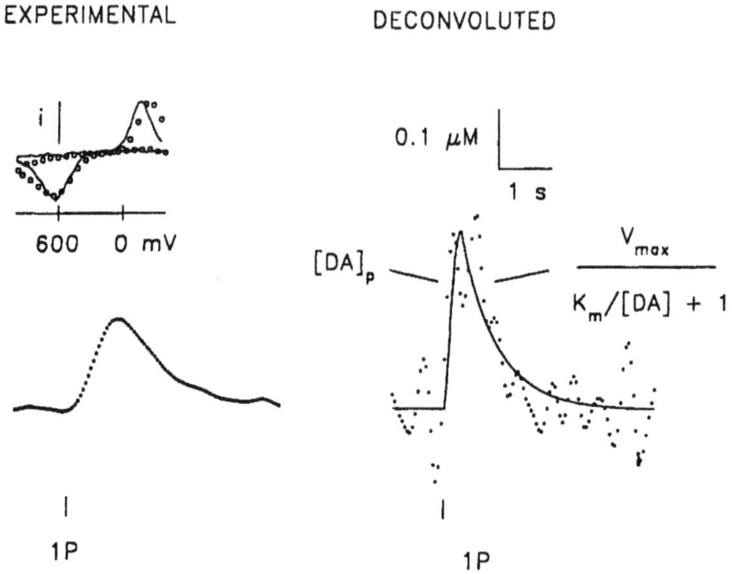

Fig. 4. Modeling of evoked DA overflow. **(Left)** Experimentally observed overflow elicited by a single stimulus pulse. Inset shows background subtracted voltammograms recorded in vivo during the stimulation (solid line) and in vitro from DA (open circles). **(Right)** Dots: Deconvoluted curve and simulation. Data and simulation were digitally filtered using cutoff frequency of 2 Hz and an apodization slope of 1 s. Dots: Deconvoluted in vivo response. Solid line: modeled curve with release modeled as an instantaneous appearance of DA ($[DA]_p$) followed by uptake, which follows Michaelis-Menten kinetics.

electrode can be directly determined from the deconvoluted response. At extracellular concentrations of DA, which are at least 10 times K_m, Michaelis-Menten kinetics reduces to zero order and V_{max} is the linear slope of the clearance rate. The V_{max} measured in the caudate-putamen is typically between 2 and 4 μM/s.

3.3. Resolving the Effect of Uptake and Release

We use the term "overflow" to describe the extracellular DA observed during stimulation. The determination of the release component of these curves is complicated, because it contains information concerning release, uptake, and diffusion. If the temporal effects of diffusion are solely the result of the Nafion film, which can be removed as described above, then uptake and release can be temporally resolved. We define release as the instantaneous concentration change that results from each stimulus pulse. Thus, release is dependent on the rate of impulse flow, whereas uptake is time-dependent and occurs in the time between pulses. If the observed rate of DA overflow in the extracellular fluid is the result of the balance of release and uptake rates (Wightman et al., 1988; Wightman and Zimmerman, 1990), then we can write:

$$\{d[DA]/dt\}_{overflow} = [DA]_p \cdot f - \{d[DA]/dt\}_{uptake} \qquad (2)$$

where the release term, $[DA]_p$, is the concentration of DA released per stimulus pulse and f is the frequency of stimulation. The parameters for the uptake during the stimulation train are obtained from the Michaelis-Menten constants measured after the stimulation has ceased. The major assumption of Eq. (2) is that the value of $[DA]_p$ remains constant during the stimulus train.

Our initial studies demonstrated that Eqs. (1) and (2) do an excellent job of describing the experimental results obtained in vivo (Wightman et al., 1988). In that evaluation, the value of K_m in the striatum was assumed to be 0.2 μM, the value determined in synaptosome preparations (Near et al., 1988; Marshall et al., 1990). V_{max} was determined from the initial disappearance rate as measured from the deconvoluted response to a supraphysiological stimulation (60 Hz, 5 s) that elicits extracellular DA levels in the 5–10 μM range. The value of $[DA]_p$ was manually adjusted to obtain the best fit as indicated by the correlation coefficient. We have recently employed nonlinear regression using simplex optimization (Press et al., 1989) to determined simultaneously $[DA]_p$, K_m, and V_{max} from a series of overflow curves evoked by frequencies between 20 and 60 Hz (Jones et al., in preparation). These results confirm the results of our initial modeling, and provide a more objective estimate of the kinetic parameters for DA release and uptake. The simulated response is convoluted with the response of the Nafion film, so that the experimental and modeled data can be directly compared.

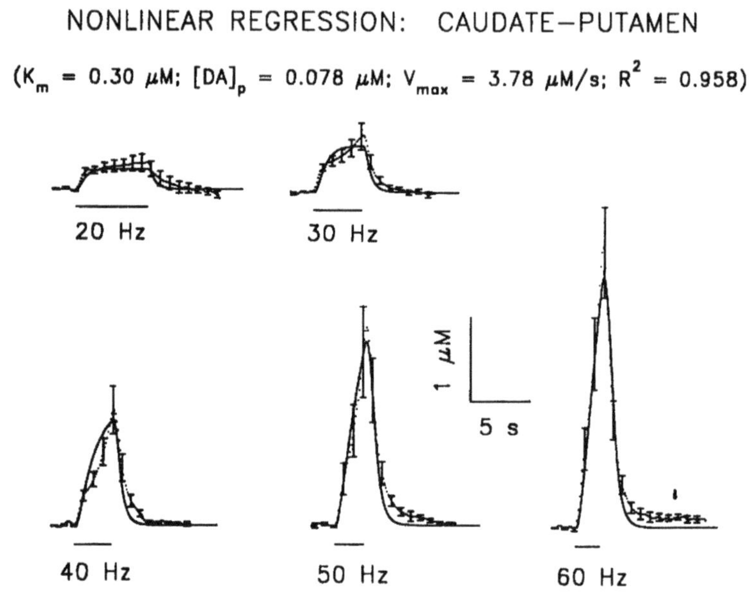

Fig. 5. Nonlinear regression of evoked DA overflow in vivo in the caudate-putamen. Data (solid circles) are the average from several animals, and simulations (solid line) are based on the best-fit parameters for DA release and uptake. Simulations were convoluted with a response function for thin-layer diffusion. The response measured for 10 Hz was excluded from curve-fitting. Open squares: the initiation and termination of the electrical stimulation, which consisted of 120 pulses for all frequencies.

Figure 5 shows DA overflow curves collected in the caudate-putamen overlaid with the resulting simulations, calculated from the best-fit parameters determined by nonlinear regression. Data and simulation agree favorably and demonstrate that the in vivo responses, which yield frequency-dependent curves with markedly different dynamics, can be described by a single set of parameters for DA uptake and release. $[DA]_p$ and V_{max} are in accord with previous estimates (Wightman et al., 1988; Wightman and Zimmerman, 1990), and K_m is similar to values obtained using classical methods in which the uptake of radiolabeled DA into synaptosomes is measured (Near et al., 1988; Marshall et al., 1990) However, the response obtained at a stimulation of 10 Hz does not agree with the simulated values. A good fit can only be

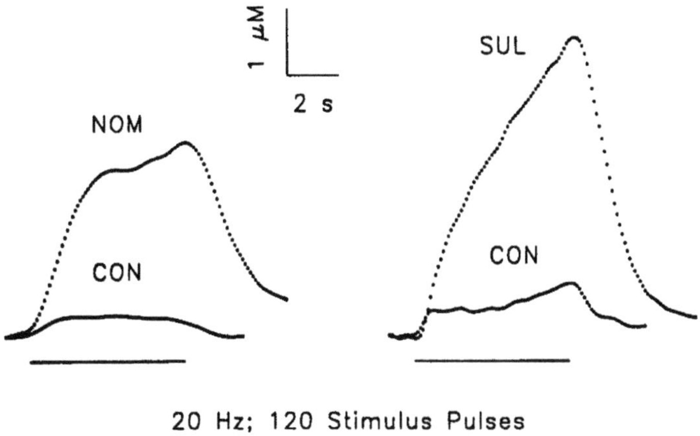

Fig. 6. Effects of dopaminergic pharmacological agents on evoked DA overflow in vivo in the nucleus accumbens. Left panel: Effects of the competitive DA uptake inhibitor nomifensine (25 mg/kg). Right panel: Effects of the D_2 receptor antagonist sulpiride (100 mg/kg). All drugs were injected ip. The duration of the electrical stimulation (6 s) is shown by the solid line directly underneath the overflow curves.

obtained if $[DA]_p$ is increased. Although this response has been seen in the striatum of every animal tested, the cause(s) is not yet understood (Kawagoe et al., 1992). The lower frequency more closely corresponds to the normal physiological firing rates and suggests that DA release caused by impulse flow becomes more tightly controlled during burst firing at higher frequencies.

The modeling of the neurochemical data is useful to evaluate the mechanism of action of pharmacological agents (Wightman and Zimmerman, 1990). For example, both sulpiride and nomifensine increase evoked extracellular concentrations of DA (Fig. 6). However, the changes in shape of the responses at different frequencies induced by each drug differ, indicating different actions of these drugs consistent with their known pharmacology. Sulpiride, a D2 receptor antagonist, blocks autoreceptors that normally inhibit release (Starke et al., 1989). In agreement, the frequency-dependent increase in extracellular DA induced by sulpiride is best modeled by an increase in $[DA]_p$ without any change in DA uptake. Conversely, nomifensine, a competitive uptake inhibitor, gives in vivo responses that are best described by a decrease in DA uptake rate with no change in DA release (Wightman et al., 1988; Wightman and Zimmerman, 1990). Furthermore, the data clearly show that the uptake rate is altered because of an increase in the apparent value of K_m, characteristic of a competitive uptake inhibitor, rather than an effect on V_{max}.

The value of $[DA]_p$ provides a fundamental quantity that allows comparison of the capacity of a particular microenvironment to support DA release. In these experiments, supramaximal stimulating currents are employed, and thus, a maximal number of terminals are forced to release. Furthermore, since the DA innervation in the striatum is $>1 \times 10^8/mm^3$ (Pickel et al., 1981), the terminals are, on average, 4 μm apart from one another. On the time scale (100 ms) of the measurements, local concentration gradients of DA are relaxed over this distance (Kawagoe et al., 1992). This means that the value of $[DA]_p$ determined with the sensor is that throughout the extracellular space adjacent to the sensor, including the synaptic–extrasynaptic interface.

3.4. Evidence for Extracellular Diffusion of Dopamine

For purposes of kinetic analysis, we normally select overflow curves that exhibit large responses and minimal temporal distortion (Kawagoe et al., 1992). Thus, our reported parameters for DA uptake and release are biased toward the regions of most active uptake and release. As shown above, the in vivo responses are well described by the kinetic equations when diffusion through the Nafion film is accounted for and there is no direct evidence from our measurements for freely diffusing DA. Rapid uptake severely limits the diffusion of DA by limiting the time of free DA in the extracellular fluid. However, significant brain diffusion could occur in regions where dense DA terminal fields are immediately adjacent to areas containing a sparse density of DA terminals.

Evoked DA overflow curves measured at different depths in the striatum reveal the chemical heterogeneity (Stamford et al., 1986b; May and Wightman, 1989) that is likely related to the documented differences within the

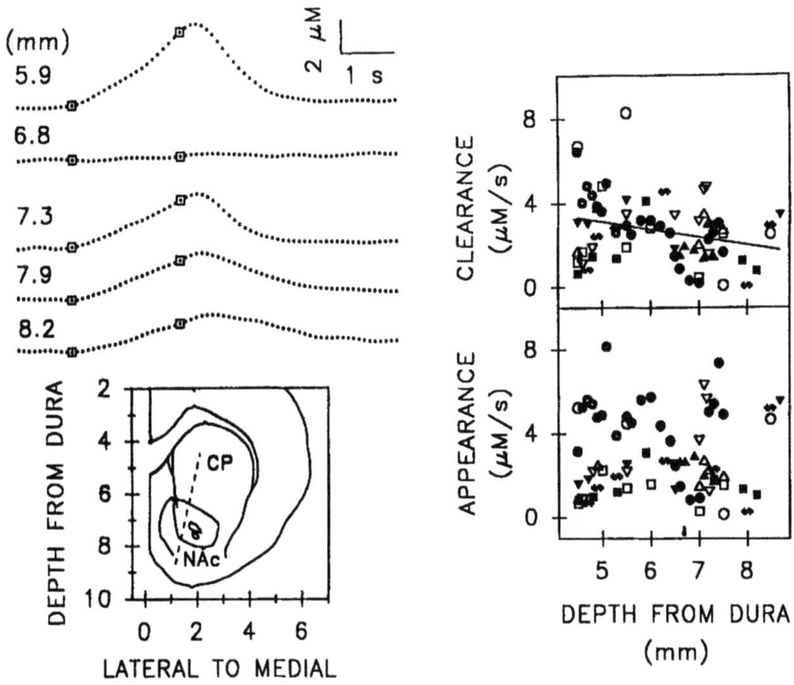

Fig. 7. Heterogeneity of evoked DA overflow in vivo in the striatum. Left panels: A diagram from a rat brain atlas (Paxinos and Watson, 1986) is found at the bottom, and the approximate locations for the measurements are represented by the dashed line. Individual overflow curves collected in a single animal using the same carbon-fiber microelectrode are shown at the top. The number above each curve corresponds to the dorsoventral position of the recording. Right panels: Compiled results of apparent DA release and uptake rates (appearance and clearance, respectively) measured in the striatum of nine different animals. Each animal is represented by a different symbol. (Reprinted from *Neuroscience* **59**, 417–427 [1994] with permission.)

mesostriatal DA innervation on the microanatomical level (Tassin et al., 1976; Graybiel, 1984; Joyce et al., 1986; Savasta et al., 1986). Individual overflow curves collected at different depths within the striatum of a single animal and measured by the same working electrode are shown in Fig. 7, left panel. Compiled results from many animals, expressed as apparent release and uptake rates, are displayed in the right panel. The heterogeneity is evident and has been observed in all regions of the striatum examined (Garris et al., 1994a). Furthermore, there exists a significant, dorsoventral gradient in DA uptake, but not DA release, which corresponds to the topography of binding sites for the DA transporter as revealed by autoradiography (Marshall et al., 1990). In the majority of these curves, any diffusional component can be accounted for by the Nafion film. Thus, normally we do not observe a diffusional component that is owing to the brain. However, diffusion can be inferred based on the lifetime of DA in the extracellular fluid.

A quite different picture can be found in the striatoamygdaloid area. Mesostriatal and mesoamygdaloid DA systems innervate regions that lie on the same dorsoventral axis, and the respective DA terminal fields exhibit markedly different dynamics for evoked extracellular DA (Garris et al., 1994a; Garris and Wightman, 1994b). Furthermore, there are sharp contrasts in the densities of the DA terminations on a microanatomical level. Interregional comparisons are simplified because a single electrode, with its associated response time, can be used to make recordings in multiple locations, negating variability caused by different Nafion films. Figure 8 shows evoked overflow curves recorded in the tail of the caudate-putamen and basolateral amygdaloid nucleus during a single pass through the striatoamygdaloid area. The response in the amygdala is atypical for any region in the telencephalon, including the amygdala, because of the pronounced time lag between the stimulus and the chemical response. When these curves are deconvoluted with thin-layer diffusion to account for the effects of Nafion, temporal distortion is completely removed from the response in the caudate-putamen, but not in the amygdala. Occasionally, such curves have been observed in the striatum, although their occurrence is quite infrequent.

The result in the amygdala seems to provide the first direct observation of diffusion of exogenous DA in the extracellular fluid. Analysis of curves that contain mixed diffusional and uptake control is complicated because of the uncertainties of the coupling of these two processes, the unknown

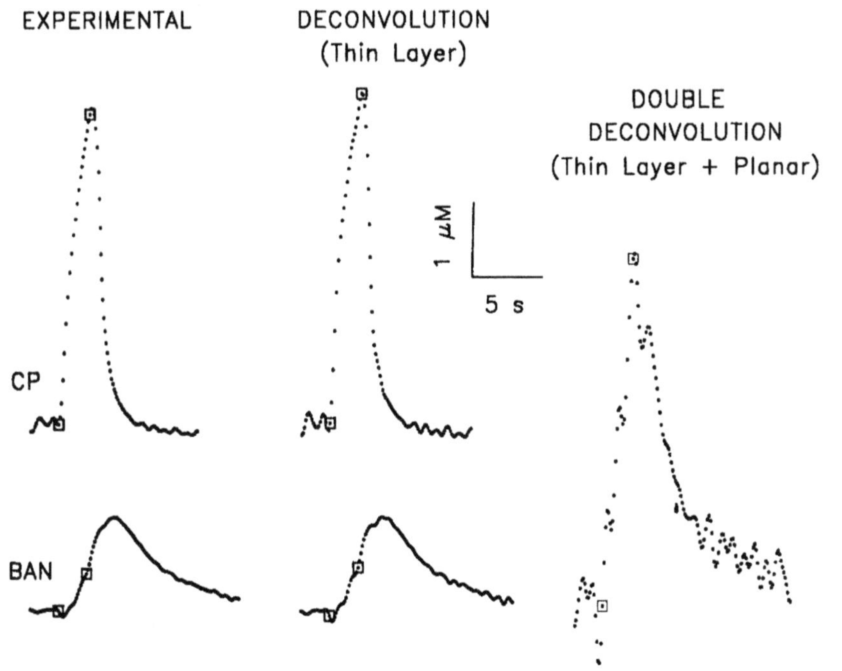

Fig. 8. Extracellular diffusion of DA in the striatoamygdaloid area. Left panels: Experimentally determined DA overflow in vivo in the tail of the caudate-putamen (CP) and basolateral amygdaloid nucleus (BAN) recorded in the same animal with a single electrode. Middle panels: Deconvolution of the measured curves using a response function calculated for thin-layer diffusion. Right panels: Double deconvolution of the measured curve recorded in the basolateral amygdaloid nucleus using response functions calculated for thin-layer and planar diffusion.

microanatomy of the DA innervation within the sensing region of the carbon-fiber microelectrode, and the complex nature of the brain extracellular microenvironment (Nicholson and Rice, 1991). To obtain a crude estimate of the average distance over which DA has diffused in this experiment, a second deconvolution was performed using the response function calculated for planar diffusion (Engstrom et al., 1988). Using a diffusion distance of 40 μm and the diffusion coefficient for free solution, the temporal distortion can be accounted for. This analysis is an oversimplification, because it treats uptake and diffusion as spatially separated processes. Furthermore, the diffusion coefficient for DA in the extracellular fluid is probably lower than in solution owing to tortuosity (Rice and Nicholson, 1991), and thus, the distance is overestimated. Nevertheless, this result clearly shows that DA can diffuse over relatively large distances in some regions of the amygdala.

4. Regional Differences in the Regulation of Extracellular Dopamine

4.1. Different Dynamics of Evoked Dopamine Overflow

The various functions of DA are associated with specific DA terminal fields (Le Moal and Simon, 1991). In addition, biochemical, pharmacological, and electrophysiological evidence indicates that mesotelencephalic DA neurons exhibit different characteristics (Bannon and Roth, 1983; Glowinski et al., 1984; Kilts et al., 1988). The regulation of extracellular levels of DA has been most extensively studied in the rat striatum. In contrast to rich terminal stores of DA (Anden et al., 1966), extracellular concentrations are in the low-nanomolar range (Parsons and Justice, 1992), and are tightly controlled by a delicate balance among release, uptake, diffusion, and degradation (Justice et al., 1988). The relationship of these factors determines, to a great extent, the nature of dopaminergic transmission. Our data show that this regulation is not uniform in all dopaminergic terminal regions.

Individual overflow curves for evoked DA collected in the medial prefrontal cortex, basolateral amygdaloid nucleus, caudate-putamen, and nucleus accumbens, and measured by fast-scan cyclic voltammetry at Nafion-coated, carbon fiber microelectrodes, are shown in Fig. 9. In all regions, electrical stimulation of ascending DA fibers results in a transient increase in extracellular DA followed by a rapid return to baseline levels. However, markedly different dynamics are clearly evident even at a single frequency. For example, steady-state concentrations develop in the caudate-putamen, but not in the other regions. A tendency toward steady state

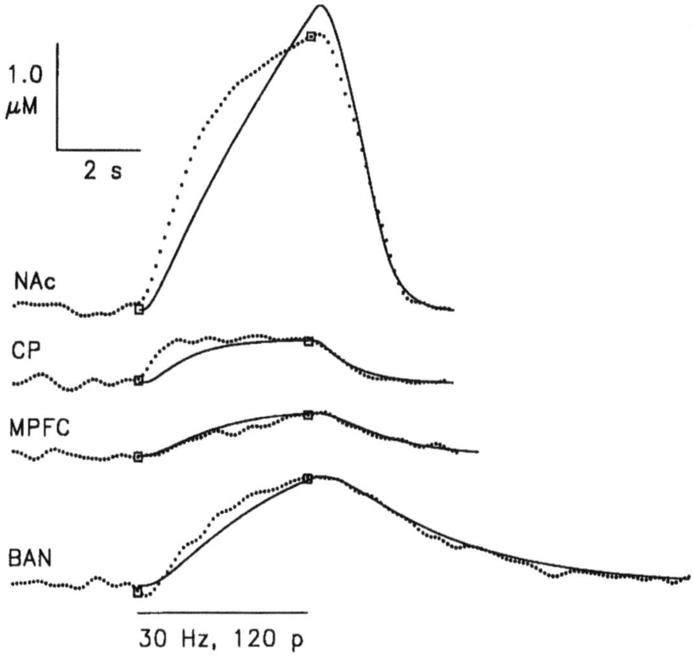

Fig. 9. Regional variation in the dynamics of evoked DA overflow. Data (solid circles) represent individual overflow curves recorded in vivo in the nucleus accumbens (NAc), caudate-putamen (CP), medial prefrontal cortex (MPFC), and basolateral amygdaloid nucleus (BAN) and evoked by electrical stimulation using 30 Hz for 120 pulses. Simulations (solid lines) are convoluted with a response function for thin-layer diffusion. (Reprinted from *J. Neurosci.* **14,** 442–450 [1994] with permission.)

is observed in the nucleus accumbens. In addition, the extracellular clearance rate of DA is much slower in the cortex and amygdala.

Differences in the regulation of extracellular DA levels are also present in the frequency response of evoked DA (Fig. 10). In contrast to the caudate-putamen and nucleus accumbens, where DA concentrations sharply increase with the frequency of stimulation, evoked DA is almost frequency independent in the amygdala and cortex. An interesting observation is that maximum overflow is similar in all regions at 10 and 20 Hz, frequencies near the physiological range (Chiodo, 1988). This is despite a huge disparity in DA tissue content and terminal density (Palkovits et al., 1979; Doucet et al., 1986; Kilts

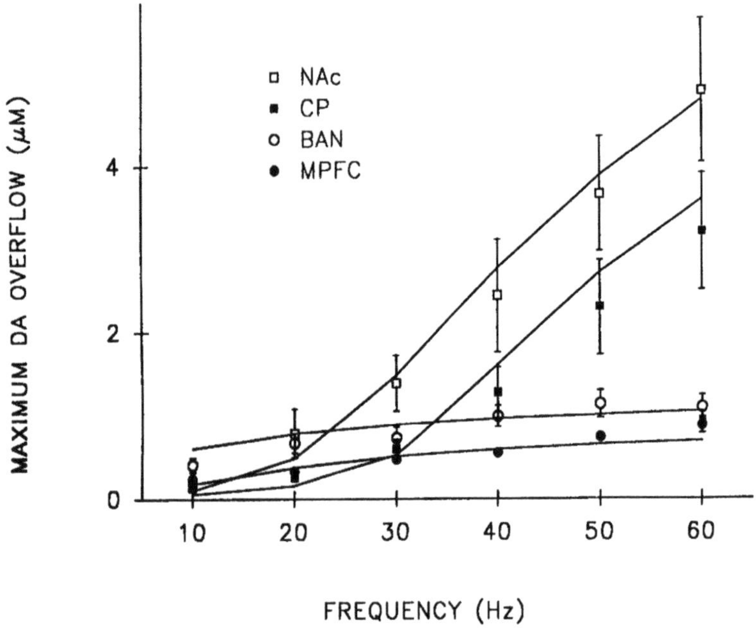

Fig. 10. Regional variation in the frequency response of evoked DA overflow. The maximum concentration of extracellular DA evoked as a function of stimulus frequency is shown in the nucleus accumbens (NAc), caudate-putamen (CP), medial prefrontal cortex (MPFC), and basolateral amygdaloid nucleus (BAN). Data are the mean ± SEM for 5 rats, 120 stimulus pulses were used for all frequencies. Calculated values, using best-fit parameters of DA release and uptake as determined for each region, are shown by the solid lines. (Reprinted from *J. Neurosci.* **14,** 442–450 [1994] with permission.)

and Anderson, 1987; Descarries et al., 1987). For example, the medial prefrontal cortex and basolateral amygdaloid nucleus contain 90- and 6-fold less DA, respectively, than the striatum. The results are consistent with the concept that regional concentrations of extracellular DA in the brain cannot be predicted solely on the basis of tissue levels and are, in fact, higher than that expected for sparsely innervated regions.

4.2. Dopamine Uptake

To examine whether the differences in the evoked overflow of DA are the result of differences in DA release and uptake, quantitative values for these parameters were determined using the kinetic analysis described above. Since K_m is not known for DA uptake in the cortex and amygdala, DA uptake was approximated using an exponential decay:

$$\{d[DA]/dt\}_{uptake} = k[DA] \tag{3}$$

where k is the first-order rate constant. For comparison, a k was calculated for DA uptake in the striatum using V_{max}/K_m. This is a valid estimate, because at the low concentrations of extracellular DA *in situ* (where $K_m >>$ than [DA]), Michaelis-Menten kinetics become first-order with a rate constant described by this ratio.

Best-fit rate constants for DA uptake in the different regions are found in Table 1. Absolute DA uptake is similar in the amygdala and cortex, but approx 50 times less compared to the striatum. This is consistent with the extracellular clearance rates of evoked DA as evident in individual overflow curves (Fig. 9). There is also a significant difference between uptake rate constants in the caudate-putamen and nucleus accumbens, and they are the result of a lower V_{max} in the latter (Stamford et al., 1988a; May and Wightman, 1989; Marshall et al., 1990). A unique picture emerges when uptake is normalized to the density of DA terminals, as reflected by the tissue content of DA: Compared to all other regions, relative uptake is 10 times less in the amygdala.

The uptake of DA in the striatum is characterized as high affinity with a K_m of around 0.2 μM (Near et al., 1988; Horn, 1990; Marshall et al., 1990). Whether the slow rates of DA uptake in the cortex and amygdala represent low-affinity uptake similar to that found in the neuroendocrine tubero-infundibular DA system (Demarest and Moore, 1979; Annunziato et al., 1980) and/or a reduced number of uptake sites cannot at present be ascertained. However, uptake in the amygdala is not related to terminal density, as is the case in the cortex and striatum, and must be distinct.

The differences in uptake can exert profound effects on the fate of extracellular DA. The half-life ($t_{1/2}$) of evoked DA is estimated by $0.693/k$, and is approx 2.0 s in the cortex and amygdala. In sharp contrast, the $t_{1/2}$ for extracellular DA in the striatum is considerably less at 0.04 s. Thus, DA has the opportunity to diffuse a distance sixfold farther in the cortex

Table 1
Values of $[DA]_p$, $[NE]_p$, k, and $t_{1/2}$ in Different Regions[a]

Brain region[c]	DA content[b], µg/g	$[DA]_p$, nM	k, s^{-1}	$t_{1/2}$, s	Normalized[d] values $[DA]_p$	Normalized[d] values k
	Absolute values					
MPFC	0.1	10.7	0.53	1.3	107	53
BAN	1.5	13.8	0.32	2.2	9.2	2.0
NAc	9.0	67.5	13.9	0.05	7.5	15
CP	9.0	89.3	22.7	0.03	9.9	25
	NE content, µg/g	$[NE]_p$ nM	k, s^{-1}	$t_{1/2}$, s	$[NE]_p$	k
ACg_{upper}	0.27	0.89	0.07	9.9	3.5	2.5
ACg_{lower}	0.36	1.68	0.27	2.6	4.7	4.6
DLG	0.41	1.66	0.19	3.6	4.0	4.6
AV	2.10	7.27	0.65	1.1	3.5	3.1

[a]From Garris and Wightman, 1994a and Mitchell et al., 1994.

[b]Original units are ng/mg protein. Converted by dividing by 10.

[c]MPFC, medial prefrontal cortex; BAN, basolateral amygdaloid nucleus; CP, caudate-putamen; NAc, nucleus accumbens; ACg_{upper}, anterior cingulate, upper; ACg_{lower}, anterior cingulate, lower; DLG, dorsal lateral geniculate; AV, anterior ventral nucleus of thalamus (Paxinos and Watson, 1986).

[d]Normalized by dividing absolute values by tissue content. Resulting units are s^{-1} µg/g for k and nM/µg/g for $[CA]_p$.

and amygdala than in the striatum. In addition, higher extracellular DA concentrations as a result of release at physiologically relevant frequencies are maintained for longer periods of time in the sparsely DA-innervated regions. This does not appear to be a function of anatomical differences between mesotelencephalic DA neurons, since these terminals make true synaptic contacts and have a similar synaptic architecture (Bouyer et al., 1984; Freud et al., 1984; Sequela et al., 1988).

4.3. Dopamine Release

Table 1 shows the release parameter, $[DA]_p$, obtained from the neurochemical model for the four different regions examined. Absolute DA release is similar in the cortex and amygdala, but approximately eight times less than in the striatum. As is the case with DA uptake, a very different picture is revealed when $[DA]_p$ is normalized to the density of the DA innervation: Compared to all other regions, relative release was 10 times greater in the cortex.

The high rate of cortical DA release may explain the usually high levels of extracellular DA as measured by dialysis here (Sharp et al., 1986), and is in agreement with in vitro studies comparing the evoked efflux of radiolabeled DA from slices of the cortex and striatum (Hoffmann et al., 1988). Since mesoprefrontal DA neurons fire at a faster mean rate compared to mesostriatal DA neurons (Chiodo, 1988), the former may have developed a more proficient secretory machinery to sustain high release. Differences in autoregulation are unlikely to be responsible, since all of the DA systems examined exhibit autoreceptors governing release (Bannon and Roth, 1983; Talmaciu et al., 1986; Altar et al., 1987; Moghaddam and Bunney, 1990; Bean and Roth, 1991; Bull et al., 1991).

4.4. Comparison to Norepinephrine

Norepinephrine (NE) neurons originate in the locus ceruleus, ascend through the medial forebrain bundle, and terminate in different regions of the brain (Ungerstedt, 1971; Moore and Card, 1984). Extracellular NE levels in the thalamus, elicited by electrical stimulation of the dorsal tegmental bundle, have been monitored with rapid-sampling chronoamperometry at Nafion-coated, carbon-fiber microelectrodes (Capella et al., 1993; Ghasemzadeh et al., 1993). This technique has an equivalent temporal resolution to fast-scan cyclic voltammetry and, thus, provides similar kinetic information concerning extracellular dynamics for NE.

Recently, Adams and coworkers (Mitchell et al., 1994) have applied the neurochemical model of evoked overflow successfully to evaluate NE release and uptake parameters in different regions of the cortex and thalamus. The results are found in Table 1 and afford a unique opportunity to compare the factors that control extracellular DA and NE levels. Several generalizations can be stated:

1. Tissue stores of DA tend to be much greater than for NE;
2. There is much less heterogeneity in the NE than in the DA systems;
3. Relative NE uptake and release rates are less than relative DA uptake and release rates, respectively; and
4. Based on the $t_{1/2}$ for evoked concentrations, the overall regulation of extracellular NE levels is similar to DA in the cortex and amygdala, and is well suited for generating extracellular concentrations over a widespread area surrounding release sites.

A useful way of characterizing the regulation of extracellular neurotransmitter levels is the ratio of release to uptake (Garris and Wightman, 1994a). This ratio for DA in the striatum is about five times less than in the cortex and amygdala. Thus, the regulation of extracellular DA in the striatum is considered "uptake-dominated" and is responsible for the low, steady-state concentrations of evoked DA during stimulation at frequencies near physiological (i.e., 10 and 20 Hz). Steady state develops from a balance between release and uptake (Wightman et al., 1988), and the large $[DA]_p$ in the striatum, which is highest in the brain, is kept in check by the potent DA uptake. Only when the time for uptake is reduced at high frequencies do excessive concentrations of extracellular DA appear.

The regulation of DA levels in the cortex and amygdala is considered "release-dominated" since the release-to-uptake ratio is greater than for DA in the striatum. This is achieved differently since relative DA release is unusually high in the cortex and relative DA uptake is unusually low in the two regions. As a direct result, these sparsely DA-innervated areas can produce identical concentrations of DA to the DA-rich striatum when stimulated by frequencies near physiological. The regulation of extracellular NE concentrations is similar, but not identical. The ratio of release to uptake is greater than for DA in the striatum, but less than for DA in the cortex and amygdala. Although $t_{1/2}$ for evoked concentrations of NE is the largest in the brain (e.g., 10 s in the anterior $cingulate_{upper}$), absolute release rates for NE are lower than for DA.

The results suggest fundamentally different modes of chemical communication for DA and NE in the brain. Neurotransmission is a function of extracellular concentrations in space and time (Nicholson and Rice, 1991). Based on the dynamics of extracellular concentrations dynamics alone, the extreme systems are the mesostriatal and mesoamygdaloid DA neurons. In the striatum, potent uptake limits the spatial domain of extracellular DA. In the amygdala, unusually low uptake combined with relative high release rates results in an expanded area of influence for release DA. Mesoprefrontal DA neurons and NE neurons lie somewhere in between, but are more closely related to mesoamygdaloid DA neurons. However, the functional significance of concentration dynamics is wholly dependent on the affinities and locations of the appropriate receptors, and further study is required for confirmation of these ideas.

The concept of release vs uptake domination may be directly relevant in understanding Parkinson's disease, where clinical symptoms are not presented until after a loss of at least 80% of the nigrostriatal DA tract (Hornykiewicz and Kish, 1986). The reduced DA uptake, owing to a loss of DA terminals, permits the high, absolute release rates for DA in the striatum to be expressed and to dominate. Thus, DA diffuses farther, and significant extracellular concentrations can be maintained, which provide a viable dopaminergic tone in the compromised striatum. According to this interpretation, the large value of $[DA]_p$, which is unique to the striatum, is a buffer against the development of Parkinson's disease. As such, it would be a trait selected for during evolution.

5. Extrapolation to the Processes in the Synaptic Cleft

5.1. Factors Regulating Dopamine Efflux from the Synapse

The evidence provided from in vivo measurements of stimulated DA overflow shows that the extracellular concentration of DA is regulated by the balance between uptake and release. Recently, however, we have become interested in extrapolating our measurements into the synaptic region. Direct measurements in the synapse are not yet possible because of its small size. For this reason, current views of chemical neurotransmission in the CNS (Cooper et al., 1991) are based on observations for acetylcholine at peripheral synapses. Acetylcholine released into the synaptic cleft is followed by diffusion, activation of receptors, and clearance by a degradative esterase (Eccles and Jaeger, 1958; Katz and Miledi, 1973; Armstrong and Lester, 1979; Wathey et al., 1979; Bartol et al., 1991). Rapid transmitter-receptor binding, which buffers the free diffusion of acetylcholine, as well as postsynaptic invaginations, slows the mass transport of acetylcholine and allows time for its degradation within the cleft. Thus, chemical communication at the neuromuscular junction is restricted to the spatial domain of the synapse. Observations of extracellular DA, which originates from synaptic DA, should be useful to test whether these concepts also apply to DA synapses.

A synaptic mode for dopaminergic transmission in the CNS is suggested by the presence of anatomical features for specific dopaminergic synapses (Pickel et al., 1981; Doucet et al., 1986, Descarries et al., 1987) and the low (nanomolar), steady-state concentration of DA in the extracellular fluid of regions with a dense dopaminergic innervation (Parsons and Justice, 1992). The high-affinity DA transporter is proposed to terminate the effects of synaptic DA (Horn, 1990; Giros and Caron, 1993), and thus play a role similar to acetylcholine esterase in the neuromuscular junction. However, physiological and behavioral experiments have suggested a role for extrasynaptic communication by DA (Johnson et al., 1986; Schenk and Bunney, 1987; Zigmond et al., 1990; Fuxe and Agnati, 1991; Gonon et al., 1991).

We have tested these concepts for DA neurotransmission in the nucleus accumbens (Garris et al., 1994b). To evaluate the likelihood for DA to leave the synapse, a single impulse is preferred (Valenta et al., 1988; Palij et al., 1990; Limberger et al., 1991; Kennedy et al., 1992). This is because synaptic efflux during stimulus trains could be enhanced because of

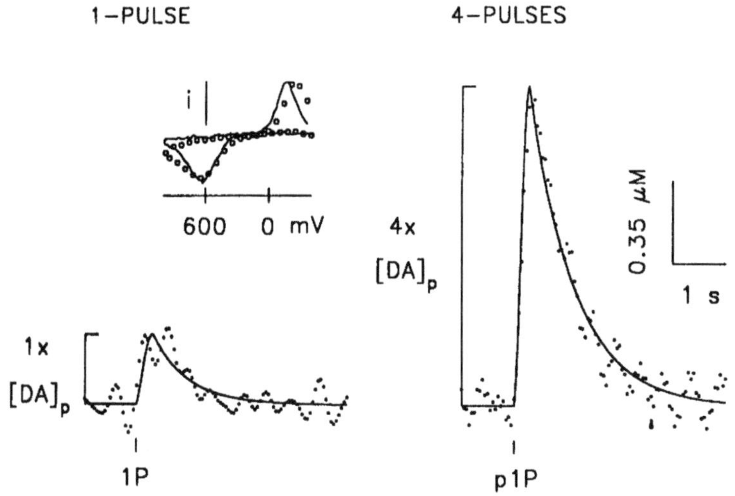

Fig. 11. Dopamine overflow curves (deconvoluted) elicited by one and four stimulus pulses. Curves were recorded in vivo in the nucleus accumbens after administration of the DA uptake inhibitor nomifensine 25 mg/kg). **(Left)** Dopamine overflow evoked by a single stimulus pulse (1P). **(Right)** Dopamine overflow evoked by four stimulus pulses at 100 Hz (p1P).

occupancy of uptake and receptor binding sites within the synapse by previously released DA. To slow the rate of removal of DA from the extracellular space by cellular uptake, rats were treated with the DA uptake inhibitor, nomifensine, and efflux elicited by a single stimulation pulse was measured (Fig. 4). The DA concentration exhibits a transient increase as a result of the stimulation, which is quite large compared to the concentration required to bind DA uptake (Scatton et al., 1985; Marshall et al., 1990; Boja et al., 1992; Mennicken et al., 1992) and receptor (Boyson et al., 1986; Dawson et al., 1986; Richfield et al., 1989) sites. Under the same conditions, four stimulus pulses delivered within 30 ms elicit a concentration increase that is exactly four times that elicited by one pulse (Fig. 11). This "pseudo-one-pulse" (p1P) stimulation (Mayer et al., 1988; Singer, 1988) provides insufficient time for the DA uptake transporters or receptors to function between pulses, although binding to these sites is not precluded.

That the observed concentration of DA is linear with the number of impulses strongly implies that binding to sites in the synapse does not restrict DA efflux from the synapse. Indeed, efflux of a substance from a circular, disk-shaped space with dimensions of the DA synapse, radius = 150 nm and width = 15 nm (Pickel et al., 1981), should occur in <1 ms if diffusion is unimpaired (Eccles and Jaeger, 1958). Note further that, if the efflux of DA from the synapse is unobstructed, as suggested by the above results, then the DA concentration measured in the extracellular space is the average synaptic concentration at the time when the concentration gradients are relaxed.

We further tested the hypothesis that DA can freely diffuse from the synapse with well-characterized pharmacological agents. The amplitude of the transient extracellular increase in DA concentration evoked by p1P stimulation was not altered by the absence of nomifensine (Fig. 12). In contrast, the efficacy of uptake inhibition is demonstrated by the dramatic differences in DA levels during 20-Hz stimulations before and after nomifensine (Fig. 6). Like nomifensine, neither the D_1 receptor antagonist, SCH 23390, nor the D_2 receptor antagonist, sulpiride, significantly increased extracellular DA levels evoked by p1P stimulation (Fig. 13). The lack of effect of DA receptor antagonists on p1P stimulation strongly suggests that binding of DA to receptors does not appreciably buffer the exit of DA from the synapse. In contrast to the p1P stimulation, sulpiride markedly increases DA efflux during a 20-Hz stimulation (Fig. 6). During the longer train, released DA has sufficient time to inhibit subsequent release by the activation of autoreceptors that are antagonized by sulpiride (Starke et al., 1989). The only treatment found to alter DA efflux evoked by a p1P stimulation is L-DOPA. This drug acts intracellularly, however, to increase DA stores (May et al., 1988).

5.2. Synaptic and Extrasynaptic Concentrations of Dopamine

If every dopaminergic terminal around the electrode tip is activated by the stimulation, then, based on the known

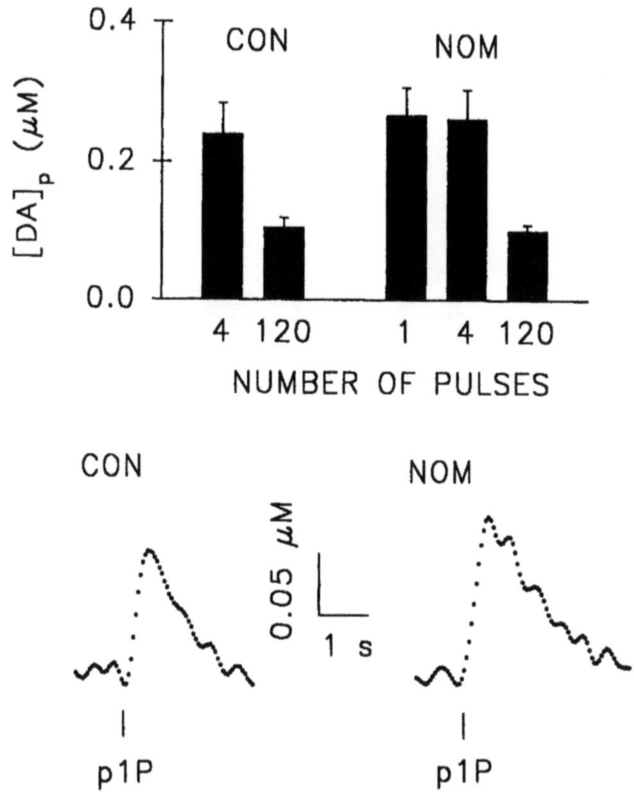

Fig. 12. Effects of DA uptake blockade on evoked DA overflow. **(Top)** The calculated parameter for DA release ($[DA]_p$) in control (CON) and after administration of nomifensine (NOM, 25 mg/kg). Recordings were made in the nucleus accumbens. Values for 120 pulses are the average for curves elicited by frequencies between 20 and 60 Hz. Data are the mean ± SEM of 5–7 animals. **(Bottom)** Individual overflow curves in control and after administration of nomifensine, and elicited by four stimulus pulses at 100 Hz (p1P).

density of the of dopaminergic terminals, we can calculate that each pulse causes an average of 1000 DA molecules per terminal to appear in the extracellular space. For this calculation, the volume of extracellular space between individual DA terminals is approximated by a sphere with a radius of 2 μm, one-half the distance between DA terminals in the striatum (Pickel et al., 1985), reduced by an extracellular volume fraction of 0.2 (Rice and Nicholson, 1991). The measurements show that $[DA]_p$ for a single stimulus pulse is 250 n*M* on aver-

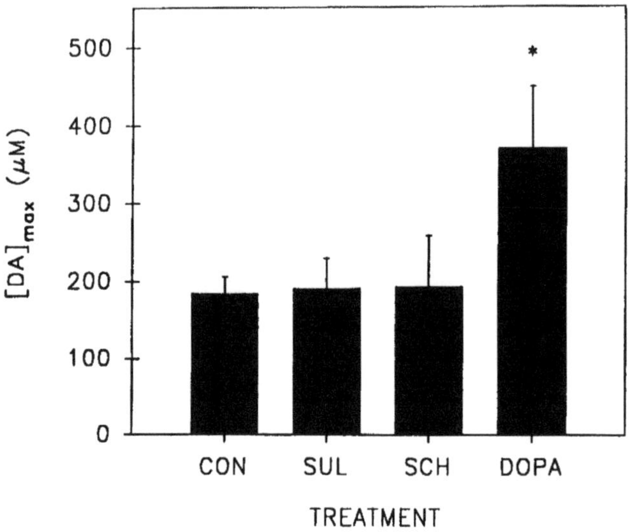

Fig. 13. Effects of DA receptor blockade on evoked DA overflow measured in vivo in the nucleus accumbens and elicited by four stimulus pulses at 100 Hz (p1P). The maximum DA concentration ($[DA]_{max}$) was recorded before (con) and after drug administration. The D_2 antagonist sulpiride (SUL, 100 mg/kg) and the D_1 antagonist SCH 23390 (SCH, 0.1 mg/kg) caused no significant effect. L-DOPA (DOPA, 250 mg/kg) administered in the presence of the peripheral decarboxylase inhibitor Ro 4-4062 (60 mg/kg) did cause an increase in stimulated overflow. The results are given as the average from four animals.

age (Fig. 12). Since this is a measure of the DA concentration in all terminal "spheres" adjacent to the sensor immediately after synaptic and extrasynaptic concentration gradients are relaxed, and before uptake occurs, we can calculate the average amount contributed per terminal. Note that 1000 DA molecules contained in a space with the dimensions of a DA synapse would lead to a concentration of 2.5 m*M*. Thus, microseconds after exocytosis, this is the concentration that we predict to be in the synapse.

Dilution into the extrasynaptic space results in the maximal concentrations observed with the sensor, which are subsequently lowered by cellular uptake. In the unstimulated animal, terminals are more likely to fire asynchronously, such that further extrasynaptic diffusion will dilute the released

DA. The net result is that the temporally and spatially averaged concentration of extracellular DA in the striatum *in situ* is predicted to be in the low nanomolar range (Kawagoe et al., 1992), a result that agrees with in vivo measurements by microdialysis (Parsons and Justice, 1992). Thus, depending on the temporal and spatial domain of a measurement, DA concentrations in the nucleus accumbens vary over six orders of magnitude.

5.3. Comparison of Synaptic Release with Uptake and Receptor Sites

The results suggest that binding to DA uptake and receptor sites does not affect the synaptic efflux of DA. Two possibilities to explain these observations are (1) the number of DA molecules released per stimulus pulse is much greater than the available synaptic binding sites and, therefore, sites are easily saturated, or (2) uptake and receptor sites are located extrasynaptically. Quantitative anatomical estimates indicate that there are approx 2000 DA uptake sites/DA terminal in the rat striatum (Scatton et al., 1985; Marshall et al., 1990; Boja et al., 1992; Mennicken et al., 1992). Similar data for DA receptor sites (Boyson et al., 1986; Dawson et al., 1986; Richfield et al., 1989) suggest there are 2000/DA terminal in this region. If 1000 DA molecules are released/stimulus pulse, there exists a sufficient number of binding sites to prevent synaptic efflux if all sites are located in the synapse. Therefore, it appears that these sites are either extrasynaptic or bind slowly compared to the time for efflux from the synapse. However, these predictions await experimental proof at the level of individual synapse.

The present interpretation is speculative because the carbon-fiber microelectrode samples from at least 30 DA terminals in vivo (Stamford et al., 1986a), and all synapses may not be activated at any given time (Cunnane and Stjarne, 1984) Nevertheless, these results show that DA concentrations near the reported K_d values of DA receptors (Boyson et al., 1986; Dawson et al., 1986; Richfield et al., 1989) can exist transiently in the extrasynaptic fluid and that the lifetime of extrasynaptic DA is regulated by uptake. The physiological consequences of these dynamic DA concentration gradients can only be established when the location and kinetics of DA binding to receptors are elucidated. These results unequivocally establish that DA can be an extrasynaptic messenger in the nucleus accumbens. Based on our measurements during stimulus trains in other regions, it is likely that extrasynaptic DA plays a key role in these areas as well.

6. Conclusions

Carbon-fiber microelectrodes provide a new way to view dopaminergic neurotransmission. The measurements are localized in space, and the measurements show that dramatic differences in extracellular DA can occur across dimensions of a few microns. The measurements are also on a time domain that is close to the firing rate of individual neurons. The rapid measurements show that high-nanomolar concentrations can exist in the extrasynaptic fluid for transient periods. Perhaps of greatest importance, the measurements show that the rates of uptake and release are finely tuned in each brain region where DA is a transmitter. Alteration of this balance by pharmacological means, lesions, or neuronal activity can profoundly alter this balance, resulting in dramatic changes in the extracellular concentration of DA. Since our evidence suggests an important role in communication for extrasynaptic DA, these imbalances could profoundly affect behavior. Future experiments with voltammetric measurements in freely moving animals will further probe this hypothesis in the spatially and temporally resolved domains.

Acknowledgments

Research in these areas was supported by the National Science Foundation (IBN) and the National Institutes of Health. The aid of Karin Pihel in preparation of Fig. 1 is gratefully acknowledged.

References

Adams R. N. (1990) In vivo electrochemical measurements in the CNS. *Prog. Neurobiol.* **35**, 297–311.

Altar C. A., Boyar W. C., Oei E., and Wood P. L. (1987) Dopamine autoreceptors modulate the in vivo release of dopamine in the frontal, cingulate and entorhinal cortices. *J. Pharmacol. Exp. Ther.* **242,** 115–120.

Anden N. E., Fuxe K., Hamberger B., and Hokfelt T. (1966) A quantitative study of the nigroneostriatal dopamine neuron system in the rat. *Acta Physiol. Scand.* **67,** 306-312.

Annunziato L., Leblanc P., Kordan C., and Weiner R. I. (1980) Differences in the kinetics of dopamine uptake in synaptosome preparations of the median eminence relative to other dopaminergically innervated brain regions. *Neuroendocrinology* **31,** 316–320.

Armstrong D. L. and Lester H. A. (1979) The kinetics of tubocurarine action and restricted diffusion within the synaptic cleft. *J. Physiol. (Lond.)* **294,** 365–386.

Bannon M. J and Roth R. H. (1983) Pharmacology of mesocortical dopamine neurons. *Pharmacol. Rev.* **35,** 53–68.

Bartol T. M., Land B. R., Salpeter E. E., and Salpeter M. M. (1991) Monte Carlo simulation of miniature endplate current generation in the vertebrate neuromuscular junction. *Biophys. J.* **59,** 1290–1307.

Baur J. E., Kristensen E. W., May L. J., Wiedemann D. J., and Wightman R. M. (1988) Fast-scan voltammetry of biogenic amines. *Anal. Chem.* **60,** 1268–1272.

Bean A. J. and Roth R. H. (1991) Extracellular dopamine and neurotensin in rat prefrontal cortex in vivo: effects of median forebrain bundle stimulation frequency, stimulation pattern, and dopamine autoreception. *J. Neurosci.* **11,** 2694–2702.

Boja J. W., Mitchell W. M., Patel A., Kopajtic T. A., Carroll F. I., Lewin A. H., Abraham P., and Kuhar M. J. (1992) High-affinity binding of [^{125}I]RTI-55 to dopamine and serotonin transporters in rat brain. *Synapse* **12,** 27–36.

Boyson S. J., McGonigle P., and Molinoff P. B. (196) Quantitaive autoradiographic localization of the D1 and D2 subtypes of dopamine receptors in rat brain. *J. Neurosci.* **6,** 3177–3188.

Bjorklund A. and Lindvall O. (1984) Dopamine-containing systems in the CNS, in *Handbook of Chemical Neuroanatomy* (Bjorklund A. and Hokfelt T., eds.), Elsevier, New York, pp. 55–122.

Bouyer J. J., Joh T. H., and Pickel V. M. (1984) Ultrastructural localization of tyrosine hydroxylase in rat nucleus accumbens. *J. Comp. Neurol.* **227,** 92–103.

Bracewell R. N. (1986) *The Fourier Transform and Its Applications.* McGraw-Hill, New York, pp. 345–355.

Bull D. R., Bakhtiar R., and Sheehan M. J. (1991) Characterization of dopamine autoreceptors in the amygdala: a fast cyclic voltammetric study in vitro. *Neurosci. Lett.* **134,** 41–44.

Capella P., Ghasemzadeh M. B., Adams R. N., Wiedemann D. J., and Wightman R. M. (1993) Real-time monitoring of electrically stimulated norepinephrine release in rat thalamus. II. Modeling of release and reuptake characteristics of stimulated norepinephrine overflow. *J. Neurochem.* **60,** 442–453.

Chiodo L. A. (1988) Dopamine-containing neurons in the mammalian central nervous system: electrophysiology and pharmacology. *Neurosci. Biobehav. Rev.* **12,** 49–91.

Cooper J. C., Bloom F. E., and Roh R. H. (1991) *The Biochemical Basis of Neuropharmacology*. Oxford University Press, New York.

Crank J. (1975) *The Mathmatics of Diffusion*. Oxford University Press, London, pp. 47–53.

Cunnane T. C. and Stjarne L. (1984) Transmitter secretion from individual varicosities of guinea-pig and mouse vas deferens: highly intermittent and monoquantal. *Neuroscience* **13,** 1–20.

Dawson T. M., Gehlert D. R., McCabe R. T., Barnett A., and Wamsley J. K. (1986) D-1 dopamine receptors in the rat brain: a quantitative autoradiographic analysis. *J. Neurosci.* **6,** 2352–2365.

Deakin M. R., Koavach P. M., Stutts K. J., and Wightman R. M. (1986) Heterogeneous mechanisms of the oxidation of catechols and ascorbic acid at carbon electrodes. *Anal. Chem.* **58,** 1474–1480.

Demarest K. T. and Moore K. E. (1979) Lack of a high affinity transport system for dopamine in the median eminence and posterior pituitary. *Brain Res.* **171,** 545–551.

Descarries L., Lemay B., Doucet G., and Berger B. (1987) Regional and laminar density of the dopamine innervation in adult rat cerebral cortex. *Neuroscience* **21,** 807–824.

Doucet G., Descarries L., and Garcia L. (1986) Quantification of the dopamine innervation in adult rat neostriatum. *Neuroscience* **19,** 427–445.

Eccles J. C. and Jaeger J. C. (1958) The relationship between the mode of operation and the dimensions of the junctional regions at synapses and motor end-organs. *Proc. R. Soc. Lond. Biol. Sci.* **148,** 38–56.

Eldon S. E., Horst K. C., Ely T., and Kilts C. D. (1994) Functional mapping of the dopamine transport complex in the rat brain: presence of low affinity transporters in the amygdaloid complex and prefrontal cortex. *Synapse* (in press).

Engstrom R. C., Wightman F. M., and Kristensen E. W. (1988) Diffusional distortion in the monitoring of dynamic events. *Anal. Chem.* **60,** 652–656.

Ewing A. G., Bigelow J. C., and Wightman R. M. (1983) Direct for in vivo monitoring of dopamine released from two striatal compartments in the rat. *Science* **221,** 169–171.

Ewing A. G. and Wightman R. M. (1984) Monitoring the stimulated release of dopamine with in vivo voltammetry. II. Clearance of released dopamine from extracellular fluid. *J. Neurochem.* **43,** 570–577.

Freund T. F., Powell J. F., and Smith A. D. (1984) Tyrosine hydroxylase-immunoreactive boutons in synaptic contact identified striatonigral neurons, with particular reference to dendritic spines. *Neuroscience* **13,** 1189–1215.

Fuxe K. and Agnati L. F. (1991) Two principal modes of electrochemical communication in the brain: volume versus wiring transmission, in *Volume Transmission in the Brain: Novel Mechanisms for Neural Transmission.* (Fuxe K. and Agnati L. F., eds.), Raven, New York, pp. 1–9.

Garris P. A., Collins L. B., Jones S. R., and Wightman R. M. (1993) Evoked extracellular dopamine in vivo in the medial prefrontal cortex. *J. Neurochem.* **61,** 637–647.

Garris P. A. and Wightman R. M. (1994a) Different kinetics govern dopaminergic transmission in the amydala, prefrontal cortex, and striatum: an in vivo voltammetric study. *J. Neurosci.* **14,** 442–450.

Garris P. A. and Wightman R. M. (1994b) In vivo voltammetric measurement of evoked extracellular dopamine in the rat basolateral amygdaloid nucleus. *J. Physiol.* **478,** 239–249.

Garris P. A. and Wightman R. M. Pharmacology of evoked dopamine overflow in the basolateral amygdaloid nucleus and comparison to the striatum. (submitted)

Garris P. A., Ciolkowski E. L., and Wightman R. M. (1994a) Heterogeneity of evoked dopamine overflow in striatal and striatomygdaloid regions. *Neuroscience* **59,** 417–427.

Garris P. A., Ciolkowski E. L., Pastore P., and Wightman R. M. (1994b) Efflux of synaptic dopamine in the nucleus accumbens of the rat brain. *J. Neurosci.* **14,** 6084–6093.

Gerhardt G. A., Oke A. F., Nagy G., Moghaddam B., and Adams R. N. (1984) Nafion-coated electrodes with high selectivity for CNS electrochemistry. *Brain Res.* **290,** 390–395.

Ghasemzadeh M. B., Capella P., Mitchell K., and Adams R. N. (1993) Real-time monitoring of electrically stimulated norepinephrine release in rat thalamus. I. Resolution of transmitter and metabolite signal component. *J. Neurochem.* **60,** 442–453.

Giros B. and Caron M. G. (1993) Molecular characterization of the dopamine transporter. *TIPS* **14,** 43–49.

Glowinski J., Tassin J. P., and Thierry A. M. (1984) The mesocortico-prefrontal dopaminergic neurons. *TINS* **7,** 414–418.

Gonon F. G., Suaud-Chagny M. F., Mermet C. C., and Buda M. (1991) Relation between impulse flow and extracellular catecholamine levels as studied by in vivo electrochemistry in CNS, in *Volume Transmission in the Brain: Novel Mechanisms for Neural Transmission.* (Fuxe K. and Agnati L. F., eds.), Raven, New York, pp. 337–350.

Graybiel A. M. (1984) Correspondence between the dopamine islands and striosomes of the mammalian striatum. *Neuroscience* **13,** 1157–1187.

Hoffmann I. S., Talmaciu R. K., Ferro C. P., and Cubeddu L. X. (1988) Sustained high release at rapid stimulation rates and reduced functional autoreceptors characterize prefrontal cortex dopamine terminals. *J. Pharmacol. Exp. Ther.* **245,** 761–772.

Horn A. S. (1990) Dopamine uptake: a review of progress in the last decade. *Prog. Neurobiol.* **34,** 387–400.

Hornykiewicz O. and Kish S. J. (1986) Biochemical pathophysiology of Parkinson's disease, in *Advances in Neurology* (Yahr M. D. and Bergmann K. J., eds.), Raven, New York, pp. 19–34.

Johnson S. W., Hoffer B. J., and Freedman R. (1986) Investigation of the failure of parenterally administered haloperidol to antagonize dopamine released from micropipettes in the caudate. *J. Neurosci.* **6,** 572–580.

Jones S. R., Michelson G. E., Collins L. B., Kawagoe K. T., and Wightman R. M. (1994) Interference by pH and Ca^{2+} ions during measurements of catecholamine release in slices of rat amygdala with fast-scan cyclic voltammetry. *J. Neurosci. Methods* **52,** 1–10.

Jones S. R., Garris P. A., Wightman R. M., and Kilts C. D. Comparison of dopamine uptake in the basolateral amygdaloid nucleus and striatum of the rat. (submitted)

Joyce J. N., Sapp. D. W., and Marshall J. F. (1986) Human striatal dopamine receptors are organized in compartments. *Proc. Natl. Acad. Sci. USA* **83,** 8002–8006.

Justice J. B., Nicolaysen L. C., and Michael A. C. (1988) Modeling the dopaminergic nerve terminal. *J. Neurosci. Methods* **22,** 239–252.

Katz B. and Miledi R. (1973) The binding of acetylcholine to receptors and its removal from the synaptic cleft. *J. Physiol. (Lond.)* **231,** 549–573.

Kawagoe K. T., Garris P. A., and Wightman R. M. (1992) Regulation of transient dopamine concentration gradients in the microenvironment surrounding nerve terminals in the rat striatum. *Neuroscience* **51,** 55–64.

Kawagoe K. T., Garris P. A., and Wightman R. M. (1993) pH-dependent processes at Nafion-coated carbon-fiber microelectrode. *J. Electroanal. Chem. Interfacial Electrochem.* **359,** 193–207.

Kawagoe K., Zimmerman J. B., and Wightman R. M. (1993) Principles of voltammetry and microelectrode surface states. *J. Neurosci. Methods* **48,** 225–240.

Kennedy R. T., Jones S. R., and Wightman R. M. (1992) Dynamic observation of dopamine autoreceptor effects in rat striatal slices. *J. Neurochem.* **59,** 449–455.

Kilts C. D. and Anderson C. M. (1987) Mesoamygdaloid dopamine neurons: differential rates of dopamine turnover in discrete amygdaloid nuclei of the rat brain. *Brain Res.* **416,** 402–408.

Kilts C. D., Anderson C. M., Ely T. D., and Mailman R. B. (1988) The biochemistry and pharmacology of dopamine neurons. *Ann. NY Acad. Sci.* **537,** 173–187.

Kristensen E. W., Kuhr W. G., and Wightman R. M. (1987) Temporal characterization of perfluorinated ion exchange coated microvoltammetric electrodes for *in vivo* use. *Anal. Chem.* **59,** 1752–1757.

Kuhr W. G., Wightman R. M., and Rebec G. V. (1987) Dopaminergic neurons: simultaneous measurements of dopamine release and single-unit activity during stimulation of the medial forebrain bundle. *Brain Res.* **419,** 122–128.

Kuhr W. G., Ewing A. G., Caudill W. L., and Wightman R. M. (1984) Monitoring the stimulated release of dopamine with in vivo voltammetry. I. Characterization of the response observed in the caudate nucleus of the rat. *J. Neurochem.* **43,** 560–569.

Le Moal M. and Simon H. (1991) Mesocorticolimbic dopamine network: functional and regulatory roles. *Physiol. Rev.* **71,** 155–234.

Limberger N., Trout S. J., Kruk Z. L., and Starke K. (1991) "Real time" measurement of endogenous dopamine release during short trains of pulses in slices of rat neostriatum and nucleus accumbens: role of autoinhibition. *Naunyn-Schmiedebergs Arch. Pharmacol.* **344,** 623–629.

Marsden C. A., Joseph M. H., Kruk Z. L., Maidment N. T., O'Neill R. D., Schenk J. O., and Stamford J. A. (1988) In vivo voltammetry: present electrodes and methods. *Neuroscience* **25,** 389–400.

Marshall J. F., O'Dell S. J., Navarrete R., Rosenstein A. J. (1990) Dopamine high-affinity transport site topography in rat brain: major differences between dorsal and ventral striatum. *Neuroscience* **37,** 11–21.

May L. J. and Wightman R. M. (1989) Heterogeneity of stimulated dopamine overflow within the rat striatum as observed with in vivo voltammetry. *Brain Res.* **487,** 311–320.

May L. J., Kuhr W. G., and Wightman R. M. (1988) Differentiation of dopamine overflow and uptake processes in the extracellular fluid of the rat caudate nucleus with fast-scan *in vivo* voltammetry. *J. Neurochem.* **51,** 1060–1069.

Mayer A., Limberger N., and Starke K. (1988) Transmitter release patterns of noradrenergic, dopaminergic and cholinergic axons in rabbit brain slices during short pulse trains, and the operation of presynaptic autoreceptors. *Naunyn-Schmiedebergs Arch. Pharmacol.* **338,** 632–643.

Mennicken F., Savasta M., Peretti-Renucci R., Feuerstein C. (1992) Autoradiographic localization of dopamine uptake sites in the rat brain with ^{3}H-GBR 12935. *J. Neural Transm.* **87,** 1–14.

Michael A. C., Ikeda M., and Justice J. B. (1987) Mechanisms contributing to the recovery of striatal releasable dopamine following MFB stimulation. *Brain Res.* **421,** 325–335.

Michael A. C., Wightman R. M., and Amatore C. A. (1989) Microdisk electrodes. I. Digital simulation with a conformal map. *J Electroanal. Chem. Interfacial Electrochem.* **267,** 33–45.

Millar J., Stamford J. A., Kruk Z. L., and Wightman R. M. (1985) Electrochemical, pharmacological and electrophysiological evidence of rapid dopamine release and removal in the rat caudate nucleus following electrical stimulation of the medial forebrain bundle. *Eur. J. Pharmacol.* **109,** 341–348.

Mitchell K., Oke A. F., and Adams R. N. (1994) In vivo dynamics of norepinephrine release-reuptake in multiple terminal field regions of rat brain. *J. Neurochem.* **63,** 917–926.

Moghaddam B. and Bunney B. S. (1990) Acute effects of typical and atypical antipsychotic drugs on the release of dopamine from prefrontal cortex, nucleus accumbens and striatum of the rat: an in vivo microdialysis study. *J. Neurochem.* **54,** 1755–1760.

Moore R. Y. and Card J. P. (1984) Noradrenaline-containing neuron systems, in *Classical Transmitters in the CNS,* Part I (Bjorklund A. and Hokfelt T., eds.), Elsevier, New York, pp. 123–156.

Nagy G., Gerhardt G. A., Oke A. F., Rice M. E., Adams R. N., Moore R. B., Szentirmay M. N., and Martin C. R. J. (1985) Ion exchange and transport of neurotransmitters in Nafion films on conventional and microelectrode surfaces. *J. Electroanal. Chem. Interfacial Electrochem.* **188,** 85–94.

Near J. A., Bigelow J. C., and Wightman R. M. (1988) Comparison of uptake of dopamine in rat striatal chopped tissue and synaptosomes. *J. Pharmacol. Exp. Ther.* **245,** 921–927.

Nicholson C. and Rice M. E. (1991) Diffusion of ions and transmitters in the brain cell microenvironment, in *Volume Transmission in the Brain: Novel Mechanisms for Neural Transmission.* (Fuxe K. and Agnati L. F., eds.), Raven, New York, pp. 279–294.

Palij P., Bull D. R., Sheehan M. J., Millar J., Stamford J., Kruk Z. L., and Humphrey P. A. (1990) Presynaptic regulation of dopamine release in corpus striatum monitored in vitro in real time by fast cyclic voltammetry. *Brain Res.* **509,** 172–174.

Palkovits M., Zaborszky L., Brownstein M. J., Fekete M. I. K., Herman J. P., and Kanyicska B. (1979) Distribution of norepinephrine and dopamine in cerebral cortical areas of the rat. *Brain Res. Bull.* **4,** 593–601.

Parsons L. H. and Justice J. B. (1992) Extracellular concentration and in vivo recovery of dopamine in the nucleus accumbens using microdialysis. *J. Neurochem.* **58,** 212–218.

Paxinos G. and Watson C. (1986) *The Rat Brain in Stereotaxic Coordinates.* Academic, New York.

Pickel V. M., Beckley S. C., Joh T. K., and Reis B. J. (1981) Ultrastructural and immunochemical localization of tyrosine hydroxylase in the neostriatum. *Brain Res.* **225,** 373–385.

Press W. H., Flannery B. P., Teukolsky S. A., and Fetterling W. T. (1989) *Numerical Recipes in Pascal.* Cambridge Llniversity Press, Cambridge, UK.

Rice M. E. and Nicholson C. (1989) Measurement of nanomolar dopamine diffusion using low-noise perfluorinated ionomer coated carbon fiber microelectrodes and high-speed cyclic voltammetry. *Anal. Chem.* **61,** 1805–1810.

Rice M. E. and Nicholson C. (1991) Diffusion characteristics and extracellular volume fraction during normoxia and hypoxia in slices of rat neostriatum. *J. Neurophysiol.* **65,** 264–272.

Richfield E. K., Penney J. B., and Young A. B. (1989) Anatomical and affinity state comparisons between dopamine D_1 and D_2 receptors in the rat central nervous system. *Neuroscience* **30,** 767–777.

Savasta M., Dubois A., and Scatton B. (1986) Autoradiographic localization of D1 dopamine receptors in the rat brain with [3H]SCH 23390. *Brain Res.* **375,** 291–301.

Scatton B., Dubois A., Dubocovich M. L., Zahniser N. R., and Fage D. (1985) Quantitative autoradiography of ^{3}H-nomifensine binding sites in rat brain. *Life Sci.* **36,** 815–822.

Schenk J. O. and Bunney B. S. (1987) Quantitative interpretations of in vivo chronoamperometry, in *Voltammetry in the Neurosciences* (Justice J. B., ed.), Humana, Clifton, NJ, pp. 139–160.

Sequela P., Watkins K. C., and Descarries L. (1988) Ultrastructural features of dopamine axon terminals in the anteromedial and suprarhinal cortex of adult rat. *Brain Res.* **442,** 11–22.

Sharp T., Zetterstrom T., and Ungerstedt U. (1986) An in vivo study of dopamine release and metabolism in rat brain regions using intracerebral dialysis. *J. Neurochem.* **47,** 113–122.

Singer E. A. (1988) Transmitter release from brain slices elicited by single pulses: a powerful method to study presynaptic mechanisms. *TINS* **9,** 274–276.

Stamford J. A., Kruk Z. L., and Millar J. (1986a) Sub-second striatal dopamine release measured by *in vivo* voltammetry. *Brain Res.* **381,** 351–355.

Stamford J. A., Kruk Z. L., and Millar J. (1986b) In vivo voltammetric characterization of low affinity striatal dopamine uptake: drug inhibition profile and relation to dopaminergic innervation density. *Brain Res.* **373,** 85–91.

Stamford J. A., Kruk Z. L., Palij P., and Millar J. (1988a) Diffusion and uptake of dopamine in rat caudate and nucleus accumbens compared using fast cyclic voltammetry. *Brain Res.* **448,** 381–385.

Stamford J. A., Kruk Z. L., and Millar J. (1988b) Stimulated limbic and striatal dopamine release measured by fast cyclic voltammetry: anatomical, electrochemical and pharmacological characterisation. *Brain Res.* **454,** 282–288.

Starke K., Gothert M., and Kilbinger H. (1989) Modulation of neurotransmitter release by presynaptic autoreceptors. *Physiol. Rev.* **69,** 864–989.

Suaud-Chagny M. F., Ponec J., and Gonon F. (1991) Presynaptic autoinhibition of the electrically evoked dopamine release studied in the rat olfactory tubercle by *in vivo* electrochemistry. *Neuroscience* **45,** 641–652.

Talmaciu R. K., Hoffmann I. S., and Cubeddu L. X. (1986) Dopamine autoreceptors modulate dopamine release from the prefrontal cortex. *J. Neurochem.* **47,** 865–870.

Tassin J. P., Cheramy A., Blanc G., Thierry A. M., and Glowinski J. (1976) Topographical distribution of dopaminergic innervation and of dopaminergic receptors in the rat striatum. I. Microestimation of [3H]dopamine uptake and dopamine content in microdiscs. *Brain Res.* **107,** 291–301.

Ungerstedt U. (1971) Stenotaxic mapping of the monamine pathways in the rat brain. *Acta Physiol. Scand.* **367s,** 1–48.

Valenta B., Drobny H., and Singer E. A. (1988) Presynaptic autoinhibition of central noradrenaline release *in vitro*: operational characteristics and effects of drugs acting at *alpha*-2 adrenoceptors in the presence of uptake inhibition. *J. Pharmacol. Exp. Ther.* **245,** 944–949.

Wathey J. C., Nass M. M., and Lester H. A. (1979) Numerical reconstruction of the quantal event at nicotinic synapses. *Biophys. J.* **27,** 145–164.

Wiedemann D. J., Kawagoe K. T., Kennedy R. T., Ciolkowski E. L., and Wightman R. M. (1991) Strategies for low detection limit measurements with cyclic voltammetry. *Anal. Chem.* **63,** 2965–2970.

Wightman R. M. and Zimmerman J. B. (1990) Control of dopamine extracellular concentration in rat striatum by impulse flow and uptake. *Brain Res. Rev.* **15,** 135–144.

Wightman R. M., Brown D. S., Kuhr W. G., and Wilson R. L. (1987) Molecular specificity of *in vivo* electrochemical measurements, in *Voltammetry in the Neurosciences* (Justice J. B., ed.), Humana, Clifton, NJ, pp. 103–138.

Wightman R. M., Amatore C., Engstrom R. C., Hale P. D., Kristensen E. W., Kuhr W. G., and May L. J. (1988) Real-time characterization of dopamine overflow and uptake in the rat striatum. *Neuroscience* **25,** 513–523.

Zigmond M. J., Abercrombie E. D., Berger T. W., Grace A. A., and Stricker E. M. (1990) Compensations after lesions of central dopaminergic neurons: some clinical and basic implications. *TINS* **13,** 290–295.

The Measurement of Brain Ascorbate In Vivo and Its Link with Excitatory Amino Acid Neurotransmission

Robert D. O'Neill

1. Introduction

L-Ascorbic acid (AA, vitamin C) is found throughout the plant and animal kingdoms (Davies et al., 1991). A major function of AA is as an antioxidant, protecting tissues from harmful free radicals and maintaining certain enzymes in their reduced form (Padh, 1990). Although mechanisms of its roles are often unknown or only poorly understood, details of specific interactions of AA with a variety of molecular complexes, such as calcium channels (Parsey and Matteson, 1993) and neurotransmitter receptors (Majewska et al., 1990), have recently been revealed.

Humans and guinea pigs are among the species that require AA as a nutrient, since they lack L-gulono-γ-lactone oxidase (EC 1.1.3.8), the final enzyme in the synthesis of AA from glucose. Although other mammalian species, such as the rat, synthesize AA in the liver, the concentration of blood AA in these two groups (with or without EC 1.1.3.8) is similar at ca. 40–80 μ*M* (Grunewald et al., 1983; Jacques, 1992; Deutsch and Kolhouse, 1993). AA concentrates in many organs, including the eye (Taylor et al., 1991), adrenal cortex (Arad et al., 1985), and brain (Milby et al., 1982). Initial evidence suggested that the major site of AA uptake into brain with a K_m of ca. 50 μ*M* is the choroid plexus at the blood–CSF inter-

From: *Neuromethods, Vol. 27: Voltammetric Methods in Brain Systems*
Eds: A. Boulton, G. Baker, and R. N. Adams © 1995 Humana Press Inc.

face (Spector and Lorenzo, 1974; Spector et al., 1977) and onward to the extracellular fluid (ECF) by flow and/or diffusion. More recent data, however, suggest that although the density of AA uptake sites is greatest at the choroid plexus, the bulk of brain AA is acquired at the BBB (Lam and Daniel, 1986; Pratt and Greenwood, 1986). AA appears to pass through cerebral capillaries both by a carrier-mediated saturable process (K_m 125 μ*M*) and by simple diffusion. Normally the BBB route represents ca. 90% of AA uptake into brain because of its greater surface area (×5000) compared with the blood–CSF barrier, and ca. 10% of this is by diffusion. When the concentration of blood AA is raised to 1 m*M*, however, as much 50% of AA penetration into brain can be via the diffusion route (Lam and Daniel, 1986). Thus, the greater surface area and direct contact with the high flow rate of blood make the BBB a much more efficient route for delivery of AA directly to the ECF.

1.1. Brain Ascorbate

The overall content of AA in brain tissue has been estimated at ca. 2 μmol/g wet wt (Kuo et al., 1978; Milby et al., 1982) and shows a high degree of heterogeneity in its distribution. The main compartmentalization occurs between the cytoplasm and ECF with a gradient of ca. 10:1 between the intracellular and extracellular spaces (Todd and Bauer, 1988; Rice and Cammack, 1991). This results from active transport across neuronal and glial cell membranes that appears to be Cl^- (Wilson and Dixon, 1989) and Na^+ (Wilson, 1989; Wilson et al., 1991) dependent, and may involve hexose transporter sites (Milosevic et al., 1989; Vera et al., 1993). AA content (Milby et al., 1982) and ECF levels (Stamford et al., 1984; Basse-Tomusk and Rebec, 1991; Ghasemzedah et al., 1991) also show uneven distribution between and within brain regions, with no clear correlation between the concentrations in the intracellular and extracellular compartments (Milby et al., 1982; Ghasemzedah et al., 1991). The concentration of AA in the ECF in vivo has been estimated using a variety of voltammetric techniques at 200–500 μ*M* (Gonon et al., 1981a; Amatore et al., 1986; Yount et al., 1991), a value that depends on brain region, state of anesthesia, and the extent of depletion caused by the measurement.

Recent reviews of in vivo voltammetric principles and techniques are available by Adams (1990), Kawagoe et al., (1993), and O'Neill (1994). AA in the ECF caused many difficulties in the early development of voltammetry in vivo (Marsden et al., 1988) by masking a variety of signals. It has continued to be treated merely as an interferant in many in vivo voltammetric studies, where the monoamines are generally the target molecules. A growing number of research groups, however, are now studying brain AA in its own right. The purpose of this chapter is to describe the *in situ* voltammetric measurement of extracellular AA in discrete brain areas, with particular emphasis on the use of carbon-paste electrodes, and to discuss the evidence linking extracellular AA with excitatory amino acid neurotransmission. Possible roles for brain AA have been discussed recently by Grunewald (1993).

1.2. Redox Chemistry of Ascorbate

The redox chemistry of AA in solution phase is extremely complicated, involving numerous radical, ionic, and complexation species (Hand and Greisen, 1942; Taqui Khan and Martell, 1967; Shin and Feather, 1990; Bansch et al., 1991; Agrawal et al., 1993). Even the thermodynamics of this redox reaction are uncertain because of the instability of the AA oxidation product—dehydroascorbic acid—in aqueous environments (*see* Fig. 1). Thus, the formal electrode potential ($E^{o'}$) of AA with respect to the saturated calomel electrode (SCE) can be estimated in a range from –190 mV (Ormonde and O'Neill 1990), through –100 mV (Falat and Cheng, 1982), to +50 mV (Wehmeyer and Wightman, 1985), although adsorption of AA on the electrode might distort such estimates. It is clear, however, that the $E^{o'}$ for this couple at pH 7.4 is significantly lower than that of other detectable redox species present in the ECF, such as catechols and 5-hydroxyindoles. The thermodynamic ground is therefore set for resolved voltammetric AA peaks at low potentials, provided the kinet-

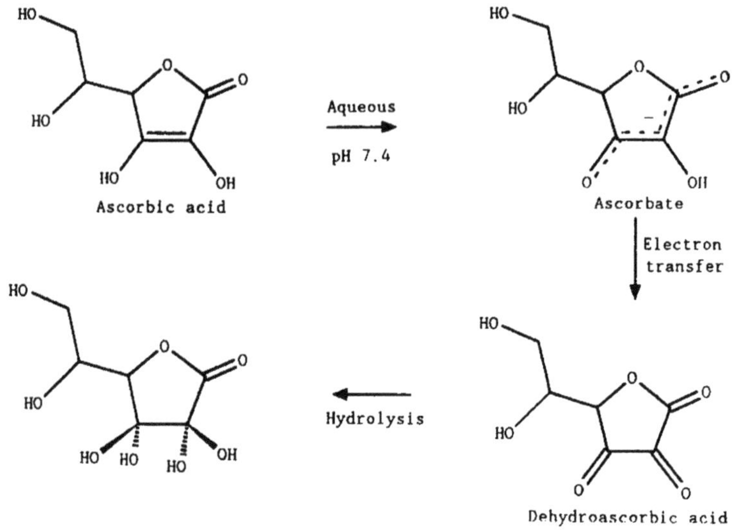

Fig. 1. Ionization, electron transfer, and hydration reactions for the ascorbic acid system in aqueous solution at pH 7.4. (*see* Section 1.2.).

ics of electron transfer at the electrode surface are sufficiently fast.

The electrochemical oxidation of AA, illustrated in Fig. 1, has been studied extensively at different electrode materials and is considered to follow an EC mechanism (electron transfer followed by a chemical reaction). Much of the early work involved mercury electrodes (Karabinas et al., 1985) at which a reversible two-electron transfer occurs; the following chemical reaction therefore distorts the shape and position of the polarographic wave (Nicholson and Shain, 1964). However, the large overpotential needed to drive AA oxidation on carbon electrodes (Ueda et al., 1982; Hu and Kuwana, 1986; Coury and Heineman, 1988) means that the electron-transfer step at these surfaces can be considered irreversible (Matsuda and Ayabe, 1955), and thus, the succeeding chemical reaction (hydration; *see* Fig. 1) has no effect on the oxidation phase of the stationary electrode voltammogram (Nicholson and Shain, 1964).

At pH 7.4, AA, with dissociation constants of pK_1 4.25 and pK_2 11.79, is a monoanion. The loss of the second acidic proton associated with the transfer of two electrons to the electrode yields dehydroascorbic acid (DHAA), which is rapidly hydrated to a species that undergoes lactone ring opening to form 2,3-diketogulonic acid (Lechien et al., 1982). Rapid oxidation-reduction techniques, such as fast cyclic voltammetry (FCV), at 1000 V/s have allowed detection of some reduction of DHAA at low overpotential mercury electrodes (Wehmeyer and Wightman, 1985). However, even FCV (Stamford, 1986) and other fast techniques involving rapid rereduction, such a reverse-pulse chronoamperometry (Kiyatkin et al., 1993), do not detect the reduction of DHAA at the high overpotential carbon electrodes usually used for in vivo voltammetry. In the present context, therefore, the voltammetric behavior of AA can be considered both electrochemically irreversible (slow electron transfer) and chemically irreversible (following homogeneous reaction).

2. Working Electrodes for Ascorbate Detection

The vast majority of voltammetric experiments in vivo involve the use of three forms of carbon electrode—carbon-epoxy electrodes, carbon-fiber electrodes (CFEs), and carbon-paste electrodes (CPEs). Both CFEs and CPEs are widely used to monitor AA in the ECF. Each has advantages and disadvantages, and the electrode of choice depends on the intended application.

2.1. Carbon Fiber Electrodes

Prepared by sealing single (Armstrong-James and Millar, 1979; Gonon et al., 1981b) or multiple (Forni and Nieoullon, 1984; Rivot et al., 1988) strands of carbon fiber into glass capillaries, CFEs are the most commonly used electrodes for in vivo voltammetry. Introduced by Armstrong-James and Millar and Ponchon et al. in 1979, the usefulness of CFEs increased dramatically when it was demonstrated that resolution of a pure AA signal (peak 1) and a mixed catechol signal (peak 2) could be achieved with differential pulse voltammetries (Gonon et al., 1980, 1981a; Marcenac and Gonon, 1985) and staircase voltammetry (Oh et al., 1989) following modification of the surface by application of either a triangular potential (Gonon et al., 1981b) or constant current (Peng et al., 1992). Recent electrochemical and spectroscopic studies indicate that these treatments are likely to improve the performance of CFEs by increasing electron transfer rates (Strein and Ewing, 1991; Swain and Kuwana, 1992) and adsorption of substrate (Kovach et al., 1986; Sujaritvanichpong et al., 1986) on the surface, and have been explained in terms of a combination of electrode cleaning, increased area of active graphite edges from cracking, and oxidation of the surface (Wang et al., 1987; Swain and Kuwana, 1991; Desimoni et al., 1992). It now appears that the important parameter of the treatment is not the applied potential, but the current density passed, and the most reproducible results have been obtained under galvanostatic rather than potentiostatic conditions (Xie and Sherwood, 1991; Peng et al., 1992). This factor may explain the difficulty some laboratories have experienced in producing good electrical treatment of CFEs, especially if the potentiostat used was not designed for the relatively high currents associated with this process. Other important factors shaping the effectiveness of these modifications include the pH (Hahn et al., 1985; Hulthe et al., 1989) and the buffer ions (Strein and Ewing, 1991; Nakahara and Shimizu, 1992) of the treatment solution, phosphate buffer pH 7.4 being the most common medium. Details of construction, properties, and monoaminergic applications of CFEs are provided elsewhere in this volume.

A main advantage of CFEs for in vivo voltammetry is their small cross-section (down to 5 μm) (Amatore et al., 1986; Tanaka and Kashiwagi, 1989), minimizing tissue damage on implantation and allowing studies of small brain regions, such as the locus ceruleus (Brun et al., 1991), raphe nuclei (Houdouin et al., 1991), suprachiasmatic nuclei (O'Connor and Kruk, 1991), amygdala (Bull et al., 1991) and in subregions of the hypothalamus (Mermet and Gonon, 1986), brainstem (Hong et al., 1992), and spinal cord (Puig et al., 1992). A major drawback of electrochemically pretreated CPEs is the loss of selectivity between AA and catechols after several hours in brain tissue, limiting the majority of studies to

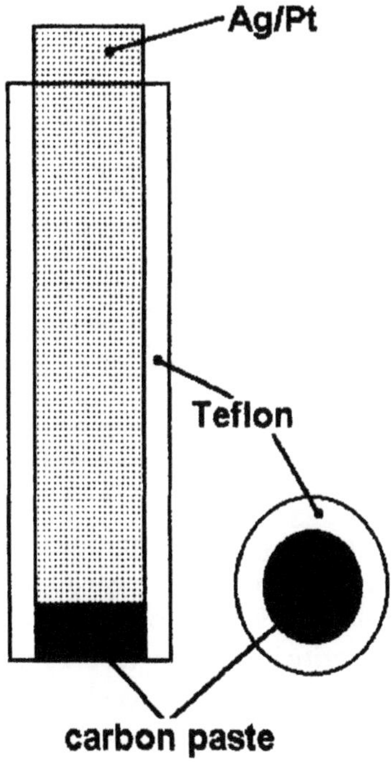

Fig. 2. Schematic representation of a CPE showing a longitudinal cross-section (left) and an end-on view of the active electrode surface—a carbon-paste disk. The outer diameter of CPEs used for voltammetry in vivo ranges from 320 to 160 µm (*see* Section 2.2.).

anesthetized preparation. The problem of loss of resolution has been addressed, however, by the use of a removable electrode system that facilitates easy replacement of the CFE in the same subject (Louilot et al., 1987; Mas et al., 1990; Rebec et al., 1993), allowing recording with acutely implanted CFEs in freely moving animals.

2.2. Carbon-Paste Electrodes

Carbon-paste, prepared by thoroughly mixing carbon powder with either Nujol (Kissinger et al., 1973; Broderick, 1988) or silicone oil (O'Neill et al., 1982b), can be easily packed into cavities, made by pulling the insulation over fine metal wires (*see* Fig. 2). This produces carbon-paste disk electrodes of diameters as small as 160 µm (Duff and O'Neill, 1994),

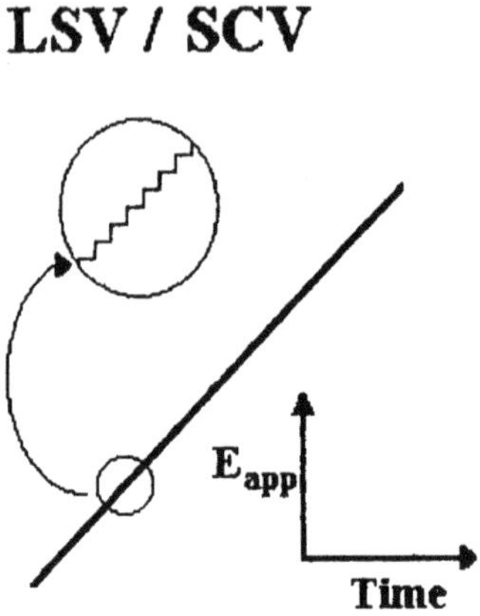

Fig. 3. Whereas differential pulse techniques are usually used with CFEs for voltammetric studies in vivo, linear ramp forms are generally used with CPEs. (*See* Adams [1990], Kawagoe et al. [1993], and O'Neill [1994] for reviews on the principles of in vivo voltammetry). When analog signal generators are used, true LSV is obtained. More often digital computer-controlled equipment is used, and this generates finite voltage steps to produce SCV. For a step size of 1 mV usually used for in vivo applications with CPEs, LSV and SCV may be regarded as equivalent.

although diameters of ca. 320 μm have generally been used for in vivo voltammetry. Cyclic voltammetry (CV) at 100 mV/s with a carbon/Nujol-paste electrode was used in the first report of voltammetry in brain tissue for neurotransmitter analysis (Kissinger et al., 1973), where a single irreversible peak was observed in rat striatum attributed mainly, but not exclusively, to the oxidation of AA. Decreasing the scan rate to 10 mV/s using linear sweep voltammetry (LSV; *see* Fig. 3) and semidifferentiation of the current later revealed three rather broad waves on the striatal voltammogram (Lane et al., 1979). Further developments, including the use of computer-controlled equipment (O'Neill et al., 1983b), silicone oil paste (O'Neill et al., 1982b), an additional reduction of the sweep rate to 5 mV/s, removal of semidifferentiation, and the subtraction of the background current recorded *in situ*, yielded CPE voltammograms (*see* Fig. 4) consisting of three well-separated peaks in striatal tissue (O'Neill et al., 1983a). The evidence discussed below (Section 3.2.) suggests

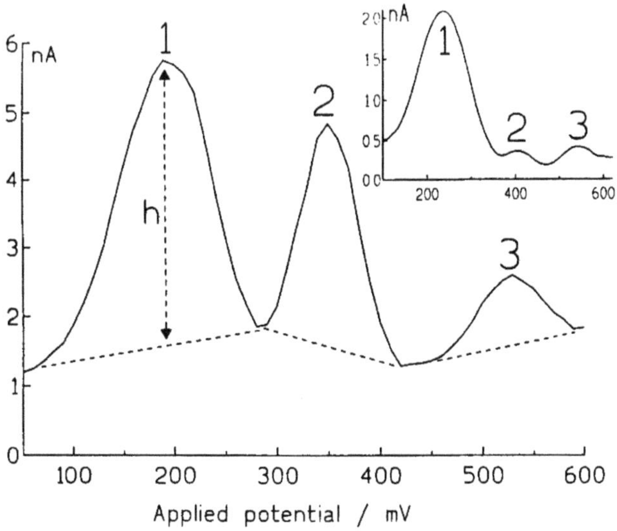

Fig. 4. Staircase voltammogram recorded at 5 mV/s in vivo with a 320-µm diameter CPE in the striatum of a freely moving rat several months after implantation. Peak 1 is caused by AA, peak 2 by ca. 90% urate and 10% 5-HIAA, and peak 3 by the DA metabolite HVA (*see* Section 3.2.). Increases in the extracellular concentration of catechols, DA, and DOPAC increase the current between peaks 1 and 2, considerably changing the potential of these peaks for large increases. Changes in the height, *h*, of peak 1, when there is no significant change in potential, reflect variations in the concentration of AA in the ECF (*see* Section 3.2.1.). Inset: When 160-µm CPEs are used under the same conditions, the voltammetric current is smaller, as predicted from Eqs. (1) and (2). Surprisingly, the height of peak 2 is smaller than expected compared with peaks 1 and 3. Evidence suggests that for 320-µm probes (main figure) the concentration of urate is anomalously high (ca. 50 µ*M*) owing to glial proliferation, but that with 160-µm CPEs, urate is below the limit of detecon (ca. 1 µ*M*) and peak 2 is owing almost entirely to 5-HIAA (*see* Section 3.2.2.).

that peak 1 recorded in this way can be used under specific conditions to monitor AA in the ECF over several months.

Unless otherwise stated, the CPE data discussed here have been obtained using electrodes made from carbon-paste prepared (O'Neill et al., 1982b) by mixing 2.8 g carbon powder (UCP-1-M, Ultra Carbon Corporation, Bay City, MI) with 1 mL silicone oil (Aldrich [Gillingham UK], 17,563-3). Carbon-powder electrodes (CPWEs) can be prepared in a simi-

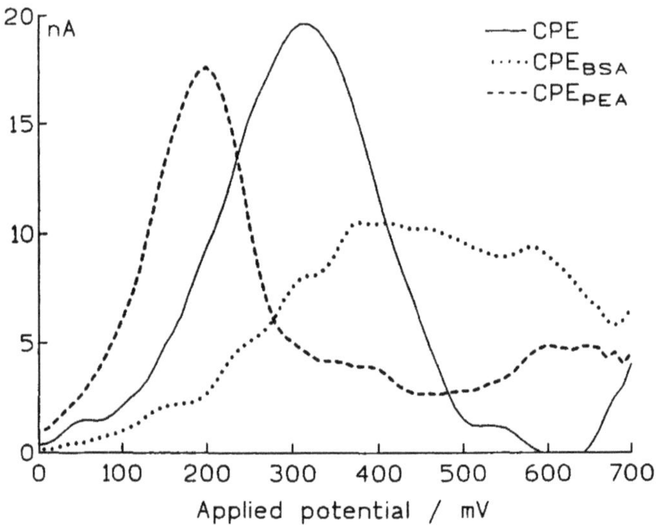

Fig. 5. Differential SCV at 50 mV/s recorded in vitro with a CPE in a 1-m*M* AA solution in PBS. The AA peak potential for fresh CPEs at ca. 300 mV was shifted to ca. 450 mV following immersion of the electrode in a 10% protein (BSA) solution. This treatment also significantly reduced peak height, increased peak width, and generally downgraded the CPE's response to AA. When the same electrode was then treated with a 10% lipid (PEA) suspension, the BSA fouling was reversed, and the electrochemical parameters of AA oxidation were similar to that for CPEs treated with lipid only, with a peak potential of ca. 200 mV. It is likely, therefore, that lipids present in brain tissue protect CPEs from protein fouling (*see* Section 2.2.1.).

lar way (Ormonde and O'Neill, 1990) by filling the Teflon™ cavity with the pure carbon powder.

2.2.1. Surfactant Modification of CPEs

To understand the behavior of CPEs in the protein- and lipid-rich environment of brain tissue, it is necessary to analyze the interaction of paste surfaces with protein and surfactant media. The low position of AA oxidation for implanted CPEs (peak 1, Fig. 4) is unexpected, since AA oxidizes at much higher sheer potentials at fresh CPEs in vitro (O'Neill et al., 1982b; Ormonde and O'Neill, 1990) (*see* Fig. 5). Furthermore, the stability of the voltammograms recorded in vivo over several months in the presence of electrode fouling agents, such as proteins, is surprising. Since surfactants reduce the overpotential of a variety of oxidations at CPEs (Albahadily and Mottola, 1987), the effects of lipids, which constitute ca. 40% dry wt of brain tissue, an other surface-active agents on the kinetics of AA oxidation at CPEs have been investigated in vitro in phosphate-buffered saline (PBS) (Ormonde and O'Neill, 1990).

The position of the AA oxidation wave on the voltage axis (quantified as the potential of maximum slope) for CPEs treated with either Triton-X solution, lipid (phosphatidylethanolamine, PEA) suspension, or brain tissue was shifted to potentials lower than that for untreated CPEs and toward the value for CPWEs, indicating an increase in the rate of charge transfer (Ormonde and O'Neill, 1990). This suggests that each treatment removes insulating pasting oil from the active sites on the electrode surface, making it more powder-like (Rice et al., 1983; Albahadily and Mottola, 1987). Although it is possible that the shift in the AA wave following treatments might be the result of a catalytic step introduced into the overall reaction, such as the interaction of AA with lipid or Triton-X adsorbed onto the electrode, surface resistance and capacitance values for CPEs (ca. 150 kΩ and 40 nF), surfactant-treated CPEs (ca. 3 kΩ and 500 nF), brain-treated CPEs (ca. 2 kΩ and 800 nF), and CPWEs (ca. 6 kΩ and 900 nF) suggest that removal of pasting oil from the surface is the mechanism of altered response. The finding that surfactant solutions, lipid suspensions, brain homogenate (O'Neill et al., 1982b), and whole tissue have similar effects on the kinetics of AA oxidation at CPEs suggests that its modification by brain tissue in vivo does not depend on either brain structure/function or on any mechanical perturbation of the electrode surface by the tissue.

In addition to lowering the overpotential for AA oxidation at CPEs, lipids appear to fulfil another important function in vivo—protection from electrode fouling caused by proteins. This phenomenon can be illustrated in vitro using differential SCV to mimic the shape of the in vivo voltammograms (*see* Section 3.1.); an example is shown in Fig. 5.

The AA peak potential for fresh CPEs at ca. 300 mV was shifted to ca. 450 mV following immersion of the electrode in a 10% protein (bovine serum albumin, BSA) solution. This treatment also significantly reduced peak height and increased peak width, characteristic of electrode fouling. When the same electrode was then treated with a 10% lipid (PEA) suspension, the BSA fouling was reversed and the electrochemical parameters of AA oxidation were similar to that for CPEs treated with lipid only, with a peak potential of ca. 200 mV. In spite of facilitated electron transfer at lipid-treated CPEs, the sensitivity of these electrodes is less than at fresh CPEs (Ormonde and O'Neill, 1990), which may be the result of a blocking effect of lipids adsorbed on the surface (Nelson and Auffret, 1988).

2.2.2. Stearate-Modified CPEs for DA Detection In Vivo

Modification of the surface of CPEs by tissue (Section 2.2.1.) has a number of analytical implications. First, electrocatalytic interaction observed between AA and dopamine (DA) at carbon electrodes in vitro (Dayton et al., 1980; Gelbert and Curran, 1986) is greatly reduced for CPEs in vivo by the shift of the AA wave following contact with tissue, since the peak potential for AA is moved below that for DA (Stamford, 1986) (*see* Section 3.2.). Second, chemical modifications of paste electrodes using lipophilic substances, such as stearic acid (stearate at pH 7.4), used to block oxidation of anionic species at low potentials (Blaha and Lane, 1983; Glynn and Yamamoto, 1989; Phillips et al., 1993), are rendered ineffective in the brain because the chemical modifier is removed together with the oil from the electrode surface (Lyne and O'Neill, 1989, 1990). The separation of the AA (anion) and DA (cation) signals, measured as the difference in the peak potentials for DA and AA ($E_p^{AA} - E_p^{DA}$), recorded separately in PBS using differential SCV with stearate-modified Nujol paste electrodes (SMEs, $E_p^{AA} - E_p^{DA}$ ca. 600 mV) is significantly reduced following contact with brain tissue (ca. 200 mV) and other surface-active media (Lyne and O'Neill, 1990), resulting in substantial overlap of the two signals (*see* Fig. 6). Additional studies using capacitance measurements and the ferrocyanide redox anion to probe the surface of SMEs following surfac-

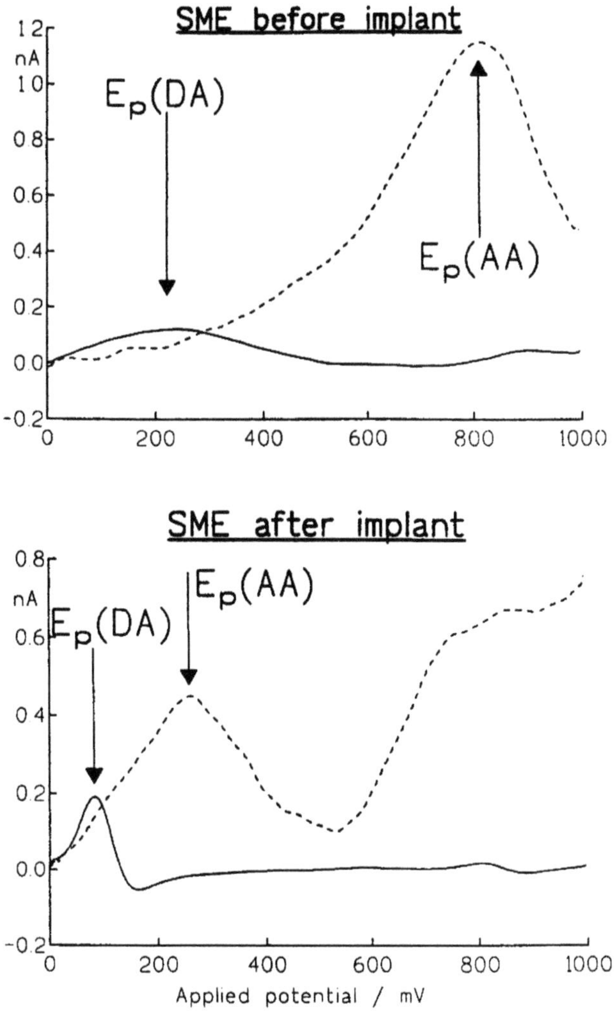

Fig. 6. The effect of brain tissue on the selectivity of a stearate-modified CPE (SME) for DA and AA. The separation of the AA and DA signals, measured as the difference in the peak potentials for DA and AA ($E_p^{AA} - E_p^{DA}$), recorded separately in PBS at 50 mV/s using differential SCV with SMEs ($E_p^{AA} - E_p^{DA}$ ca. 600 mV) is significantly reduced following contact with brain tissue (ca. 200 mV) and other surface-active media, resulting in substantial overlap of the two signals (*see* Section 2.2.2.). Even at the concentation ratio AA:DA used here of 100:1, there is signficant interference by AA in the DA signal after contact of the SME with brain tissue. Since the ratio in vivo in rat striatum appears to be of the order of 10,000:1 (500 µ*M* AA and 50 n*M* DA; Smith et al., 1992), these data indicate that SMEs cannot be used to monitor DA unambiguously in the brain (*see* Section 2.2.2.). Data presentation modified from Lyne and O'Neill (1989).

tant and tissue treatments (Lyne and O'Neill, 1990) support these conclusions. These data, together with studies of the electrocatalytic interactions of AA and DA at SMEs (Gelbert and Curran, 1986), indicate that SMEs cannot be used to monitor DA unambiguously in the brain.

It has been suggested more recently (Blaha and Jung, 1991) that reliable DA detection in brain tissue is possible with SMEs provided that fast (short time scale) voltammetric techniques, such as chronoamperometry, are used. Although the arguments leading even to this circumspect conclusion are questionable, it is surprising that putative DA signals recorded in vivo with SMEs and slow sweep voltammetry are still reported (Broderick et al., 1993; Phillips et al., 1993).

3. Voltammetry In Vivo with CPEs

To obtain stable voltammograms over several months in vivo with CPEs (*see* Fig. 2), the carbon paste should be packed carefully into a cavity of reasonable depth. If the paste is too loose or the cavity too short, the ECF will soon penetrate to the metal wire producing large background currents. If packed too tightly, the Teflon™ may crack and allow ECF into the cavity; cracking is more difficult to avoid with long cavities. CPEs of different sizes can be prepared from Teflon™-coated wire in diameters ranging from 10 T Ag (320 μm external, 250 μm internal) to 5 T Pt (160 μm, 125 μm) as described before (O'Neill et al., 1982b). Briefly, cavities 1–1.5 mm long are packed firmly with the paste mixture after clamping a stainless-steel collar around the wire to prevent slipping of the Teflon™. Pt wire is used for the 5 T electrodes for extra rigidity. The CPEs were implanted in male Sprague-Dawley rats (initial weight, 270–350 g) under chloral hydrate anesthesia (7 mL/kg ip, 5% solution in normal saline) as described in detail (O'Neill et al., 1982b). The coordinates, with the head level between bregma and lambda, were striatum: AP –0.5 (from bregma), L 3.0 (from bregma), and DV4.8 (from skull); and hippocampus AP –3.5, L 2.2, and DV 4.0. Silver wires (200–250 μm) placed in the cortex served as reference and auxiliary electrodes. In some animals implanted with bilateral striatal electrodes, a stainless-steel cannula (600 µm) was attached to the left striatal electrode 1 mm above the tip to avoid electrical contact. In others, a bipolar stainless-steel stimulating electrode was placed in the perforant path at AP –0.2, L 4.7, and DV 3.3–3.8 (O'Neill et al., 1984b).

Staircase voltammograms were recorded at 12- or 6-min intervals at a scan rate of either 5, 10, 25, or 50 mV/s using microcomputer-controlled equipment (O'Neill et al., 1983b). The potential was increased from 0–600 mV in small equal steps; the step size depended on the scan rate and was in the range 1–5 mV (*see* Fig. 3). The background current for each electrode was measured *in situ* before each experiment (O'Neill et al., 1983a) and was subtracted from voltammograms recorded at 12-min intervals for graphical representations, but not for quantitative analysis. Current-voltage plots show three distinct peaks between 0 and 600 mV at a scan rate of 5 mV/s (Fig. 4): peak 1 (centered at ca. 20 mV), peak 2 (ca. 350 mV), and peak 3 (ca. 550 mV). The height of each peak *h*, was measured by constructing a baseline between the two minima and measuring the distance from this to the current maximum. In some experiments, currents were recorded amperometrically at 2-s intervals at the potential of peak 1.

3.1. Voltammogram Shape

A number of different models have been reported to describe the behavior of electrodes implanted in brain tissue (Cheng et al., 1979; Albery et al., 1984; Amatore et al., 1986). To understand the shape of the AA oxidation signal (*see* Fig. 4), the diffusion environment of CPEs in brain tissue has been investigated by analyzing the effect of scan rate on peak height using a theory developed for implanted electrodes (Albery et al., 1984). For species electrolyzed in a restricted compartment by a disk electrode, using either LSV or SCV, the current is peak-shaped and peak height, *h*, is directly proportional to the scan rate, ν (*see* Eq. [1]).

$$h = K[\alpha n V c^{*} \nu] \qquad (1)$$

K (also K' and K'' below) represents a conglomeration of thermodynamic constants whose value depends on the reversibility of the redox system and is not important in this analysis; α is the transfer coefficient (equal to n for the reversible case); n is the number of electrons involved in the electrooxidation; V is the volume of the compartment; and c* is the initial concentration of the substrate. For species adsorbed on the surface, the LSV/SCV current is also peak-shaped, and the dependence of h on scan rate has the same form as Eq. (1) (*see* Eq. [2]),

$$h = K'[\alpha n A \Gamma^* \nu] \quad (2)$$

where $V \times c^*$ in Eq. (1) is replaced by the electrode area $A \times \Gamma^*$, the initial surface coverage. Finally, a third contribution to h may arise from molecules arriving at the electrode surface by diffusion. The shape of the LSV/SCV voltammogram depends on the scan rate, ranging from sigmoidal at low values of ν to peak-shaped at high ν values. Here, in contrast to the compartment and adsorption case, there are two terms, one for linear diffusion varying with $(\nu)^{1/2}$ and the other being a scan-rate-independent term for contributions from spherical diffusion at the edge of the electrode (*see* Eq. [3])

$$h = K''[nAc^*(\alpha D\nu)^{1/2}] + (GnAFDc^*/r) \quad (3)$$

where h in this context is the difference between the peak and baseline currents, D is the diffusion coefficient of the substrate, G is an electrode geometric factor, F the Faraday constant, and r the radius of the disk.

Thus, the slope of ln(h) plotted as a function of ln(ν) indicates the contribution to the current of unrestricted diffusing species to the implanted electrode. Since the slope was not significantly different from unity for both 10 T (Albery et al., 1984) and 5 T (Duff and O'Neill, 1994) CPEs, this contribution was negligible for chronically implanted CPEs. It appears that both adsorbed (Ormonde and O'Neill, 1990) and compartment (Albery et al., 1984) molecules are involved in producing the in vivo voltammogram (Fig. 4). The transfer coefficient, α, for AA measured under these conditions is very close to the value of 0.5 for an ideal irreversible electrooxidation (Albery et al., 1984). Thus, the AA peak is asymmetric, with the current before the peak rising less steeply than the current falls after the peak potential is reached.

3.2. Signal Identification

Linear staircase voltammograms recorded with CPEs in rat striatum consist of three well-resolved peaks even several months following electrode implantation (Fig. 4). The identity of the compounds responsible for these signals has been investigated using a wide variety of techniques, including: theoretical analysis (Albery et al., 1984); electrochemical studies in vitro (O'Neill et al., 1982b; Ormonde and O'Neill, 1990); microinfusion of candidate substances (O'Neill et al., 1983b; Joseph et al., 1989) and metabolizing enzymes (Brazell and Marsden, 1982; O'Neill et al., 1984a; Boutelle et al., 1989) beside electrodes implanted in brain tissue; lesions to striatal afferent pathways (O'Neill et al., 1983c; O'Neill and Fillenz, 1985a; Haskett and Mueller, 1987); pharmacological perturbations using drugs with well-characterized actions (Clemens and Phebus, 1983; Mueller et al., 1985); and comparison with other in vivo techniques, such as microdialysis (Justice et al., 1983; Joseph and Young, 1991; O'Neill et al., 1991). Although peak 1 is the focus of the present discussion, it will be useful to consider this signal in the context of the other peaks present in the voltammogram since each reveals an interesting aspect of the problems associated with interpreting and ascribing significance to data recorded in the ECF.

3.2.1. Peak 1—The Ascorbate/Catechol Problem

Microinfusion into brain tissue of AA beside the electrode leads to an increase in the height of peak 1, centered at ca. 200 mV, with no detectable change in peak potential (O'Neill et al., 1983b). Similar experiments infusing DA or the DA metabolite, 3,4-dihydroxyphenylacetic acid (DOPAC), show that the oxidation peak potential of these catechols in vivo is ca. 60 mV greater than peak 1, shifting peak 1 to more positive potentials (O'Neill et al., 1982b, 1983b). More importantly, infused ascorbate oxidase causes an effective collapse in peak 1 (Boutelle et al., 1989). These results have led to the conclusion that peak 1 is caused mainly by the oxidation of AA (O'Neill et al., 1982b, 1983b; Boutelle et al., 1989), which has an extracellular concentration of about 200–500 μM (Gonon et al., 1981a; Amatore et al., 1986; Yount et al., 1991). Moreover, changes in the height of peak 1 result from variations in the extracellular concentration of AA, provided there is no shift in peak potential. When such potential shifts occur, they are unlikely the result of electrocatalysis, since AA oxidizes at potentials below that of DA at these electrodes (Stamford, 1986), but instead are likely to be caused by the algebraic combination of the catechol and AA currents, and difference voltammograms can be used to analyze the changes in terms of fluctuations in the catechol concentration (O'Neill et al., 1983b). Under most circumstances (some extreme exceptions being stroke [Slivka et al., 1988] and death [Vulto et al., 1988; Gonzalez-Mora et al., 1989]), these latter changes are mainly owing to DOPAC with a baseline concentration of ca. 5 μ*M* (Zetterstrom et al., 1983; Ballarin et al., 1987), since the level of DA in the ECF is of the order of 5–50 n*M* (Church and Justice, 1987; Kawagoe et al., 1992; Smith et al., 1992).

The use of peak 1 to monitor AA has been questioned by Joseph and Young (1991) on the basis of a comparison between results obtained with in vivo microdialysis (2–3 d after implantation) and in vivo voltammetry (2–3 mo after implantation) in "parallel" experiments to determine the substances that contribute to the voltammograms recorded with CPEs. Many of their conclusions were based on the assumption that there is a quantitative correspondence between changes observed with the two techniques. However, one would not expect these two very different methodologies to give quantitatively similar results under all conditions. This is especially true for measurements made at very different times after implantation, since there is growing evidence of marked time-dependent histological and neurochemical changes in the brain following probe implantation (Cespuglio et al., 1986; Ruggeri et al., 1990; O'Neill et al., 1991; Woodroofe et al., 1991). In addition, there is evidence that the brain contains nondialyzable agents that normally protect species from oxidation in the ECF (Kovachich and Mishra, 1984), and that AA oxidation can be catalyzed by trace amounts of heavy metal ion impurities in the buffer (Davies, 1992). These factors, together with the continuous depletion caused by perfusion and the difficulty in separating AA successfully from the solvent front at the HPLC stage, may explain the very low estimates of AA in the ECF determined by microdialysis, e.g., ca. 70 μ*M* (Joseph and Young, 1991) and 25 μ*M* (Slivka et al., 1988), compared with voltammetric values of 300 μ*M* (not corrected for relatively mild voltammetric depletion; Gonon et al., 1981a) and 570 μ*M* (corrected for depletion, Amatore et al., 1986). In addition, the inability of their microdialysis experiments to detect an increase in striatal AA following pargyline administration, when such an increase has been demonstrated using CFEs (Gonzalez-Mora et al., 1988, 1991), questions the reliability of microdialysis for measurement of AA.

3.2.2. Peak 2–The 5-HIAA/Urate Problem

Peak 2 recorded in vivo with CPEs was originally thought to be the result of the oxidation of 5-hydroxyindoles (Joseph and Kennett, 1981; O'Neill et al., 1983b). Later, the failure of serotonergic lesion of the medial forebrain bundle to decrease the height of peak 2 recorded in striatum, and the almost total collapse of peak 2 following microinjection of uricase led to the conclusion that peak 2 was essentially the result of uric acid oxidation (O'Neill et al., 1984a). This deduction was supported by additional manipulations of peak 2 (Mueller et al., 1985). However, more detailed analysis showed that there was a small (ca. 10%) contribution from 5-HIAA to striatal peak 2 recorded with chronically implanted 320-μm diameter CPEs (O'Neill et al., 1991). It now appears that the concentration of urate in the ECF is anomalously high (ca. 50 μ*M*) when electrodes or microdialysis probes of this size are implanted for more than 24 h, a phenomenon that may be linked to a glial reaction. The problem of urate production by tissue reaction has been reduced, however, by the use of 160-μm diameter CPEs (Duff and O'Neill, 1994). Here, peak 2 is the result of 5-HIAA oxidation, urate is not detected in the ECF, and the glial reaction is minimal even several weeks after implantation (*see* Fig. 4 inset).

A possible explanation has thus emerged for the apparently paradoxical results for urate and 5-HIAA obtained with CPEs and CFEs. For several hours after an electrode is implanted, the concentrations of 5-HIAA and urate are similar in the low-micromolar range (O'Neill et al., 1991). Since both CPEs (O'Neill et al., 1991) and CFEs (Crespi et al., 1983) are twice as sensitive to 5-HIAA as urate, and neither electrode can resolve the two substrates, their respective in vivo voltammetric peaks (peak 2 and peak 3) recorded under acute conditions have ca. 70% (5-HIAA) and 30% (urate) contributions. However, the vast majority of CPE data have been recorded in awake, freely moving animals several days following implantation. With large (320-μm diameter) CPEs, the concentration of urate increases to ca. 50 μ*M* by 24 h after surgery and remains high for months; peak 2 is much larger compared with the acute condition and reflects mostly (ca. 90%) urate. When chronically implanted small electrodes are used, such as CFEs (Cespuglio et al., 1986) or 160-μm diameter CPEs (Duff and O'Neill, 1994; *see* Fig. 4), the low level of urate observed under acute conditions actually decreases to below the limit of detection (ca. 1 μ*M*). This phenomenon appears to be linked to the lesser extent of glial reaction to the smaller electrodes (Duff and O'Neill, 1994), and suggests that the low-micromolar level of urate detected in the ECF under acute conditions may arise from leakage across the damaged BBB.

3.2.3. Peak 3—The Release/Metabolism Problem

Changes in the height of peak 3 recorded in striatum, and other DA-rich forebrain regions, are directly proportional to changes in the extracellular concentration of methylated metabolites of the neuromodulator DA. Normally this is homovanillic acid (HVA) (Lane et al., 1978; O'Neill and Fillenz, 1985a; Joseph et al., 1989), but when monoamine oxidase is inhibited pharmacologically, 3-methoxytyramine may contribute (Joseph and Young, 1991). In either case, variations in this peak reflect changes in the metabolism of DA, and peak 3 has been used to study the relationship between diurnal changes in extracellular HVA levels and motor activity patterns (Fillenz and O'Neill, 1986; Brose et al., 1987, 1988), and the response of the nigrostriatal DA system in trained responses (Joseph et al., 1989). The link, however, between DA release and HVA remains unclear, as illustrated by experiments using microdialysis probes in striatal tissue of awake rats where the following divergent data have been reported, even in the absence of drugs: diurnal increases in extracellular DA with no concomitant change in extracellular HVA (Smith et al., 1992); increases in HVA with no change in DA following exposure to food (Inoue et al., 1993); and parallel increases in DA and HVA in the ECF of males exposed to receptive females (Wenkstern et al., 1993). The relationship between synaptic DA release and its concentration in the ECF is an additional, poorly understood, dimension to this problem and beyond the scope of the present discussion.

4. Link Between Extracellular Ascorbate and Excitatory Amino Acid Neurotransmission

The concentration of AA in the ECF of awake, freely moving animals shows wide fluctuations. In brain areas associated with locomotion, such as nucleus accumbens and striatum, AA correlates with motor activity levels, and in all regions thus far monitored over 24-h periods, diurnal variations have been observed (O'Neill and Fillenz, 1985b; Brose et al., 1989). An example of the diurnal pattern of striatal AA concentration and simultaneously recorded motor activity over a 48-h period is shown in Fig. 7. During the dark, active phase of the rat's circadian cycle, the AA signal increases on average by ca. 50%. (The term AA signal refers to peak 1 recorded with CPEs and SCV [*see* Fig. 4] unless otherwise stated.) Changes of similar magnitude have been reported for frontal cortex and nucleus accumbens, with significantly smaller increases observed in hippocampus (O'Neill and Fillenz, 1985b). A wide variety of physiological and pharmacological stimuli also give rise to changes in the concentration of AA in brain ECF (Boutelle et al., 1989; Cammack et al., 1991; Ghasemzedah et al., 1991; Pierce and Rebec, 1992; Svensson et al., 1992a,b). Two aspects associated with this phenom-

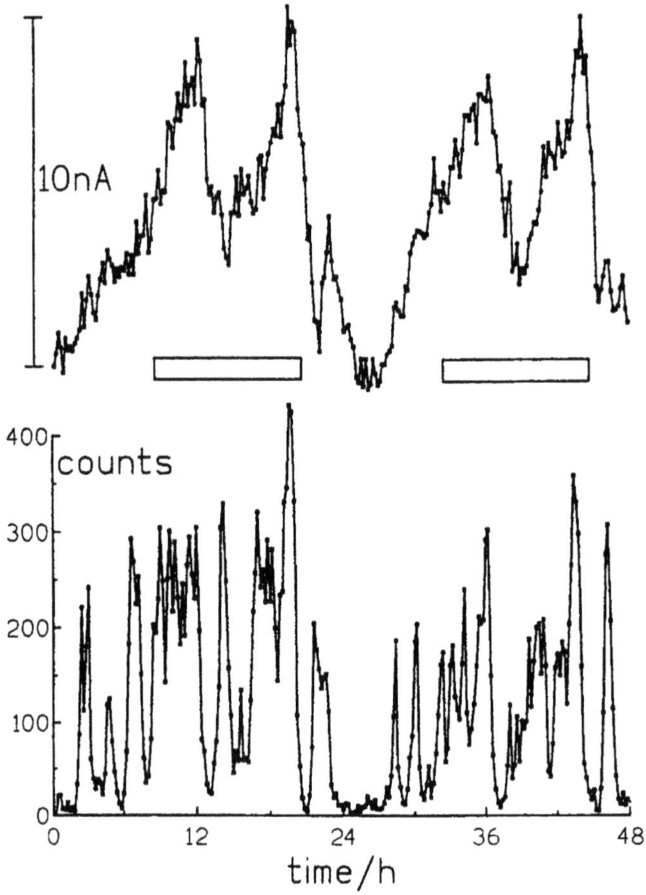

Fig. 7. Example of diurnal changes in the CPE AA signal (peak 1, Fig. 4) recorded in rat striatum **(Top)** and simultaneously monitored motor activity **(Bottom)** over a 48-h period at 12-min intervals. Bars indicate periods of darkness.

enon, namely, the mechanism of AA release and its function in neurotransmission, will now be considered.

4.1. Neuropil Origin of Extracellular Ascorbate Fluctuations

Since AA is concentrated into the brain from blood, changes observed in the level of AA in the ECF could arise from changes in the supply of blood AA. The finding that systemic administration of amphetamine caused a decrease in the AA signal recorded in rat hippocampus at the same time as an increase was observed in the striatum indicates that changes in extracellular AA do not reflect variations in the concentration of AA in blood (Grunewald et al., 1983). Furthermore, the failure of amphetamine to affect the distribution of iv [^{14}C]AA to different brain regions suggests that changes in brain AA are also not owing to modification by the drug of AA uptake from blood. It appears, therefore, that mechanisms exist whereby the concentration of AA in the ECF can change as a result of alterations in the release or uptake of AA by neuropil. In addition to these drug-related studies, the finding that the diurnal variations in the AA signal (O'Neill et al., 1982a; O'Neill and Fillenz, 1985b) do not correlate with blood AA (Grunewald et al., 1983) suggests that spontaneous fluctuations in brain AA are also of neuropile origin.

4.2. Experiments In Vitro with Synaptosomes

To investigate the cellular mechanisms associated with AA release in vivo, its efflux from synaptosomes has been studied in detail. AA is released from synaptosomes by high extracellular levels of K^+ in a concentration-dependent, Ca^{2+}-independent manner (Milby et al., 1981; Grunewald and Fillenz, 1984). This mechanism of AA release appears to be of minimal significance in vivo, however, since microinfusions of K^+ beside implanted electrodes do not affect extracellular AA (Cammack et al., 1991). Depolarization-related AA release is discussed in Section 4.3.1.

4.2.1. Effect of Ions and Neuroactive Substances

None of the following agents affected the efflux of AA from synaptosomes prepared from rat cortex, striatum, hippocampus, or cerebellum (Grunewald and Fillenz, 1984): GABA, glycine, acetylcholine, noradrenaline, DA, amphetamine, tyrosine, baclofen, or choline. However, incubation of synaptosomes with 50 μM glutamate increased the release of AA from all but cerebellar preparations. This effect was dose-dependent, saturable, and stereo-selective. Aspartate caused a similar, but nonstereo-selective, enhancement of release. The release of AA induced by the excitatory amino acids (EAAs) glutamate and aspartate was significantly reduced in Na^+-free media, but was unaffected by the absence of Ca^{2+} or the presence of Mg^{2+}, Mn^{2+}, or Co^{2+}.

4.2.2. Effect of Glutamate-Related Drugs

A variety of EAA receptor antagonists had no effect on glutamate-evoked AA release from synaptosomes (Grunewald and Fillenz, 1984), indicating that it is not receptor mediated. Receptor agonists kainate and NMDA (50 μ*M*) failed to mimic the ability of glutamate to release AA from all regions studied. However, kainate at 5 m*M* did release AA from cerebellar synaptosomes, but not from those prepared from cortex, striatum, or hippocampus. This kainate-induced AA efflux could be blocked by EAA receptor antagonists, suggesting that the cerebellum possesses a unique receptor-mediated mechanism of AA release.

The majority of glutamate uptake inhibitors, such as aspartate-β-hydroxamate and cysteate, are themselves substrates for the uptake site, and these drugs actually increased AA release from synaptosomes (Grunewald and Fillenz, 1984). However, high concentrations of kainate that block glutamate uptake without acting as a substrate blocked the effect of glutamate on AA release. On the basis of these in vitro results, a heteroexchange model of AA release in vivo has been proposed (O'Neill et al., 1984; *see* Fig. 8) based on the observations that AA release is coupled to EAA, mainly glutamate, uptake. Glutamate released by nerve terminals following the arrival of an action potential acts at a variety of receptor sites associated with its neurotransmission function. Inactivation of the chemical signal occurs by removal through a number of glutamate uptake systems that can be distinguished on the basis of their kinetic parameters, localization, ionic dependence, substrate specificity, and susceptibility to inhibitors. The relevance of the different uptake systems to AA release has been discussed previously (Fillenz et al., 1986), and it appears that the Na^+-dependent high-affinity sites located on EAA nerve terminals and on glial cells play a dominant role.

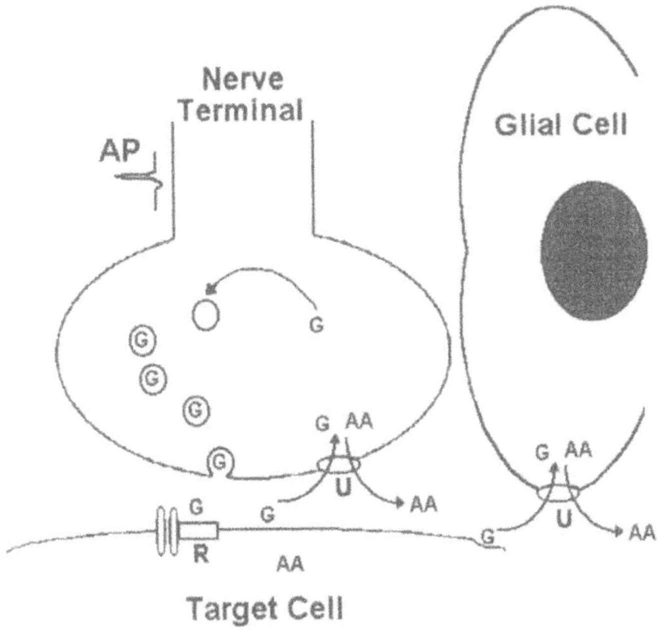

Fig. 8. Representation of the heteroexchange model proposed to explain changes in AA concentration in the ECF. Glutamate (G), released by nerve terminals following the arrival of an action potential (AP), acts at a variety of receptor sites (R) associated with its neurotransmission function. Inactivation of the chemical signal occurs by removal through a number of glutamate uptake systems (U) that can be distinguished on the basis of their kinetic parameters, localization, ionic dependence, substrate specificity, and susceptibility to inhibitors (*see* Section 4.2.2.). Both heteroreceptors and autoreceptors on the nerve terminal have been omitted for clarity.

4.3. Manipulation of Ascorbate Signal *In Vivo*

A number of voltammetric techniques have been used with both CPEs and CFEs to detect AA in vivo. The principal types of AA signal that have been generated electrochemically are: peak 1 obtained using differential pulse voltammetries (Gonon et al., 1981a) or SCV (Oh et al., 1989) with electrochemically treated CFEs; peak 1 with CPEs and either SCV (*see* Fig. 4) or LSV (O'Neill et al., 1982b); the amperometric currents at the potential of peak 1 recorded with CPEs (Boutelle et al., 1989);

and high-speed chrono-coulometric data at low potentials for treated CFEs (Ghasemzedah et al., 1991). The applicability to AA release in vivo of the model represented in Fig. 8 that is based on the behavior of AA release from synaptosomes can now be tested with results available from these in vivo AA signals. Although work with synaptosomes has been very useful, leading directly to the heteroexchange hypothesis, one would not expect the effect of agents to be necessarily the same in vivo, even if injected directly into the brain. Factors giving rise to differences include the presence in vivo of action potentials, polysynaptic feedback loops involving many neurotransmitter pathways, and the greater significance of AA uptake.

4.3.1. Microinfusion Experiments

Microinjection of glutamate beside implanted electrodes causes a stereo-selective (O'Neill et al., 1984b) and dose-dependent (Cammack et al., 1991) increase in brain extracellular AA. This increase in AA could be completely blocked by glutamate uptake inhibitors that acted on both neuronal and glial carrier sites, but only partially by blockers specific to glial cells, such as 4-acetamido-4'-isothiocyano-2,2'-disulfonic acid stilbene (Cammack et al., 1991). This agrees with earlier conclusions based on synaptosomal work that uptake sites on both neurons and glia are involved in releasing AA (Fillenz et al., 1986). It is somewhat surprising, however, that competitive uptake inhibitors enhanced glutamate-evoked AA release in vitro (Grunewald and Fillenz, 1984) while blocking it in vivo (Cammack et al., 1991).

Locally applied glutamate receptor ligands appear to play a more important role in AA release in vivo compared with that in vitro, presumably because of the greater number of possible feedback mechanisms available in vivo. For example, glutamate enhances EAA release (Ferkany et al., 1982; Collins et al., 1983) and EAA terminal excitability (Garcia-Munoz et al., 1991) through presynaptic receptors. Thus, it is consistent with the heteroexchange model that infused NMDA antagonists suppress (Pierce and Rebec, 1993), whereas NMDA itself enhances (Cammack et al., 1991), AA release in vivo, possibly by acting through these positive feedback mechanisms. Autofeedback in the glutamate system is complicated, however, differing considerably from region to region (Lovinger et al., 1993), and may involve both positive (Ferkany et al., 1982) and negative (Lovinger et al., 1993) presynaptic feedback.

It is somewhat surprising that K^+ infused at high concentrations does not release AA in vivo (Cammack et al., 1991) at least by directly releasing EAAs (O'Shea et al., 1992; Yamamoto and Davy, 1992). Cammack et al. (1991) have argued, however, that this is consistent with the heteroexchange model (Fig. 8), since high K^+ levels inhibit Na^+-dependent EAA uptake.

The inhibitory amino acid GABA has been implicated in AA release on the basis of experiments with brain slices where GABA released AA under depolarization conditions (Biglow et al., 1984). This may not be important in vivo, since infused GABA failed to affect AA release in the presence or absence of high K^+ levels (Cammack et al., 1991). The presence of anesthesia in the latter report may have interfered with the GABA-mediated mechanism, however, since the widely distributed GABA-receptor/chloride-channel complex may be a common mediator in some aspects of anesthetic action (Moody et al., 1988; Mihic et al., 1992; Tanelian et al., 1993). Indeed, specific examples of different (sometimes opposite) results in the presence and absence of anesthetics have been reported for in vivo sampling techniques (Clemens and Phebus, 1983; O'Neill et al., 1984b; Crespi et al., 1986; Hamilton et al., 1992). Receptor-mediated GABA enhancement of K^+-stimulated glutamate release (Mitchell, 1980) may be involved, therefore, in the effect of GABA on AA release.

It is possible that depolarization of cell membranes in vivo might lead to a change in extracellular AA concentration, if only by interfering with AA uptake into cells. This appears not to be significant, however. For example, the coinfusion with glutamate into brain tissue of either glutamate uptake inhibitors (enhancing glutamate-evoked depolarization) or glutamate receptor antagonists (depressing glutamate-evoked depolarization) does not affect glutamate-induced AA release (Cammack et al., 1991).

4.3.2. Lesion Experiments

Systemic administration of the catecholamine mimetic drug, amphetamine, has long been known to release AA from rat striatum in vivo (Dayton et al., 1981; Gonon et al., 1981a; O'Neill et al., 1982b), and for sometime the release of AA was linked with DA neurotransmission within the striatum—either release or receptor activation. However, the failure of nigrostriatal lesions to block amphetamine-evoked AA release in the striatum (Dayton et al., 1981; Gonon et al., 1981a; Kamata et al., 1986) and the inability of the drug to affect AA when infused directly in the striatum (O'Neill et al., 1982b; Wilson and Wightman, 1985), together with a variety of other pharmacological manipulations (Basse-Tomusk and Rebec, 1990; Zetterstrom et al., 1991; Pierce and Rebec, 1993) have led to the conclusion that activation of DA receptors outside the striatum, possibly in the substantia nigra, is responsible for amphetamine-evoked striatal AA release by a mechanism that does not significantly involve the nigrostriatal DA system. Further studies involving lesions to the crus cerebri (Wilson et al., 1986), thalamus (Basse-Tomusk and Rebec, 1990), and excitotoxic lesions of the striatum itself (Pierce et al., 1992) suggest that systemically administered DA mimetics act through a striato-nigro-thalamo-cortico-striatal feedback loop (Carlsson and Carlsson, 1990) to affect striatal AA.

The heteroexchange model (Fig. 8) predicts that AA release in any brain area should depend on the integrity of the EAA inputs to that region. Since the striatum receives a major EAA input from a glutamatergic corticostriatal pathway (Walaas, 1981; Slater et al., 1992), the effects on striatal AA of lesions to this system have been investigated (O'Neill et al., 1983c; Basse-Tomusk and Rebec, 1990). Figure 9 illustrates the effect of such lesions on the baseline and diurnal change in striatal AA over a 54-h period. The finding that the baseline AA level was reduced by over 50% and the diurnal increase by ca. 80% on the lesioned compared with the intact side suggests that the corticostriatal pathway is intimately involved with spontaneous striatal AA release. A reduction in the amphetamine-evoked striatal AA release by

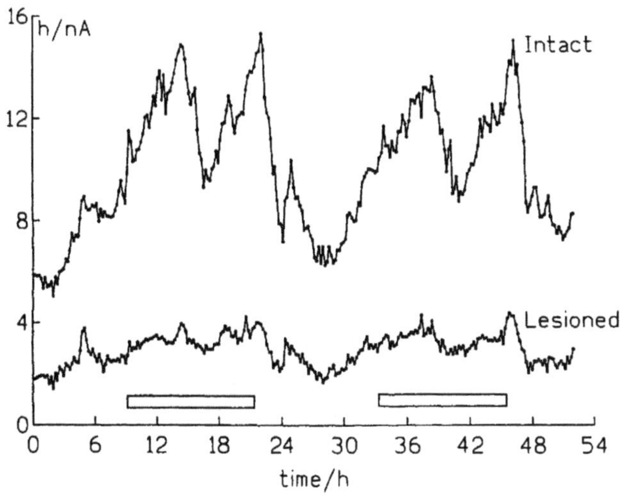

Fig. 9. Effect of unilateral cortical lesion on the diurnal changes in striatal extracellular AA monitored bilaterally with CPEs at 12-min intervals over a 54-h period. Both the baseline level and diurnal increase have been significantly reduced by lesioning the cortico-striatal glutamatergic pathway. Bars indicate periods of darkness. Based on data from O'Neill et al. (1983c) (*see* Section 4.3.2.).

50% caused by this lesion suggests that drug-induced changes in striatal AA are also related to release from this projection (Fillenz et al., 1986).

4.3.3. Stimulation of Ascorbate Release

A direct prediction of the EAA/AA heteroexchange model shown in Fig. 8 is that electrical stimulation of EAA pathways should lead to increases in extracellular AA concentrations in the terminal region. Figure 10 shows the effect of stimulation of the perforant pathway on the AA signal recorded in the dentate gyrus of the hippocampus. The increase in AA caused by the stimulus was reversibly blocked by the anesthetic chloral hydrate, which in addition decreased the baseline AA level (O'Neill et al., 1984b). These findings are consistent with the proposed model, since there is evidence that anesthetics reduce EAA neurotransmission (El-Beheiry and Puil, 1989) and release (Kendall and Minchin, 1982). The effect showed good anatomical specificity in that the response

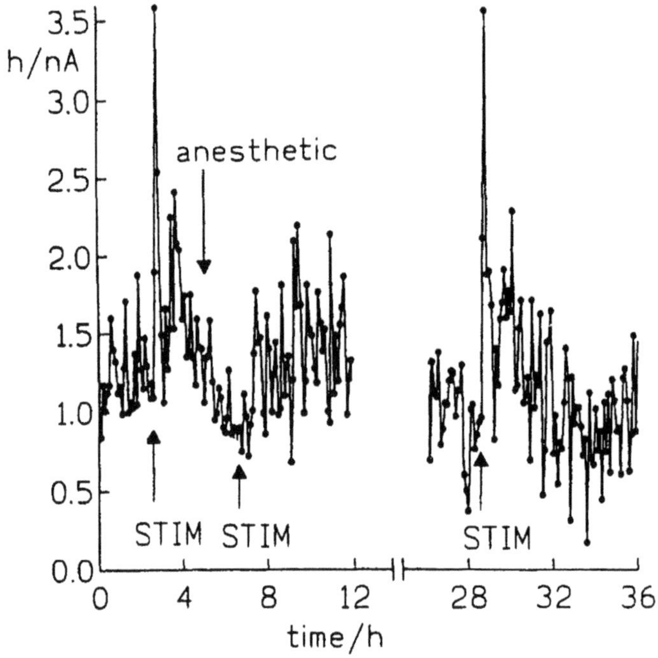

Fig. 10. Effect of electrical stimulation of the perforant pathway (10 V, 200 Hz, 2.5 s duration) on the AA signal recorded at 6-min intervals with a CPE implanted in the dentate gyrus of the hippocampus. The anesthetic (chloral hydrate, 7 mL/kg 5% solution ip) reversibly blocked the increase normally observed in hippocampal AA. Based on data from O'Neill et al. (1984b) (*see* Section 4.3.3.).

to electrical stimulation was very sensitive to the location of the CPE tip with respect to the terminal field. More recently, Cammack et al. (1991) have supported the original findings by showing that AA monitored with CFEs in the dentate gyrus was not increased following electrical stimulation of the perforant pathway in deeply anesthetized rats, but did respond when the animals were close to recovery.

Since EAAs are the principal excitatory neurotransmitter systems in the brain (Lovinger et al., 1993), one might expect on the basis of the heteroexchange model that behavioral stimulation would lead to AA release in brain areas associated with behavioral output. Figure 11 shows an example of a mild stressor (10-s tail pinch) on the amperometric current recorded at 2-s intervals with a CPE at the potential

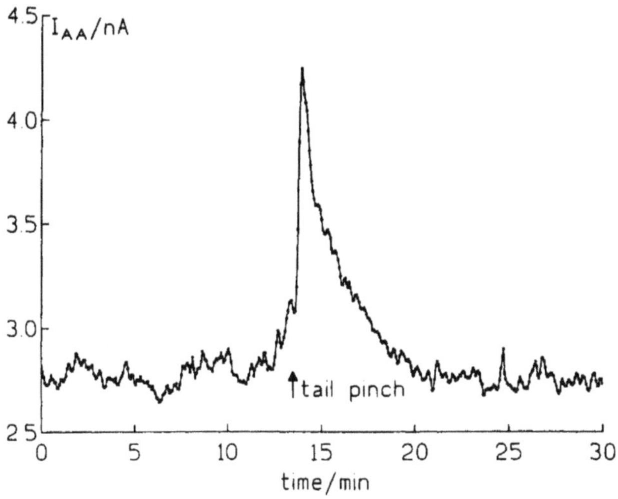

Fig. 11. Example of the effect of a 10-s tail pinch (using a plastic clothes peg) on the amperometric current recorded at 2-s intervals at the peak potential of peak 1 (*see* Fig. 4) for a CPE in the striatum of a freely moving rat (*see* Section 4.3.3.). Data from Lowry and O'Neill (unpublished); similar data reported in Boutelle et al. (1990b).

of peak 1 (*see* Fig. 4). These results are similar to those published (Boutelle et al., 1989, 1990a,b) and, positing the model shown in Fig. 8, indicate that it may be possible to monitor EAA release with subsecond time resolution (Boutelle et al., 1990b) over extended periods using stable CPEs in behaving animals. Indeed, recent data from tail-pinch experiments (Miele et al., 1994) provide further support for some aspects of the heteroexchange model. The increase in the AA signal produced by this behavioral activation is blocked by either Cd^{2+} or tetrodotoxin, indicating that physiologically-induced release of AA in rat brain is dependent on both impulse traffic and transmitter release. The finding that glutamate uptake inhibitors also blocked the tail-pinch induced AA release represents the strongest evidence to date for the model.

5. Neurophysiological Effects of Ascorbate

The thesis presented here is concerned primarily with the origin of changes in the concentration of AA in brain ECF. An obvious progression from this is to ask what, if any, are the neurophysiological effects of AA? A number of reports are now available on the effects of AA on such parameters as receptor binding and membrane properties, through to animal behavior. The space available here, however, permits only an introductory glance at possible roles of AA in brain function.

AA has been found to affect the binding of ligands to a variety of receptors in vitro. Since the binding parameters of receptors for DA (Heikkila et al., 1982; Wiener et al., 1989; Tolbert et al., 1992), serotonin (May et al., 1988; Todd and Bauer, 1988; Britt et al., 1992), adenosine (Fonseca and Ribeiro, 1991), and acetylcholine (Lotwick and Haynes, 1989; Ghosh et al., 1993) have been observed to change in the presence of AA, these effects may be through a nonspecific redox mechanism. Indeed, AA has both antioxidant and pro-oxidant properties depending on conditions. In general, AA is considered to play an antioxidant, radical scavenger role in biological systems (Padh, 1990). In the presence of heavy metal ions as catalysts, however, AA can induce lipid peroxidation (May et al., 1988; Gosh et al., 1993). May et al. (1988) suggest that AA should be used in binding studies to protect ligands from oxidation, but that EDTA should be incorporated in the medium to prevent oxidation of the membrane and consequent artifacts. The problem of lipid peroxidation may not be present in vivo to any great extent, however, since the concentration of peroxidation catalysts present in in vitro buffers is not realized in the ECF in vivo (Lowry et al., 1994). Thus, the effect, if any, of AA on the properties of these receptors in vivo remains unclear.

A special interaction between AA and EAA receptors has been reported. AA can protect neurons from injury caused by glutamate and NMDA (Majewska and Bell, 1990) at concentrations of AA of the order of that normally present in the ECF. It appears that the NMDA receptor contains one or more redox-sensitive centers where AA can downregulated ligand binding and thus the inward current associated with receptor activation (Majewska et al., 1990). The possibility exists, therefore, that the sensitivity of EAA receptors can be fine-tuned by the level of release of the neurotransmitter through the heteroexchange process (Fig. 8). This mechanism may be the basis of AA effects observed in vivo on electrophysiological parameters (Gardiner et al., 1985; Sutor and Ten Bruggencate, 1990) and on behavior itself (Pierce et al., 1991).

6. Summary and Perspectives

With its high concentration and ease of oxidation, AA is probably the simplest molecule to detect and monitor in brain ECF using in vivo voltammetric techniques. Microdialysis methodologies, however, are not as fortunate in that they appear to have difficulty recovering and detecting changes in brain AA. This may be because of the presence in brain tissue of nondialyzable agents that normally protect species from oxidation in the ECF, and the oxidation of AA in the buffer, catalyzed by trace amounts of heavy metal ion impurities. These factors, together with the continuous depletion caused by perfusion and the difficulty in separating AA successfully from the solvent front at the HPLC stage, may explain these analytical difficulties.

Voltammetric signals like that represented in Fig. 4 have revealed that the extracellular concentration of AA in discrete brain areas of awake animals change with the animal's behavior and respond to the administration of drugs. On the basis of experiments in vitro with synaptosomes, a model (Fig. 8) relating AA in the ECF with the uptake of EAAs through a heteroexchange mechanism has been proposed to explain these variations. The in vivo applicability of the model has been tested by a number of independent laboratories using a variety of lesion, stimulation, microinfusion, and other pharmacological experiments, and so far has not been invalidated. It remains to be seen, however, whether this idea will survive more detailed testing in vivo. These tests, together with experiments designed to elucidate the function of brain AA, should provide a greater understanding of the phenomenon first observed in the earliest report using an in vivo voltammetric technique—the significance of changing concentrations of AA in the ECF.

The true value of in vivo voltammetric signals, including those for AA, and microdialysis and cannot be fully appreciated until there is a clearer understanding of the relationships between levels of neurochemicals in the ECF and the synaptic events associated with them. In particular, greater recognition of the significance of extrasynaptic vs synaptic neurotransmitter function is needed for a more comprehensive view of chemical signalling in the CNS. Maybe voltammetry alone, which allows chemical analysis with enough spatial and temporal resolution to detect release from single vesicles (Chow et al., 1992; Schroeder et al., 1992), is in a position to begin to provide data on these questions.

Abbreviations

5-HIAA, 5-hydroxyindoleacetic acid; AA, ascorbic acid, ascorbate; BSA, bovine serum albumin (protein); CFE, carbon-fiber electrode; CPE, carbon-paste electrode; CPWE, carbon-powder electrode; CV, cyclic voltammetry; DA, dopamine; DHAA, dehydroascorbic acid; DOPAC, 3,4-dihydroxyphenylacetic acid; EAA, excitatory amino acid; GABA, γ-aminobutyric acid; HVA, homovanillic acid; LSV, linear sweep voltammetry; PBS, phosphate-buffered saline; PEA, phosphatidylethanolamine (lipid); SCV, staircase voltammetry; SME, stearate-modified CPE.

Acknowledgments

Space does not permit mention here of all those who have contributed to the work on CPEs in general and on the heteroexchange hypothesis in particular; the reference list acknowledges their efforts. I should, however, like to pay special thanks to Marianne Fillenz (Oxford University), who enthused a then-young thermodynamicist with her love of neurochemistry, and to John W. Albery (Oxford University), who introduced that same steady-statist to his ebullient world of dynamic electrochemistry. I also thank J. P. Lowry and M. Fillenz for helpful discussions. The work has been supported financially by the following organizations: Medical Research

Council (UK); Beit Memorial Trustees; The Royal Society (UK); Royal Irish Academy; Irish Science and Technology Agency (EOLAS); Forbairt; and University College Dublin.

References

Adams R. N. (1990) In vivo electrochemical measurements in the CNS. *Prog. Neurobiol.* **35,** 297–311.

Agrawal A., Rao I., and Sharma P. D. (1993) Trace metal-ion catalysis—kinetics and mechanism of oxidation of L-ascorbic acid by chromium(VI) in phosphate buffers. *Transit. Metal Chem.* **18,** 191–196.

Albahadily F. N. and Mottola H. A. (1987) Improved response of carbon-paste electrodes for electrochemical detection in flow systems by pretreatment with surfactants. *Anal. Chem.* **59,** 958–962.

Albery W. J., Goddard N. J., Beck T. W., Fillenz M., and O'Neill R. D. (1984) Theoretical and experimental studies of linear sweep voltammetry in the rat brain. *J. Electroanal. Chem.* **161,** 221–233.

Amatore C., Kelly S., Kristensen E. W., Kuhr W. G., and Wightman R. M. (1986) Effects of restricted diffusion at ultramicroelectrodes in brain tissue. The pool model: theory and experiment for chronoamperometry. *J. Electroanal. Chem.* **213,** 31–42.

Arad I., Sidi A., and Shohami E. (1985) Effect of acute hypoxia on ascorbate content of plasma, cerebral cortex and adrenal gland. *J. Neurochem.* **45,** 766–769.

Armstrong-James M. and Millar J. (1979) Carbon fibre micro-electrodes. *J. Neurosci. Methods* **1,** 279–287.

Ballarin M., Herrera-Marschitz M., Casas M., and Ungerstedt U. (1987) Striatal adenosine levels measured "in vivo" by microdialysis in rats with unilateral dopamine denervation. *Neurosci. Lett.* **83,** 338–344.

Bansch B., Martinez P., Uribe D., Zuluaga J., and Vaneldik R. (1991) Is the oxidation of L-ascorbic acid by aquated iron(III) ions in acidic aqueous solution substitution-transfer-controlled or electron-transfer controlled—a combined chloride, pH temperature, and pressure dependence study. *Inorg. Chem.* **30,** 4555–4559.

Basse-Tomusk A. and Rebec G. V. (1990) Corticostriatal and thalamic regulation of amphetamine-induced ascorbate release in the neostriatum. *Pharmacol. Biochem. Behav.* **35,** 55–60.

Basse-Tomusk A. and Rebec G. V. (1991) Regional distribution of ascorbate and 3,4–dihydroxyphenylacetic acid (DOPAC) in rat striatum. *Brain Res.* **538,** 29–35.

Bigelow J. C., Brown D. S., and Wightman R. M. (1984) Gamma-aminobutyric acid stimulates the release of endogenous ascorbic acid from rat striatal tissue. *J. Neurochem.* **2,** 412–419.

Blaha C. D. and Jung M. E. (1991) Electrochemical evaluation of stearate-modified graphite paste electrode: selective detection of dopamine is maintained after exposure to brain tissue. *J. Electroanal. Chem.* **310,** 317–334.

Blaha C. D. and Lane R. F. (1983) Chemically modified electrode for in vivo monitoring of brain catecholamines. *Brain Res. Bull.* **10,** 861–864.

Boutelle M. G., Svensson L., and Fillenz M. (1989) Rapid changes in striatal ascorbate in response to tail-pinch monitored by constant potential voltammetry. *Neuroscience* **30,** 11–17.

Boutelle M. G., Svensson L., and Fillenz M. (1990a) Effect of diazepam on behaviour and associated changes in ascorbate concentration in rat brain areas: striatum, n. accumbens and hippocampus. *Psychopharmacology* **100,** 230–236.

Boutelle M. G., Zetterstrom T., Pei Q., Svensson L., and Fillenz M. (1990b) In vivo neurochemical effects of tail pinch. *J. Neurosci. Methods* **34,** 151–157.

Brazell P. and Marsden C. (1982) Intracerebral injection of ascorbate oxidase–effect on in vivo electrochemical recordings. *Brain Res.* **249,** 167–172.

Britt S. G., Chiu V. W. S., Redpath G. T., and VandenBerg S. R. (1992) Elimination of ascorbic acid-induced membrane lipid peroxidation and serotonin receptor loss by Trolox-C, a water soluble analogue of vitamin E. *J. Recept. Res.* **12,** 181–200.

Broderick P. A. (1988) Distinguishing in vitro electrochemical signatures for norepinephrine and dopamine. *Neurosci. Lett.* **95,** 275–280.

Broderick P. A., Kornak E. P. Jr., Eng F., and Wechsler R. (1993) Real time detection of acute (IP) cocaine-enhanced dopamine and serotonin release in ventrolateral nucleus accumbens of the behaving Norway rat. *Pharmacol. Biochem. Behav.* **46,** 715–722.

Brose N., O'Neill R. D., Boutelle M. G., and Fillenz M. (1989) The effects of anxiolytic and anxiogenic benzodiazepine receptor ligands on motor activity and levels of ascorbic acid in the nucleus accumbens and striatum of the rat. *Neuropharmacology* **28,** 509–514.

Brose N., O'Neill R. D., Boutelle M. G., and Fillenz M. (1988) Dopamine in the basal ganglia and benzodiazepine-induced sedation. *Neuropharmacology* **27,** 589–595.

Brose N., O'Neill R. D., Boutelle M. G., Anderson S. M. P., and Fillenz M. (1987) Effects of an anxiogenic benzodiazepine receptor ligand on rat motor activity and dopamine release in nucleus accumbens and striatum. *J. Neurosci.* **7,** 2917–2926.

Brun P., Suaud-Chagny M. F., Lachuer J., Gonon F., and Buda M. (1991) Catecholamine metabolism in locus coeruleus neurons: a study of its activation by sciatic nerve stimulation in the rat. *Eur. J. Neurosci.* **3,** 397–406.

Bull D. R., Bakhtiar R., and Sheehan M. J. (1991) Characterization of dopamine autoreceptors in the amygdala: a fast cyclic voltammetric study in vitro. *Neurosci. Lett.* **134,** 41–44.

Cammack J., Ghasemzadeh B., and Adams R. N. (1991) The pharmacological profile of glutamate-evoked ascorbic acid efflux measured by in vivo electrochemistry. *Brain Res.* **565,** 17–22.

Carlsson M. and Carlsson A. (1990) Interactions between glutamatergic and monoaminergic systems with the basal ganglia—implications for schizophrenia and Parkinson's disease. *Trends Neurosci.* **13,** 272–276.

Cespuglio R, Sarda N., Gharib A., Faradji H., and Chastrette N. (1986) Differential pulse voltammetry in vivo with working carbon fibre electrodes: 5-hydroxyindole compounds or uric acid detection. *Exp. Brain Res.* **64,** 589–595.

Cheng H. Y., Schenk J., Huff R., and Adams R. N. (1979) In vivo electrochemistry: behaviour of micro electrodes in brain tissue. *J. Electroanal. Chem.* **100,** 23–31.

Chow R. H., Vonruden L., and Neher E. (1992) Delay in vesicle fusion revealed by electrochemical monitoring of single secretory events in adrenal chromaffin cells. *Nature* **356,** 60–63.

Church W. H. and Justice J. B. Jr. (1987) Rapid sampling and determination of extracellular dopamine in vivo. *Anal. Chem.* **59,** 712–716.

Clemens J. A. and Phebus L. A. (1983) Changes in brain chemistry produced by dopaminergic agents: in vivo electrochemical monitoring reveals opposite changes in anaesthetized vs unanaesthetized rats. *Brain Res.* **267,** 183–186.

Collins G. G. S., Anson J., and Surtees L. (1983) Presynaptic kainate and N-methyl-D-aspartate receptors regulate excitatory amino-acid release in the olfactory cortex. *Brain Res.* **265,** 157–159.

Coury L. A. Jr. and Heineman W. R. (1988) Electrochemical examination of unusual properties of spectroscopic graphite electrodes. *J. Electroanal. Chem.* **256,** 327–341.

Crespi F., Keane P. E., and Morre M. (1986) Anaesthesia abolishes the effect of valproate on extracellular 5HIAA, DOPAC and ascorbate as measured in rat striatum by differential pulse voltammetry. *Br. J. Pharmacol.* **87,** 279–283.

Crespi F., Sharp T., Maidment N., and Marsden C. A. (1983) Differential pulse voltammetry in vivo—evidence that uric acid contributes to the indole oxidation peak. *Neurosci. Lett.* **43,** 203–207.

Davies M. B. (1992) Reactions of L-ascorbic acid with transition metal complexes. *Polyhedron* **11,** 285–321.

Davies M. B., Austin J., and Partridge D. A. (1991) *Vitamin C. Its Chemistry and Biochemistry.* The Royal Society of Chemistry (Pubs.), Cambridge, England.

Dayton M. A., Ewing A. G., and Wightman R. M. (1980) Response of microvoltammetric electrodes to homogeneous catalytic and slow heterogeneous charge-transfer reactions. *Anal. Chem.* **52,** 2392–2396.

Dayton M. A., Ewing A. G., and Wightman R. M. (1981) Evaluation of amphetamine-induced in vivo electrochemical response. *Eur. J. Pharmacol.* **75,** 141–144.

Desimoni E., Casella G. I., Cataldi T. R. I., Salvi M., Rotunno T., and Dicroce E. (1992) Remarks on the surface characterization of carbon fibres. *Surf. Interface Anal.* **18,** 623–630.

Deutsch J. C. and Kolhouse J. F. (1993) Ascorbate and dehydroascorbate measurements in aqueous solutions and plasma determined by gas chromatography/mass spectrometry. *Anal. Chem.* **65,** 321–326.

Duff A. and O'Neill R. D. (1994) Effect of probe size on the concentration of brain extracellular uric acid monitored with carbon paste electrodes. *J. Neurochem.* **62,** 1496–1502.

El-Beheiry H. and Puil E. (1989) Anaesthetic depression of excitatory synaptic transmission in neocortex. *Exp. Brain Res.* **77,** 87–93.

Falat L. and Cheng H. Y. (1982) Voltammetric differentiation of ascorbic acid and dopamine at an electrochemically treated graphite-epoxy electrode. *Anal. Chem.* **54,** 2108–2111.

Ferkany J. W., Zaczek R., and Coyle J. T. (1982) Kainic acid stimulates excitatory amino acid neurotransmitter release at presynaptic receptors. *Nature* **28,** 757–759.

Fillenz M. and O'Neill R. D. (1986) Effects of light reversal on the circadian pattern of motor activity and voltammetric signals recorded in rat forebrain. *J. Physiol. (Lond.)* **374,** 91–101.

Fillenz M., O'Neill R. D., and Grunewald R. A. (1986) Changes in extracellular brain ascorbate concentration as an index of excitatory amino acid release, in *Measuring Peripheral and Central Neurotransmitter Release During Behaviour* (Joseph M. F., Fillenz M., MacDonald I. A., and Marsden C. A., eds.), Ellis Horwood, Chichester, pp. 144–163.

Fonseca E. and Ribeiro J. A. (1991) Ascorbate-induced lipid peroxidation and inhibition of [^{3}H] adenosine binding to rat brain synaptosomes. *Nucleosides Nucleotides* **10,** 1213–1214.

Forni C. and Nieoullon A. (1984) Electrochemical detection of dopamine release in the striatum of freely moving hamsters. *Brain Res.* **297,** 11–20.

Garcia-Munoz M., Young S. J., and Groves P. M. (1991) Terminal excitability of the corticostriatal pathway. II. Regulation by glutamate receptor stimulation. *Brain Res.* **551,** 207–215.

Gardiner T. W., Armstrong-James M., Caan A. W., Wightman R. M., and Rebec G. V. (1985) Modulation of neostriatal activity by iontophoresis of ascorbic acid. *Brain Res.* **344,** 181–185.

Gelbert M. B. and Curran D. J. (1986) Alternating current voltammetry of dopamine and ascorbate at the carbon paste electrodes and stearate modified carbon paste electrodes. *Anal. Chem.* **5,** 1028–1032.

Ghasemzedah B., Cammack J., and Adams R. N. (1991) Dynamic changes in extracellular fluid ascorbic acid monitored by in vivo electrochemistry. *Brain Res.* **547,** 162–166.

Ghosh C., Dick R. L., and Ali S. F. (1993) Iron/ascorbate-induced lipid peroxidation changes membrane fluidity and muscarinic cholinergic receptor binding in rat frontal cortex. *Nature* **23,** 479–484.

Glynn G. E. and Yamamoto B. K. (1989) In vivo neurochemical and anatomical heterogeneity of the dopamine uptake system in the rat caudate putamen. *Brain Res.* **481,** 235–241.

Gonon F., Buda M., Cespuglio R., Jouvet M., and Pujol J. F. (1980) In vivo electrical detection of catechols in the rat neostriatum: dopamine or DOPAC? *Nature* **286,** 902–904.

Gonon F., Buda M., Cespuglio R., Jouvet M., and Pujol J. F. (1981a) Voltammetry in the striatum of chronic freely moving rats: detection of catechols and ascorbic acid. *Brain Res.* **223,** 69–80.

Gonon F., Fombarlet C. M., Buda M., and Pujol J. F. (1981b) Electrochemical treatments of pyrolytic carbon fibre electrodes. *Anal. Chem.* **53,** 1386–1389.

Gonzalez-Mora J. L., Sanchez-Bruno J. A., and Mas M. (1988) Concurrent on-line analysis of striatal ascorbate, dopamine and dihydroxyphenylacetic acid concentrations by in vivo voltammetry. *Neurosci. Lett.* **86,** 61–66.

Gonzalez-Mora J. L., Maidment N. T., Guadalupe T., and Mas M. (1989) Post-mortem dopamine dynamics assessed by voltammetry and microdialysis. *Brain Res. Bull.* **23,** 323–327.

Gonzalez-Mora J. L., Guadalupe T., Fumero B., and Mas M. (1991) Mathematical resolution of mixed in vivo voltammetry signals. Models, equipment, assessment by simultaneous microdialysis sampling. *J. Neurosci. Methods* **39,** 231–244.

Grunewald R. A. (1993) Ascorbic acid in the brain. *Brain Res. Rev.* **18,** 123–133.

Grunewald R. A. and Fillenz M. (1984) Release of ascorbate from synaptosomal fraction of rat brain. *Neurochem. Int.* **6,** 491–500.

Grunewald R. A., O'Neill R. D., Fillenz M., and Albery W. J. (1983) The origin of circadian and amphetamine-induced changes in the extracellular concentration of brain ascorbate. *Neurochem. Int.* **5,** 773–778.

Hahn Z., Cespuglio R., Faradji H., and Jouvet M. (1985) Factors influencing the properties of voltammetric carbon fibre electrodes: the importance of the pH of the medium used for the electrical treatment and of the resin coating of the fibres. *J. Biochem. Biophys. Methods* **11,** 265–275.

Hamilton M. E., Mele A., and Pert A. (1992) Striatal extracellular dopamine in conscious vs. anesthetized rats: effects of chloral hydrate anesthetic on responses to drugs of different classes. *Brain Res.* **597,** 1–7.

Hand D. B. and Greisen E. C. (1942) Oxidation and reduction of vitamin C. *J. Am. Chem. Soc.* **64,** 358–361.

Haskett C. and Mueller K. (1987) The effects of serotonin depletion on the voltammetric response to amphetamine. *Pharmacol. Biochem. Behav.* **28,** 381–384.

Heikkila R. E., Cabbat F. S., and Manzino L. (1982) Inhibitory effects of ascorbic acid on the binding of [^{3}H]dopamine to neostriatal membrane preparations: relationship to lipid peroxidation. *J. Neurochem.* **38,** 1000–1006.
Hong M., Milne B., Loomis C., and Jhamandas K. (1992) Stereoselective effects of central Alpha2-adrenergic agonist medetomidine on in vivo catechol activity in the rat rostral ventrolateral medulla (RVLM). *Brain Res.* **592,** 163–169.
Houdouin F., Cespuglio R., Gharib A., Sarda N., and Jouvet M. (1991) Detection of the release of 5-hydroxyindole compounds in the hypothalamus and the n. raphe dorsalis throughout the sleep-waking cycle and during stressful situations in the rat: a polygraphic and voltammetric approach. *Exp. Brain Res.* **85,** 153–162.
Hu I. F. and Kuwana T. (1986) Oxidative mechanism of ascorbic acid at glassy carbon electrodes. *Anal. Chem.* **58,** 3235–3239.
Hulthe P., Johannessen K., Svensson L., and Engel J. (1989) A new pretreatment method for carbon fibre microelectrodes enhances the selectivity for dopamine. *J. Neurosci. Methods* **29,** 297.
Inoue K., Kiriike N., Okuno M., Ito H., Fujisaki Y., Matsui T., and Kawakita Y. (1993) Scheduled feeding caused activation of dopamine metabolism in the striatum of rats. *Physiol. Behav.* **53,** 177–181.
Jacques P. F. (1992) Relationship of vitamin C status to cholesterol and blood pressure. *Ann. NY Acad. Sci.* **669,** 205–214.
Joseph M. H. and Young A. M. J. (1991) Pharmacological evidence, using in vivo dialysis, that substances additional to ascorbic acid, uric acid and homovanillic acid contribute to the voltammetric signals obtained in unrestrained rats from chronically implanted carbon paste electrodes. *J. Neurosci. Methods* **36,** 209–218.
Joseph M. H. and Kennett G. A. (1981) In vivo voltammetry in the rat hippocampus as an index of drug effects on extraneuronal 5-HT. *Neuropharmacology* **20,** 1361–1364.
Joseph M. H., Hodges H., and Gray J. A. (1989) Lever pressing for food reward and in vivo voltammetry: evidence for increases in extracellular homovanillic acid, the dopamine metabolite, and uric acid in the rat caudate nucleus. *Neuroscience* **32,** 195–201.
Justice J. B. Jr., Wages S. A., and Michael A. C. (1983) Interpretations of voltammetry in the striatum based on chromatography of striatal dialysate. *J. Liq. Chromatog.* **6,** 1873–1896.
Kamata K., Wilson R. L., Alloway K. D., and Rebec G. V. (1986) Multiple amphetamine injections reduce the release of ascorbic acid in the neostriatum of the rat. *Brain Res.* **362,** 331–338.
Karabinas P., Sazou D., and Jannakoudakis D. (1985) Comparative electrochemical study of L-ascorbic acid and dihydroxyfumaric acid on a mercury electrode in neutral media. *J. Electroanal. Chem.* **192,** 469–478.
Kawagoe K. T., Zimmerman J. B., and Wightman R. M. (1993) Principles of voltammetry and microelectrode surface states. *J. Neurosci. Methods* **48,** 225–240.

Kawagoe K. T., Garris P. A., Wiedemann D. J., and Wightman R. M. (1992) Regulation of transient dopamine concentration gradients in the microenvironment surrounding nerve terminals in the rat striatum. *Neuroscience* **51,** 55–64.

Kendall T. J. G., and Minchin M. C. W. (1982) The effects of anaesthetics on the uptake and release of amino acid neurotransmitters in thalamic slices. *Br. J. Pharmacol.* **75,** 219–227.

Kissinger P. T., Hart J. B., and Adams R. N. (1973) Voltammetry in brain tissue—a new neurophysiological measurement. *Brain Res.* **55,** 209–213.

Kiyatkin E. A., Wise R. A., and Gratton A. (1993) Drug- and behavior-associated changes in dopamine-related electrochemical signals during intravenous heroin self-administration in rats. *Synapse* **14,** 60–72.

Kovach P. M., Deakin M. R., and Wightman R. M. (1986) Electrochemistry at partially blocked carbon-fiber microcylinder electrodes. *J. Phys. Chem.* **90,** 4612–4617.

Kovachich G. B. and Mishra O. P. (1984) Stabilization of ascorbic acid and norepinephrine in vitro by the subcellular fractions of rat cerebral cortex. *Neurosci. Lett.* **52,** 153–158.

Kuo C. H., Yonehara N., Hata F., and Yoshida H. (1978) Subcellular distribution of ascorbic acid in the rat brain. *Jpn. J. Pharmacol.* **28,** 789–791.

Lam D. K. C. and Daniel P. M. (1986) The influx of ascorbic acid into the rat brain. *Quart. J. Exp. Physiol.* **71,** 483–490.

Lane R. F., Hubbard A. T., and Blaha C. D. (1978) Brain dopaminergic neurons: in vivo electrochemical information concerning storage, metabolism and release processes. *Bioelectrochem. Bioenerg.* **5,** 504–525.

Lane R. F., Hubbard A T., and Blaha C. D. (1979) Application of semiderivative electroanalysis to studies of neurotransmitters in the central nervous system. *J. Electroanal. Chem.* **95,** 117–122.

Lechien A., Valenta P., Nurnberg H. W., and Patriarche G. J. (1982) Determination of ascorbic acid by differential pulse voltammetry. *Fresenius Z. Anal. Chem.* **311,** 105–108.

Lotwick H. S. and Haynes L. W. (1989) CGRP and ascorbate oppositely regulate cell-surface acetylcholine receptor number in quail myotubes. *Neurosci. Res. Commun.* **5,** 171–179.

Louilot A., Serrano A., and D'Angio M. (1987) A novel carbon fibre implantation assembly for cerebral voltammetric measurements in freely moving rats. *Physiol. Behav.* **41,** 227–231.

Lovinger D. M., Tyler E., Fidler S., and Merritt A. (1993) Properties of a presynaptic metabotropic glutamate receptor in rat neostriatal slices. *J. Neurophysiol.* **69,** 1236–1244.

Lowry J. P., McAteer K., El Atrash S. S., Duff A., and O'Neill R. D. (1994) Characterization of glucose oxidase modified poly(phenylenediamine)-coated electrodes for in vitro and in vivo: homogeneous interference by ascorbic acid in hydrogen peroxide detection. *Anal. Chem.* **66,** 1754–1761.

Lyne P. D and O'Neill R. D. (1989) Selectivity of stearate-modified carbon paste electrodes for dopamine and ascorbic acid. *J. Anal.* **61,** 2323,2324.

Lyne P. D. and O'Neill R. D. (1990) Stearate-modified carbon paste electrodes for detecting dopamine in vivo: decrease in selectivity caused by lipids and other surface-active agents. *Anal. Chem.* **62,** 2347–2351.

Majewska M. D. and Bell J. A. (1990) Ascorbic acid protects neurons from injury induced by glutamate and NMDA. *NeuroReport* **1,** 194–196.

Majewska M. D., Bell J. A., and London E. D. (1990) Regulation of the NMDA receptor by redox phenomena: inhibitory role of ascorbate. *Brain Res.* **537,** 328–332.

Marcenac F. and Gonon F. (1985) Fast in vivo monitoring of dopamine release in the rat brain with differential pulse amperometry. *Anal. Chem.* **57,** 1778,1779.

Marsden C. A., Joseph M. H., Kruk Z. L., Maidment N. T., O'Neill R. D., Schenk J. O., and Stamford J. A. (1988) In vivo voltammetry—Present electrodes and methods. *Neuroscience* **25,** 389–400.

Mas M., Gonzalez-Mora J. L., Louilot A., Sole C., and Guadalupe T. (1990) Increased dopamine release in the nucleus accumbens of copulating male rats as evidenced by in vivo voltammetry. *Neurosci. Lett.* **110,** 303–308.

Matsuda H. and Ayabe Y. (1955) Zur Theorie der Randles-Seveikschen Kathodenstrahl-Polarographie. *Z. Elektrochem.* **59,** 494–503.

May P. C., Morgan D G., Salo D., Goss J. R., and Finch C. E. (1988) Effects of radioligand oxidation and ascorbate-induced lipid peroxidation on serotonin–1 receptor assay: use of ascorbate and ethylenediaminetetraacetic acid buffers to prevent [^{3}H]-5-HT binding artifacts. *J. Neurosci. Res.* **20,** 257–262.

Mermet C. and Gonon F. (1986) In vivo voltammetric monitoring of noradrenaline release and catecholamine metabolism in the hypothalamic paraventricular nucleus. *Neuroscience* **19,** 829–838.

Miele M., Boutelle M. G., and Fillenz M. (1994) The physiologically-induced release of ascorbate in rat brain is dependent on impulse traffic calcium influx and glutamate uptake. *Neuroscience* **62,** 87–91.

Mihic S. J., Wu P. H., and Kalant H. (1992) Potentiation of gamma-aminobutyric acid-mediated chloride flux by pentobarbital and diazepam but not ethanol. *J. Neurochem.* **58,** 745–751.

Milby K., Oke A. F., and Adams R. N. (1982) Detailed mapping of ascorbate distribution in the rat brain. *Neurosci. Lett.* **28,** 15–20.

Milby K. H., Mefford L. N., Chey W., and Adams R. N. (1981) In vitro and in vivo depolarization-coupled efflux of ascorbic aid in rat brain preparations. *Brain Res. Bull.* **7,** 237–242.

Milosevic P., Jorga I., and Milosevic A. (1989) Brain capillary transport of glucose during fatiguing physical work: effect of ascorbic acid. *Med. Sci. Res.* **17,** 11,12.

Mitchell R. (1980) A novel GABA receptor modulates stimulus-induced glutamate release from corticostriatal terminals. *Eur. J. Pharmacol.* **67,** 119–122.

Moody E. J., Suzdak P. D., Paul S. M., and Skolnick P. (1988) Modulation of the benzodiazepine/gamma-aminobutyric acid receptor chloride-channel complex by inhalation anaesthetics. *J. Neurochem.* **51,** 1386–1393.

Mueller K., Palmour R., Andrews C. D., and Knott P. J. (1985) In vivo voltammetric evidence of production of uric acid by rat caudate. *Brain Res.* **335,** 231–235.

Nakahara M. and Shimizu K. (1992) Effects of electrolyte on the structure of pyrolytic graphite surfaces in anodic oxidation. *J. Mater. Sci.* **27,** 1207–1211.

Nelson A. and Auffret N. (1988) Phospholipid monolayers of di-oleoyl lecithin (di-o-PC) at the mercury water interface. Effect on faradaic reactions. *J. Electroanal. Chem.* **248,** 167–180.

Nicholson R. S. and Shain I. (1964) Theory of stationary electrode polarography. *Anal. Chem.* **36,** 706–723.

O'Connor J. J. and Kruk Z. L. (1991) Frequency dependence of 5-HT autoreceptor function in rat dorsal raphe and suprachiasmatic nuclei studied using fast cyclic voltammetry. *Brain Res.* **568,** 123–130.

Oh C., Gardiner T. W., and Rebec G. V. (1989) Blockade of both D1- and D2-dopamine receptors inhibits amphetamine-induced ascorbate release in the neostriatum. *Brain Res.* **80,** 184–189.

O'Neill R. D. (1994) Microvoltammetric techniques and sensors for monitoring neurochemical dynamics in vivo—a review. *Analyst* **119,** 767–779.

O'Neill R. D. and Fillenz M. (1985a) Detection of homovanillic acid in vivo using microcomputer-controlled voltammetry: simultaneous monitoring of rat motor activity and striatal dopamine release. *Neuroscience* **14,** 753–763.

O'Neill R. D. and Fillenz M. (1985b) Circadian changes in extracellular ascorbate in rat cortex accumbens striatum and hippocampus: correlations with motor activity. *Neurosci. Lett.* **60,** 331–336.

O'Neill R. D., Fillenz M., and Albery W. J. (1982a) Circadian changes in homovanillic acid and ascorbate levels in the rat striatum using microprocessor-controlled voltammetry. *Neurosci. Lett.* **34,** 189–193.

O'Neill R. D., Grunewald R. A., Fillenz M., and Albery W. J. (1982b) Linear sweep voltammetry with carbon paste electrodes in the rat striatum. *Neuroscience* **7,** 1945–1954.

O'Neill R. D., Fillenz M., and Albery W. J. (1983a) The development of linear sweep voltammetry with carbon paste electrodes in vivo. *J. Neurosci. Methods* **8,** 263–273.

O'Neill R. D., Fillenz M., Albery W. J., and Goddard N. J. (1983b) The monitoring of ascorbate and monoamine transmitter metabolites in the striatum of unanaesthetised rats using microprocessor-based voltammetry. *Neuroscience* **9,** 87–93.

O'Neill R. D., Grunewald R. A., Fillenz M., and Albery W. J. (1983c) The effect of unilateral cortical lesions on the circadian changes in rat striatal ascorbate and homovanillic acid levels measured in vivo using voltammetry. *Neurosci. Lett.* **42,** 105–110.

O'Neill R. D., Fillenz M., Grunewald R. A., Bloomfield M. R., Albery W. J., Jamieson C. M., Williams J. H., and Gray J. A. (1984a) Voltammetric carbon paste electrodes monitor uric acid and not 5HIAA at the 5-hydroxyindole potential in the rat brain. *Neurosci. Lett.* **45,** 39–46.

O'Neill R. D., Fillenz M., Sundstrom L., and Rawlins J. N. P. (1984) Voltammetrically monitored brain ascorbate as an index of excitatory amino acid release in the unrestrained rat. *Neurosci. Lett.* **52,** 227–233.

O'Neill R. D., Gonzalez-Mora J. L., Boutelle M. G., Ormonde D. E., Lowry J. P., Duff A., Fumero B., Fillenz M., and Mas M. (1991) Anomalously high concentrations of brain extracellular uric acid detected with chronically implanted probes: implications for in vivo sampling techniques. *J. Neurochem.* **57,** 22–29.

Ormonde D. E. and O'Neill R. D. (1990) The oxidation of ascorbic acid at carbon paste electrodes. Modified response following contact with surfactant, lipid and brain tissue. *J. Electroanal. Chem.* **279,** 109–121.

O'Shea T. J., Weber P. L., Bammel B. P., Lunte C. E., Lunte S. M., and Smyth M. R. (1992) Monitoring excitatory amino acid release in vivo by microdialysis with capillary electrophoresis-electrochemistry. *J. Chromatogr.* **608,** 189–195.

Padh H. (1990) Cellular functions of ascorbic acid. *Biochem. Cell Biol.* **68,** 1166–1173.

Parsey R. V. and Matteson D. R. (1993) Ascorbic acid modulation of calcium channels in pancreatic Beta cells. *J. Gen. Physiol.* **102,** 503–523.

Peng T. Z., Lu H. H., Liu G. Q., and Cao Y. P. (1992) An improved electrochemical procedure for treating carbon fiber electrodes for in vivo determination of some neurotransmitters. *Anal. Lett.* **25,** 795–805.

Phillips A. G., Atkinson L. J., Blackburn J. R., and Blaha C. D. (1993) Increased extracellular dopamine in the nucleus accumbens of the rat elicited by a conditional stimulus for food: an electrochemical study. *Can. J. Physiol. Pharmacol.* **71,** 387–393.

Pierce R. C., Miller D. W., Reising D. B., and Rebec G. V. (1992) Unilateral neostriatal kainate, but not 6-OHDA, lesions block dopamine agonist-induced ascorbate release in the neostriatum of freely moving rats. *Brain Res.* **597,** 138–143.

Pierce R. C. and Rebec G. V. (1992) Dopamine-, NMDA-, and sigma-receptor antagonists exert differential effects on basal and amphetamine-induced changes in neostriatal ascorbate and DOPAC in awake, behaving rats. *Brain Res.* **579,** 59–66.

Pierce R. C. and Rebec G. V. (1993) Intraneostriatal administration of glutamate antagonists increases behavioral activation and decreases neostriatal ascorbate via nondopaminergic mechanisms. *J. Neurosci.* **13,** 4272–4280.

Pierce R. C., Rowlett J. K., Bardo M. T., and Rebec G. V. (1991) Chronic ascorbate potentiates the effects of chronic haloperidol on behavioral supersensitivity but not D2 dopamine receptor binding. *Neuroscience* **45,** 373–378.

Ponchon J. L., Cespuglio R., Gonon F., Jouvet M., and Pujol J. F. (1979) Normal pulse polarography with carbon fibre electrodes for in vitro and in vivo determination of catecholamines. *Anal. Chem.* **51,** 1483–1486.

Pratt O. E. and Greenwood J. (1986) Movement of vitamins across the blood–brain barrier, in *Blood–Brain Barrier in Health and Disease* (Suckling A. J., Rumsby M. G., and Bradbury M. W. B., eds.), Ellis Horwood, Chichester, England, pp. 87–97.

Puig S., Rivot J. P., and Besson J. M. (1992) Effect of subcutaneous administration of the chemical algogen formalin, on 5-HT metabolism in the nucleus raphe magnus and the medullary dorsal horn: a voltammetric study in freely moving rats. *Brain Res.* **590,** 250–254.

Rebec G. V., Langley P. E., Pierce R. C., Wang Z., and Heidenreich B. A. (1993) A simple micromanipulator for multiple uses in freely moving rats: electrophysiology, voltammetry, and simultaneous intracerebral infusions. *J. Neurosci. Methods* **47,** 53–59.

Rice M. E. and Cammack J. (1991) Anoxia-resistant turtle brain maintains ascorbic acid content in vitro. *Neurosci. Lett.* **132,** 141–145.

Rice M. E., Galus Z., and Adams R. N. (1983) Graphite paste electrodes. Effects of paste composition and surface states on electron transfer rates. *J. Electroanal. Chem.* **143,** 89–102.

Rivot J. P., Pointis D., and Besson J. M. (1988) In vivo electrochemical detection of 5-hydroxyindole within the trigeminal nucleus caudatus of freely moving rats: the effect of morphine. *Brain Res.* **463,** 275–283.

Ruggeri M., Zoli M., Grimaldi R., Ungerstedt U., Eliasson A., Agnati L. F., and Fuxe K. (1990) Aspects of neural plasticity in the central nervous system. III. Methodological studies on the microdialysis technique. *Neurochem. Int.* **16,** 427–435.

Schroeder T. J., Jankowski J. A., Kawagoe K. T., Wightman R. M., Lefrou C., and Amatore C. (192) Analysis of diffusional broadening of vesicular packets of catecholamines released from biological cells during exocytosis. *Anal. Chem.* **64,** 3077–3083.

Shin D. B. and Feather M. S. (1990) The degradation of L-ascorbic acid in neutral solutions containing oxygen. *J. Carbohydr. Chem.* **9,** 461–469.

Slater P., Simpson M. D. C., Hunter A. J., and Cross A. J. (1992) Loss of excitatory amino acid transport sites after lesioning a glutamatergic pathway in rat brain. *Neurosci. Res. Commun.* **10,** 135–140.

Slivka A., Brannan T. S., Weinberger J., Knott P. J., and Cohen G. (1988) Increase in extracellular dopamine in the striatum during cerebral ischemia: a study utilizing cerebral microdialysis. *J. Neurochem.* **50,** 1714–1718.

Smith A. D., Olson R. J., and Justice J. B. Jr. (1992) Quantitative microdialysis of dopamine in the striatum: effect of circadian variation. *J. Neurosci. Methods* **44,** 33–41.

Spector R. and Lorenzo A. V. (1974) Specificity of ascorbic acid transport system of the central nervous system. *Am. J. Physiol.* **226,** 1468–1473.

Spector R., Spector A. Z., and Snodgrass S. R. (1977) Model of transport in the central nervous system. *Am. J. Physiol.* **232,** R73–R79.

Stamford J. A. (1986) Effect of electrocatalytic and nucleophilic reactions on fast voltammetric measurements of dopamine at carbon fiber microelectrodes. *Anal. Chem.* **58,** 1033–1036.

Stamford J. A., Kruk Z. L., and Millar J. (1984) Regional differences in extracellular ascorbic acid levels in the rat brain determined by high-speed cyclic voltammetry. *Brain Res.* **299,** 289–295.

Strein T. G. and Ewing A. G. (1991) In situ laser activation of carbon fiber microdisk electrodes. *Anal. Chem.* **63,** 194–198.

Sujaritvanichpong S., Aoki K., Tokuda K., and Matsuda H. (1986) Electrochemical behavior of dopamine at carbon fibre electrodes. *J. Electroanal. Chem.* **198,** 195–203.

Sutor B. and Ten Bruggencate G. (1990) Ascorbic acid: a useful reductant to avoid oxidation of catecholamines in electrophysiological experiments in vitro? *Neurosci. Lett.* **116,** 287–292.

Svensson L., Wu C., Hulthe P., Johannessen K., and Engel J. A. (1992a) Rapid changes in ascorbate and dopamine release in rat nucleus accumbens after intracerebroventricular administration of NMDA. *Brain Res.* **586,** 195–202.

Svensson L., Wu C., Johannessen K., and Engel J. A. (1992b) Effect of ethanol on ascorbate release in the nucleus accumbens and striatum of freely moving rats. *Alcohol* **9,** 535–540.

Swain G. M. and Kuwana T. (1991) Electrochemical formation of high surface area carbon fibers. *Anal. Chem.* **63,** 517–519.

Swain G. M. and Kuwana T. (1992) Anodic fracturing and vacuum heat treated annealing of pitch-based carbon fibers. *Anal. Chem.* **64,** 565–568.

Tanaka K. and Kashiwagi N. (1989) In vivo voltammetry with an ultramicroelectrode. *J. Electroanal. Chem.* **275,** 95–98.

Tanelian D. L., Kosek P., Mody I., and MacIver M. B. (1993) The role of the $GABA_A$ receptor/chloride channel complex in anesthesia. *Anesthesiology* **78,** 757–776.

Taqui Khan M. M., and Martell A. E. (1967) Metal ion and metal chelate catalysed oxidation of ascorbic acid by molecular oxygen. I. Cupric and ferric ion catalysed oxidation *J. Am. Chem. Soc.* **89,** 4176–4185.

Taylor A., Jacques P. F., Nadler D., Morrow F., Sulsky S. I., and Shepard D. (1991) Relationship in humans between ascorbic acid consumption and levels of total and reduced ascorbic acid in lens, aqueous humor, and plasma. *Curr. Eye Res.* **10,** 751–759.

Todd R. D. and Bauer P. (1988) Ascorbate modulates 5-[^{3}H]hydroxytryptamine binding to central 5-HT_3 sites in bovine frontal cortex. *J. Neurochem.* **50,** 1505–1512.

Tolbert L C., Morris P. E. Jr., Spollen J. J., and Ashe S. C. (1992) Stereospecific effects of ascorbic acid and analogues on D1 and D2 agonist binding. *Life Sci.* **51,** 921–930.

Ueda C., Tse D. C., and Kuwana T. (1982) Stability of catechol modified carbon electrodes for electrocatalysis of dihydronicotinamide adenine dinucleotide and ascorbic acid. *Anal. Chem.* **54,** 850–856.

Vera J. C., Rivas C. I., Fischbarg J., and Golde D. W. (1993) Mammalian facilitative hexose transporters mediate the transport of dehydroascorbic acid. *Nature* **364,** 79–82.

Vulto A. G., Sharp T., Ungerstedt U., and Versteeg D. H. G. (1988) Rapid postmortem increase in the extracellular dopamine in the rat brain as assessed by brain microdialysis. *J. Neurochem.* **51,** 746–749.

Walaas I. (1981) Biochemical evidence for overlapping neocortical and allocortical glutamate projections to the nucleus accumbens and rostral caudatoputamen in the rat brain. *Neuroscience* **6,** 399–405.

Wang J., Tuzhi P., and Villa V. (1987) Activation of carbon fiber microelectrodes by alternating current electrochemical treatment. *J. Electroanal. Chem.* **234,** 119–131.

Wehmeyer K. R. and Wightman R. M. (1985) Cyclic voltammetry and anodic stripping voltammetry with mercury ultramicroelectrodes. *Anal. Chem.* **57,** 1989–1993.

Wenkstern D., Pfaus J. G., and Fibiger H. C. (1993) Dopamine transmission increases in the nucleus accumbens of male rats during their first exposure to sexually receptive female rats. *Brain Res.* **618,** 41–46.

Wiener H. L., Lajtha A., and Sershen H. (1989) Ascorbic acid inhibits [^{3}H]SCH-23390 binding to striatal dopamine D1 receptors. *J. Recept. Res.* **9,** 331–340.

Wilson J. X. (1989) Ascorbic acid uptake by a high-affinity sodium-dependent mechanism in cultured rat astrocytes *J. Neurochem.* **53,** 1064–1071.

Wilson J. X. and Dixon S. J. (1989) Ascorbic acid transport in mouse and rat astrocytes is reversibly inhibited by furosemide, SITS and DIDS. *Neurochem. Res.* **14,** 1169–1175.

Wilson J. X, Jaworski E. M., and Dixon S. J. (1991) Evidence for electrogenic sodium-dependent ascorbate transport in rat astroglia. *Neurochem. Res.* **16,** 73–78.

Wilson R. L. and Wightman R. M. (1985) Systemic and nigral application of amphetamine both cause an increase in extracellular concentration of ascorbate in the caudate nucleus of the rat. *Brain Res.* **339,** 219–226.

Wilson R. L., Kamata K., Bigelow J. C., Rebec G. V., and Wightman R. M. (1986) Crus cerebri lesions abolish amphetamine-induced ascorbate release in the rat neostriatum. *Brain Res.* **370,** 393–396.

Woodroofe M. N., Sarna G. S., Wadhwa M., Hayes G. M., Loughlin A. J., Tinker A., and Cuzner M. L. (1991) Detection of interleukin-1 and interleukin-6 in adult rat brain, following mechanical injury, by in vivo microdialysis: evidence of a role for microglia in cytokine production. *J. Neuroimmunol.* **33,** 227–236.

Xie Y. M. and Sherwood P. M. A. (1991) X-ray photoelectron spectroscopic studies of carbon fibers. 15. Electrochemical treatment on pitch-based fibers by potentiostatic and galvanostatic methods. *Appl. Spectrosc.* **45,** 1158–1165.

Yamamoto B. K. and Davy S. (1992) Dopaminergic modulation of glutamate release in striatum as measured by microdialysis. *J. Neurochem.* **58,** 1736–1742.

Yount S. E., Kraft M. E., Pierce R. C., Langley P. E., and Rebec G. V. (1991) Acute and long-term amphetamine treatments alter extracellular ascorbate in neostriatum but not nucleus accumbens of freely moving rats. *Life Sci.* **49,** 1237–1244.

Zetterstrom T., Sharp T., Marsden C. A., and Ungerstedt U. (1983) In vivo measurement of dopamine and its metabolites by intracerebral dialysis: changes after d-amphetamine. *J. Neurochem.* **41,** 1769–1773.

Zetterstrom T., Wheeler D. B., Boutelle M. G., and Fillenz M. (1991) Striatal ascorbate and its relationship to dopamine receptor stimulation and motor activity. *Eur. J. Neurosci.* **3,** 940–946.

Voltammetric and Amperometric Probes for Single-Cell Analysis

Andrew G. Ewing, Ta-Kung Chen, and Guangyao Chen

1. Introduction

Chemical analysis of single cells is an area of great interest in the biological and medical sciences. Knowledge of the chemical composition and dynamics of single nerve cells should lead to better models of the cellular neurotransmission process. Information of this type promises to advance our knowledge of neurotransmitter storage, exocytosis, and the physiological effects of external stimuli, such as drugs and toxins. Advances in these areas require the development of analytical methods capable of monitoring neurotransmitter dynamics at the single-cell and subcellular levels.

The considerable interest in studying single-cell chemistry has resulted in the development of a number of analytical techniques. These include enzyme activity measurements (Giacobini, 1968), immunoassay (Giacobini, 1987; Ishikawa et al., 1990), microgel electrophoresis (Matioli and Niewisch, 1965), fluorescence imaging techniques (Tank et al., 1988), microscale ion-selective electrodes (Amman, 1986; Nicholson and Rice, 1988), microcolumn separation techniques (Kennedy et al., 1989; Ewing et al., 1992; Hogan and Yeung, 1992), optical and electron-microscope techniques (Sossin and Scheller, 1989; Betzig et al., 1991), and secondary ion mass spectrometry (Mantus et al., 1991). Although these methods

From: *Neuromethods, Vol. 27: Voltammetric Methods in Brain Systems*
Eds: A. Boulton, G. Baker, and R. N. Adams

have provided valuable information, they have important limitations. Most suffer from either inadequate sensitivity, poor quantitative capabilities, or an inability to monitor chemical dynamics on a time scale similar to the neurotransmission process.

Voltammetric microelectrodes (Wightman, 1981, 1988; Wightman and Wipf, 1988) are ideally suited for monitoring dynamic chemical changes resulting from discrete neurochemical events since they possess rapid response times. Furthermore, many neurochemicals are easily oxidized. To date, voltammetric microelectrodes have been used to monitor dynamics of catecholamine neurotransmitters both inside single-cell cytoplasm and outside single nerve cells (Ewing et al., 1992). The latter experiments have led to the first direct, real-time chemical monitoring of exocytosis and neurotransmitter reverse transport.

Voltammetric microelectrodes used for monitoring single-cell neurochemistry have generally been constructed using carbon as the electrode material and have ranged in tip size from 10 or 50 μm down to 400 nm. These electrodes and pretreatment procedures were discussed in Chapter 1.

This chapter describes the use of voltammetric microelectrodes for analysis of neurotransmitter dynamics in and at single nerve cells. The small size of the single-cell environment requires that extremely small electrodes be used. In addition since many of these systems involve monitoring model systems in culture, the problem of electrode fouling by components in the cellular culture medium must be addressed. Electrode characterization and application to single-cell neurochemical dynamics are discussed.

2. Intracellular Voltammetry in Invertebrates

Direct measurements of neurotransmitter and drug concentrations inside single nerve cells can provide unique information concerning the dynamics of drug action, and about neurotransmitter storage and dynamics. Intracellular measurements of cell membrane potential (Parsons et al., 1983) and inorganic ion concentration (Thomas and Moody, 1980) are possible with ultrasmall glass pipets filled with a saline solution or a liquid ion exchanger, respectively. Voltammetric electrodes small enough to be placed inside single cells while maintaining cell viability have been developed (Meulemans et al., 1986, 1987; Chien et al., 1988, 1990). In this section, the application of voltammetry to measurements in the cytoplasm of single nerve cells will be discussed.

2.1. Intracellular Voltammetry with Polished Carbon-Fiber Electrodes

Electrodes with tip diameters of 0.5–2 µm have been constructed by electropolishing carbon fibers and then sealing the polished tips in glass. These electrodes have been used to monitor metronidazole, a widely used antibiotic, and antipyrine, an antipyretic drug, in the cytoplasm of identified neurons of *Aplysia californica* (Meulemans et al., 1986). Identified neurons (150–300 µm diameter) of the buccal ganglia were used. Differential pulse voltammetry was used to monitor the concentrations of these two drugs and ascorbic acid (AA) as a function of penetration into the cell and the subsequent clearance process. These electrodes were capable of following intracellular drug concentration for periods of 30–40 min.

2.2. Intracellular Voltammetry with Platinum and Platinum-Coated Electrodes

Platinum electrodes with micrometer tip dimensions have also been constructed by electropolishing platinum microwires and sealing the tips in glass as discussed above for carbon. Differential pulse voltammetry has been used to monitor endogenous serotonin (5-hydroxytryptamine, 5-HT) injected into cells in the buccal ganglion of *A. californica* and to monitor free levels of serotonin in single-cell cytoplasm following cell stimulation or extracellular application of L-tryptophan, reserpine, or *p*-chlorophenylalanine (Meulemans et al., 1987). Differential pulse voltammograms obtained in an *Aplysia* neuron are shown in Fig. 1. Voltammograms in the background buffer and in the cytoplasm of single cholinergic

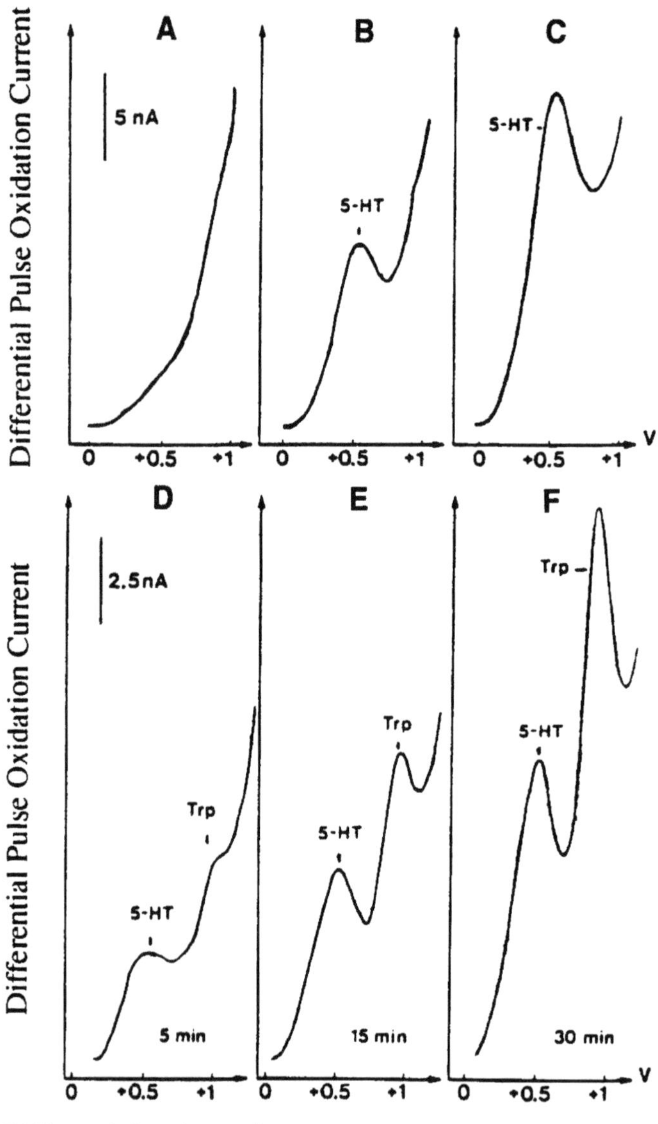

Fig. 1. Differential pulse voltammograms from artificial sea water (ASW) **(A)** and from two living metacerebral cells **(B–C and D–F)**. In the first experiment (B, C), the endogenous peak at +530 mV was increased by ionophoretic intracellular injection of 5-HT from 0.76–1.62 m*M*. In the second experiment (D–F), the cerebral ganglion was superfused by a 5 m*M* solution of L-tryptophan (Trp) in ASW. With time, L-tryptophan accumulated inside the neuron (peak at +1000 mV) and induced a synthesis of 5-HT as manifested by the increase of the peak at +530 mV. The intracellular concentrations of 5-HT and tryptophan were 0.20 and 0.09 m*M* at 5 min, 0.44 and 0.41 m*M* at 15 min, 0.74 and 1.06 m*M* at 30 min, respectively (with permission from Meulemans et al., 1987).

neurons showed no peak for serotonin, whereas a distinct serotonin peak is observed in the metacerebral neuron. This was confirmed by injecting serotonin into the cell body of this neuron and observing an increase in the voltammetric peak. The increase in serotonin persisted for approx 10 min, and the endogenous level of serotonin measurements agrees with estimates of cytoplasmic serotonin in this cell obtained with fluorescent (McCaman et al., 1973) and radioenzymatic (Brownstein et al., 1974) assays.

Intracellular voltammetry has been used in the metacerebral neuron to investigate the effect of cell stimulation on intracellular serotonin levels (Meulemans et al., 1987). Figure 2 shows the levels of serotonin recorded with differential pulse voltammetry during stimulation by direct intracellular depolarization of the soma and by application of extracellular potassium chloride. It was speculated that cell stimulation likely leads to increased serotonin synthesis, resulting in the increased cytoplasmic levels. If this is true in terminals also, this might be very important in the regulation of transmitter availability and synaptic efficacy.

2.3. Intracellular Voltammetry with Carbon-Ring Electrodes

Electrodes with tip diameters as small as 1 µm have been fabricated using hydrocarbon pyrolysis inside silica microcapillaries (Kim et al., 1986; Chien et al., 1988). Methane is pyrolyzed inside silica microcapillaries at temperatures below the silica melting point, leaving a conductive film of pyrolytic carbon inside the capillary tip. Cleaving the epoxy-filled tip with a scalpel exposes a ring of carbon.

The intracellular response of ultrasmall carbon and platinum ring electrodes has been investigated (Lau et al., 1991; Chen et al., 1992). Adsorption of high-mol-wt species on the electrode surface appears to result in deterioration of the amperometric signal. An analytical method has been developed to allow quantitative voltammetry to be obtained (Lau et al., 1991). This method assumes that there is a linear dependence between the degree of electrode fouling of carbon-

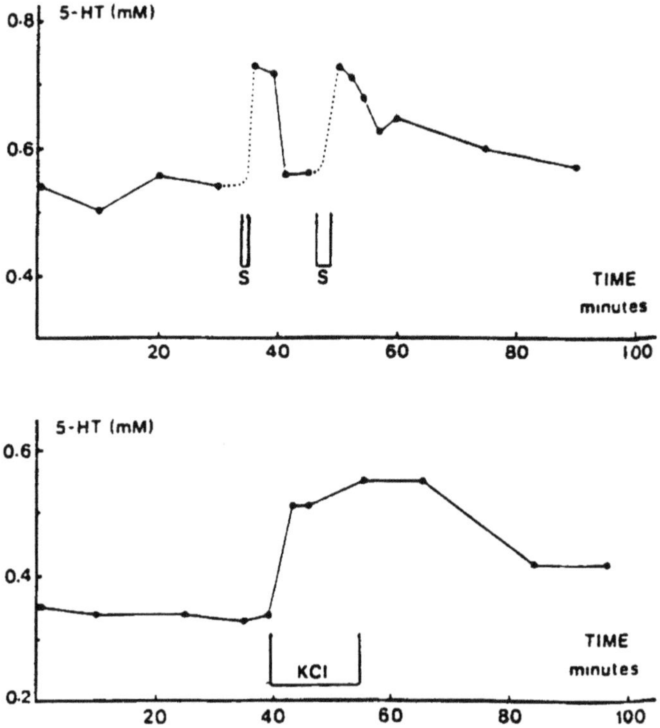

Fig. 2. Increase of intracellular 5-HT concentration owing to stimulation. The two metacerebral cells of one ganglion were successively impaled with the same platinum microelectrode. **(Top)** The soma was directly stimulated by an intracellular 3*M* KCl microelectrode (S: 3 Hz for 70 and 150 s). **(Bottom)** The cell was depolarized from –51 to –27 mV by bath application of 100 m*M* KCl for 15 min (with permission from Meuelmans et al., 1987).

ring electrodes and the number of scans taken in the neuronal microenvironment during intracellular voltammetry.

The left pedal ganglion of the pond snail *Planorbis corneus* contains a single large dopamine (DA) neuron that closely resembles mammalian neurons in several ways (Osborne et al., 1975). It displays nomifensine-sensitive DA uptake, reserpine-sensitive vesicular storage of DA, and metabolizes DA to 3,4-dihydroxyphenylacetic acid (DOPAC). However, unlike mammalian DA neurons, it is approx 200 μm in diameter facilitating single-cell analysis. In addition, this cell lacks AA, allowing the unambiguous electrochemical detection of DA.

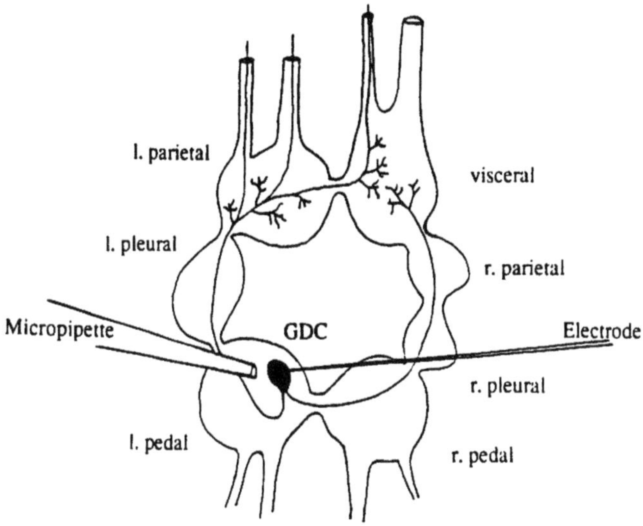

Fig. 3. Schematic of the brain of *P. corneus* showing electrode placement and microinjector relative to the giant DA cell (GDC) (with permission from Chien et al., 1988).

Carbon-ring electrodes have been used to monitor DA levels in the cytoplasm of the large DA cell located on the left pedal ganglion of the pond snail, *P. corneus* (Chien et al., 1988, 1990; Lau et al., 1991). A schematic of the system employed is shown in Fig. 3. In these experiments, endogenous levels of DA were below the detection limit of the electrodes (approx 3 μM), which is considerably below the cytoplasmic levels of serotonin observed in the metacerebral neuron of *Aplysia* (Meulemans et al., 1987). Cytoplasmic levels of DA could, however, be increased to detectable levels by external application of ethanol or exogenous DA. Figure 4 shows the electrochemical response of an electrode placed in the cytoplasm of the DA cell following external application of DA before and after nomifensine. Nomifensine, a competitive inhibitor of the DA transporter, clearly reduces the amount of DA entering the cell in this experiment.

In addition to monitoring intracellular DA, carbon-ring electrodes have been coated with platinum and platinum with glucose oxidase to monitor intracellular oxygen levels (Lau

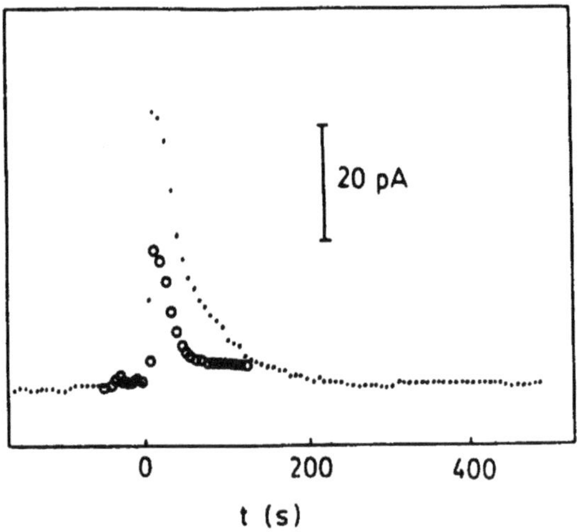

Fig. 4. Effect of 700 µL of 0.6 m*M* nomifensine on repeated DA bathings (50 µL; 0.5 m*M*): before (●) and after (○) nomifensine application. The oxidation current was monitored at 0.78 V vs a sodium-saturated calomel reference electrode with 7 s between measurements (with permission from Chien et al., 1990).

et al., 1992) and intracellular glucose transients (Abe et al., 1991, 1992), respectively. The intracellular level of oxygen was estimated at 32 µ*M* in the identified DA neuron of *P. corneus*. Changes in intracellular glucose concentration on the order of 0.8 m*M* have been observed following injection of exogenous glucose into the cell.

3. Extracellular Voltammetry at Single Cells

Direct measurements of neurotransmitters outside single neurons can provide unique information about dynamic processes, including neurotransmitter transport and exocytosis. This section deals with electrochemical measurements to monitor catecholamine dynamics outside single cells. Both operation of the DA transporter in reverse (reverse transport) and exocytosis of catecholamines have been observed with this methodology.

3.1. Reverse Transport of Dopamine from the Cell Body of Panorbis corneus

Voltammetry can be used to monitor DA levels both inside and outside single nerve cells. As shown above, intracellular voltammetry can be used to monitor DA transport into the cell body of the identified DA neuron of *P. corneus*. Likewise, a carbon-ring electrode placed immediately outside the cell body of this neuron can be used to monitor the same transporter operating in reverse. Injection of DA into the cytoplasm of the DA neuron leads to a rapid release of DA through a nomifensine-sensitive transporter (Fig. 5). Since nomifensine blocks substrate flux across the plasma membrane via the DA transporter, this suggests that most DA release following intracellular injection occurs via reverse transport.

Measurement of reverse transport has been used to study the mechanism of action of amphetamine. The actions of amphetamine at the cellular level have been controversial and have included mechanisms involving blocking catecholamine transport across the plasma membrane and also involving vesicular storage of catecholamine. Injection of amphetamine directly into the cytoplasm of the identified DA neuron on *P. corneus*, which bypasses the uptake transporter, causes DA release (Fig. 6). In these experiments, carbon-ring electrodes were used to measure DA concentrations immediately outside the cell body following an injection of 8 pL of 100 m*M* D-amphetamine into the cell. Based on an estimated cell volume of 8 nL, this resulted in an intracellular level of 100 μ*M*. Extracellular DA increased to a maximal level 14 s following the injection. The shape of the voltammogram obtained at the peak current value closely matched the voltammogram of a DA standard and showed that extracellular DA reached about 1.6 μ*M* following the amphetamine injection. Combined with evidence for specific uptake of amphetamine (Zaczek et al., 1991) and the ability of amphetamine to release catecholamines from isolated vesicles (Sulzer and Rayport, 1990), these observations indicate that, at least in part, amphetamine releases DA via an intracellular action.

3.2. Monitoring Single Exocytotic Events

Secretion of catecholamines from single exocytotic events has been monitored by use of carbon-fiber electrodes placed against the outside of the plasma membrane of bovine

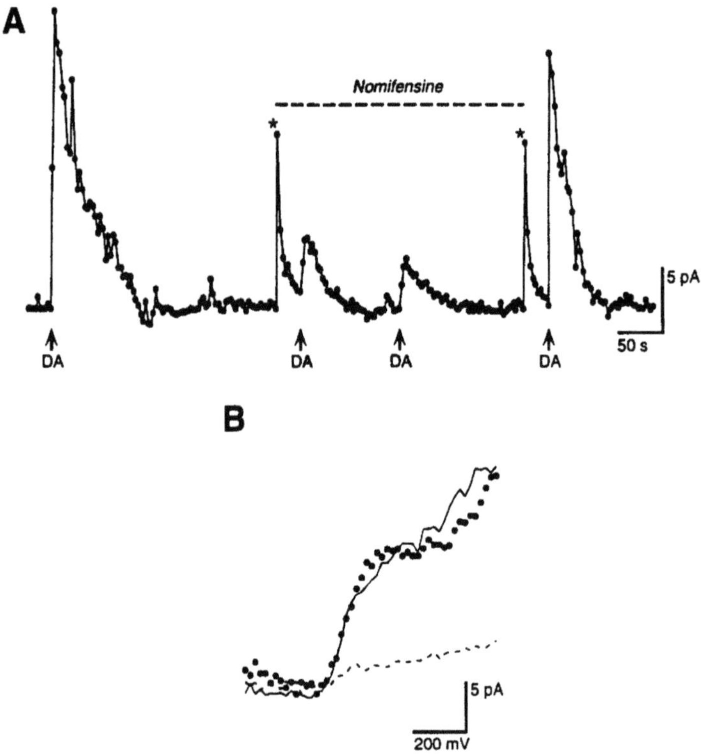

Fig. 5. Intracellular injection of DA induces reverse transport. **(A)** Intracellular injection of 4 pL of 0.5 mM DA (arrows) induced reliable increases in extracellular DA release. During extracellular perfusion with 10 μ*M* nomifensine (dashed line), the same DA injections were markedly attenuated. Perfusion of nomifensine-containing medium and its replacement by control medium produced current spikes (asterisks). Data have been baseline corrected by subtracting the curve fit to the background signal. **(B)** Background subtracted voltammograms (–0.2 to 0.8 V) for the first (—) and second (- -), and fourth (···) intracellular DA injections shown in (A). The voltammograms monitored at the face of the neuron were nearly identical to DA. All three voltammograms were sigmoidal with a half-wave potential of approx 0.2 V (with permission from Sulzer et al., 1995).

adrenal medullary chromaffin cells in primary culture (Leszczyszyn et al., 1990, 1991; Wightman et al., 1991; Chow et al., 1992). These cells were stimulated with nicotine via a microinjection pipet and a pressure injection system, while

A

B

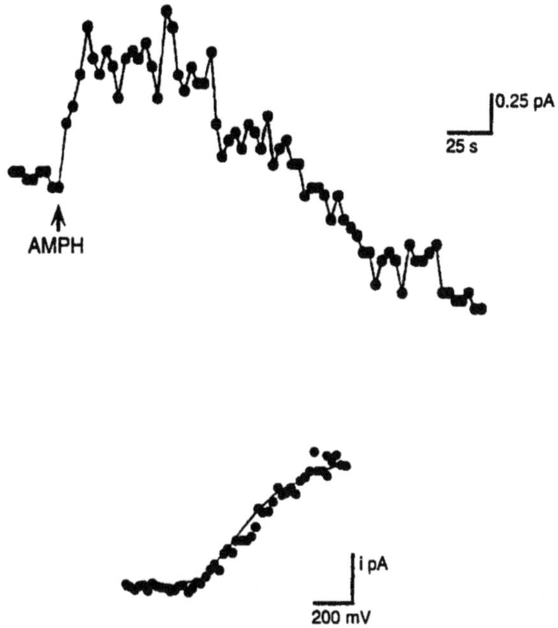

Fig. 6. **(A)** AMPH (8 pL of 100 m*M*) was injected intracellularly (arrow). The increase in extracellular DA was measured by oxidation current with a 2-µm carbon-ring ultramicroelectrode. The electrode was touching the GDC membrane and held at a steady potential of 0.8 V. **(B)** The shape of the voltammogram (–0.2 to 0.8 V) taken at the peak current value 14 s postinjection (filled circles) closely matched the voltammogram for a DA standard (solid line). Calibration: i = 0.5 for the experiment and 31.5 for the standard (with permission from Sulzer et al., 1995).

simultaneously monitoring catecholamines with the small electrode (Fig. 7). The response to this stimulation was monitored with voltammetry repeated every 100 ms (Fig. 8). When norepinephrine (NE) is ejected from the micropipet as a control, the current at the electrode increases momentarily and then returns to baseline over a period of 10–20 s (Fig. 8C). However, nicotine stimulation results in numerous current transients that are in fact chemical spikes from the cell (Fig. 8D). Voltammograms obtained for authentic NE are similar to those corresponding to the current transients observed

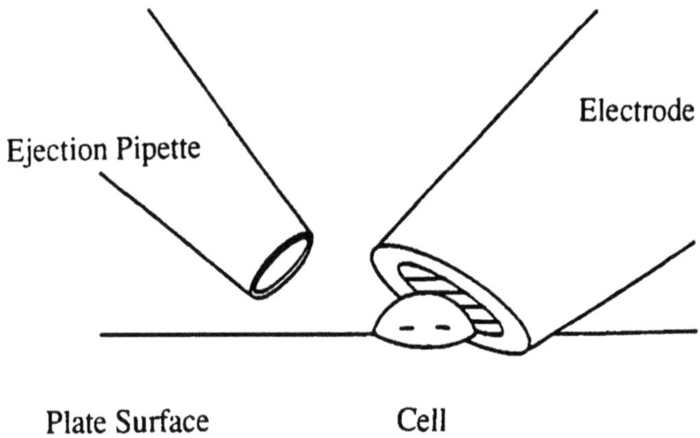

Fig. 7. Experimental arrangement for measuring secretion from single adrenal medullary chromaffin cells. Drawing is approximately to scale (with permission from Leszczyszyn et al., 1990).

(Fig. 8A,B). The current rise under the transients observed in these preliminary experiments might be the result of an overall increase in the extracellular concentration of catecholamine, or it might be an artifact of the nicotine ejection. Later experiments with lower nicotine concentrations have shown much smaller changes in the baseline current (Wightman et al., 1991). Initial estimates of the number of catecholamine molecules in each current transient have assumed that all the released catecholamine is oxidized. Using Faraday's law to calculate the amount of catecholamine released, 5–10 amol of catecholamine is released in each exocytotic event (Leszczyszyn et al., 1990, 1991; Wightman et al., 1991).

The effect of different secretagogs on the release of catecholamines from adrenal cells has been examined (Wightman et al., 1991). Ejection of nanoliter volumes of 100 μ*M* nicotine or 1 m*M* carbamoylcholine leads to release of catecholamines characterized by a broad envelope under the transients with a concentration >10 μ*M* (Wightman et al., 1991). When cells are stimulated with only 10 μ*M* nicotine or 60 m*M* KCl, this envelope is significantly less. Following 10 μ*M* nicotine or 60 m*M* KCl, the envelope is 23 or 22%, respectively, of that following 100 μ*M* nicotine. In addition, the frequency of exocytotic events decreases from 1.2 ± 0.2 Hz to

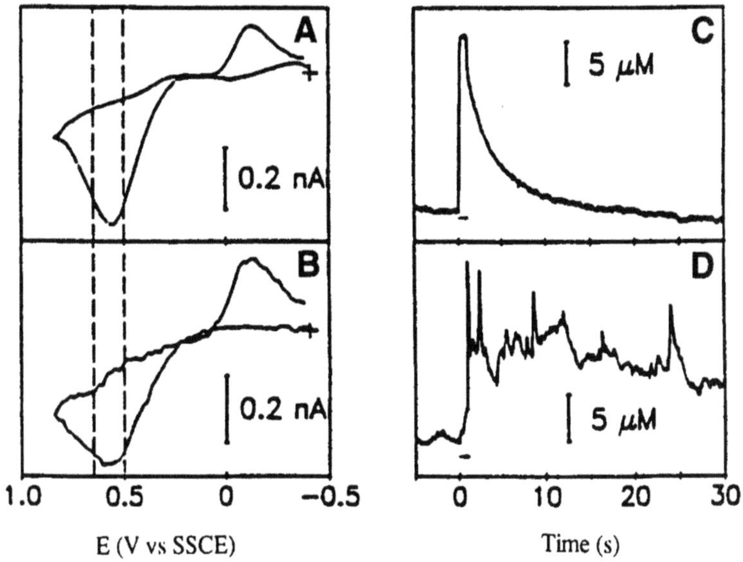

Fig. 8. Cyclic voltammetric response (200 V/s, repeated at 100-ms intervals) of a carbon-fiber electrode to NE ejection from an adjacent micropipet and to catecholamine secretion from a single chromaffin cell. **(A,B)** Averaged background-subtracted voltammograms of the substances whose concentration changed during the measurement interval of **(C)** and **(D)**. Each time-point in (C) and (D) is the integrated current recorded from 0.5–0.6 V from individual voltammograms (hatched lines in [A] and [B]); bars to the right in (C) and (D) are the conversion of current to catecholamine concentration based on calibration curves constructed with standards. (A, C) Electrode response to 1-s 3-nL ejection of 20 μ*M* NE applied at $t = 0$ with the ejection pipet 20 μm from the electrode. (B, D) Electrochemical response obtained with the electrode tip adjacent to a single-cell; at $t = 0$, a 1-s ejection of nicotine (100 μ*M*) was made 20 μm away from the cell (with permission from Leszczyszyn et al., 1990).

0.5 ± 0.2 Hz and 0.6 ± 0.2 Hz, respectively. Histograms of the areas of the electrochemical responses shown, however, are similar for all the secretagogs (Fig. 9). It is interesting to note that the distribution of spike areas is clearly skewed for all the secretagogs, suggesting that a distribution of granule content is present in each cell.

Adrenal medullary chromaffin cells in culture present an excellent model system to study neurosecretion (Trifaro

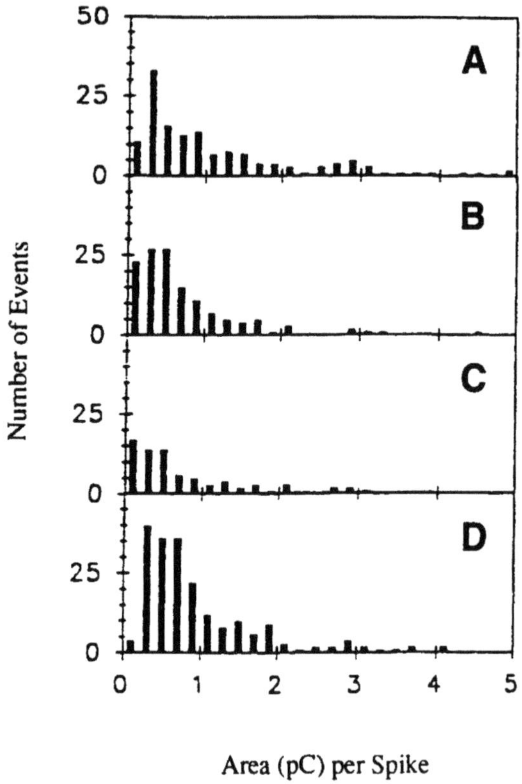

Fig. 9. Frequency distribution of individual spike areas determined for various secretagogs. The spikes were measured over a 30-s segment beginning at the half-rise time. The bins are divided into 0.2-pC intervals. **(A)** Spike distribution during exposure to ten 3-s, 60 m*M* K^+ exposures (136 spikes; the balanced salt solution in the ejection pipet contained 90 m*M* Na^+ and 60 m*M* K^+). **(B)** Spike distribution during seven exposures to 1 m*M* carbamoylcholine simulations (137 spikes). **(C)** Spike distribution during 10 exposures to 10 μ*M* nicotine (80 spikes). **(D)** Spike distribution during 10 exposures to 100 μ*M* nicotine (211 spikes) (with permission from Wightman et al., 1991).

and Lee, 1980). However, there are clearly at least two subpopulations of cells that secrete either NE or epinephrine. In fact, microcolumn liquid chromatography of single adrenal cells has shown that these cells can contain either predominantly NE, epinephrine, or a high level of both (Cooper et al., 1992). Despite their structural similarity, voltammetry at

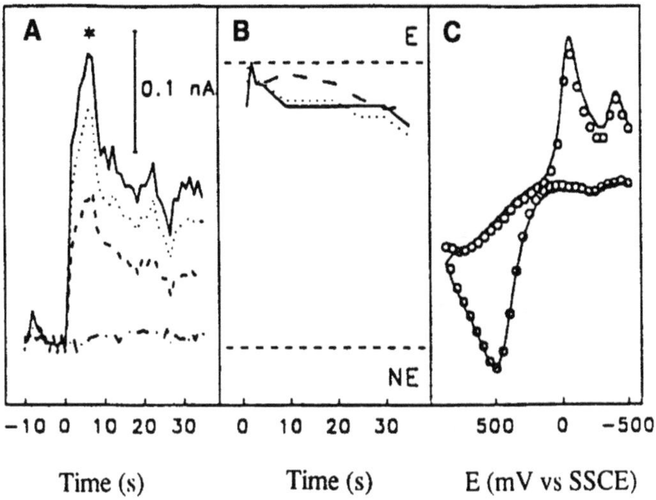

Fig. 10. Secretion of epinephrine from an individual adrenal cell. **(A)** Current versus time profile of single-cell response to 100 μ*M* nicotine introduced at 0 s: oxidative current (solid line), reductive current (dotted line), adrenochrome current (dashed line), current at 0 V (dashed-dotted line). **(B)** E_p (solid line), ΔE_p (dotted line), and i_{cyc}/i_{ox} (dashed line) vs time. The scale for each of these parameters is adjusted so that the values will fall between the limits of pure epinephrine (upper dashed line) and pure NE (lower dashed line) as determined by the postcalibrations. **(C)** Background-subtracted cyclic voltammogram (solid line) taken at the asterisk in (A). Circles represent the background-subtracted cyclic voltammogram of the epinephrine postcalibration. Voltammograms have been normalized to the oxidative peak currents (with permission from Ciolkowski et al., 1992).

microelectrodes can be used to discriminate between NE and epinephrine by monitoring the rate of internal cyclization following oxidation (Hawley et al., 1967). This cyclization reaction leads to the formation of an adrenochrome, which can be detected as an additional voltammetric wave on the negative-going scan of the cyclic voltammogram (Ciolkowski et al., 1992). This is shown in Figs. 10 and 11 for current traces obtained by voltammetry at two different adrenal cells with epinephrine released from one cell (Fig. 10) and NE released from the other cell (Fig. 11). In these experiments, approx 75% of the cells examined released only epinephrine or NE,

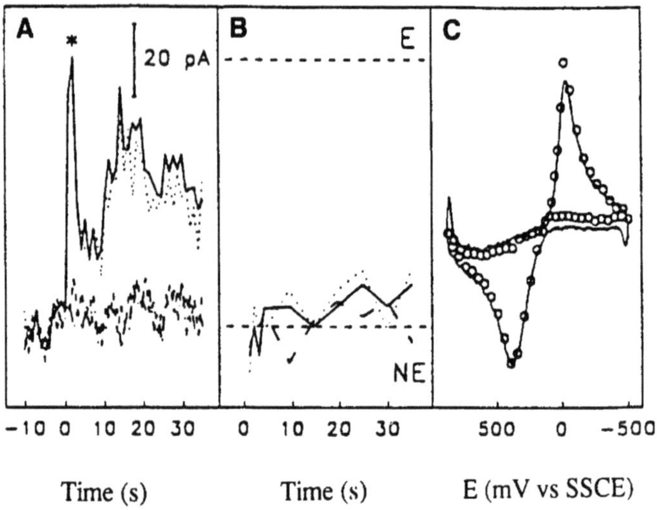

Fig. 11. Secretion of NE from an individual adrenal cell. The parameters for **(A)** and **(B)** are the same as those for Fig. 10. **(C)** is the same as Fig. 10, except circles represent the background-subtracted cyclic voltammogram of the NE postcalibration (with permission from Ciolkowski et al., 1992).

whereas 25% of the cells released both catecholamines in response to a 100-μ*M* nicotine stimulus. It appears that these cells release catecholamine in the same proportions that they store catecholamines.

Single exocytotic events can also be monitored by voltammetry at adrenal cells following permeabilization with digitonin (Jankowski et al., 1992). Digitonin is commonly used to permeabilize adrenal cells to introduce reagents into the cell (Dunn and Holz, 1983). During digitonin-induced permeabilization, catecholamine secretion can be induced by low levels of calcium ion in the medium surrounding the cells. Voltammetry has been used to monitor the exocytosis of single chromaffin granules from digitonin-permeabilized cells. Strong evidence is given that catecholamine release following digitonin is via exocytosis. This release is calcium-dependent and reaches a maximum at calcium concentrations at or above 19 μ*M*. Finally, exocytotic release can be distinguished from nonexocytotic release by analysis of the shape of the current transients observed at the electrode.

3.3. Delay in Vesicle Fusion Observed by Amperometry

Whole-cell patch-clamping has been used to examine the delay between adrenal cell stimulation and exocytosis (Chow et al., 1992). Exocytosis is thought to occur on a submillisecond time scale based on measurements of postsynaptic membrane potential following stimulation (Llinas et al., 1981; Edwards et al., 1990). In nonsynaptic cells, the absence of a postsynaptic neuron prevents this type of measurement. Amperometric measurements with microelectrodes provide a means to monitor single exocytotic events directly. When cells are depolarized for 25 ms, exocytosis is observed following highly varied delay times (Fig. 12). A histogram of the areas of the amperometric responses and the delay from the initiation of the depolarization to the observed release are shown in Fig. 13. Cells were held at –60 mV and depolarized at +10 mV for 25 ms to activate calcium channels. The average spike contained 2×10^6 catecholamine molecules, although this again follows a skewed distribution similar to that shown above for release following exposure to a secretagog. The latency period between depolarization and release is usually <100 ms, and events after this time are approximately constant. Since the exocytosis process in adrenal cells is slow following depolarization (5–100 ms), it is likely that calcium or membrane depolarization triggers a cascade of events that manifests itself later as the exocytosis process. The rationale for the wide variability of the latency times is not yet clear.

In addition to the latency time between depolarization and exocytosis, evidence has also been obtained by amperometry to demonstrate the existence of a fusion pore during exocytosis (Chow et al., 1992). This could represent a slow leakage of catecholamines through a small opening at the initiation of the exocytotic event via a structure similar to a gap junction (Thomas et al., 1988). Figure 13A shows three current transients observed at the electrode following cell depolarization with two of these showing a small rise in current prior to the larger spike. Two-thirds of all current transients observed are preceded by a prerise or "foot." Histograms for

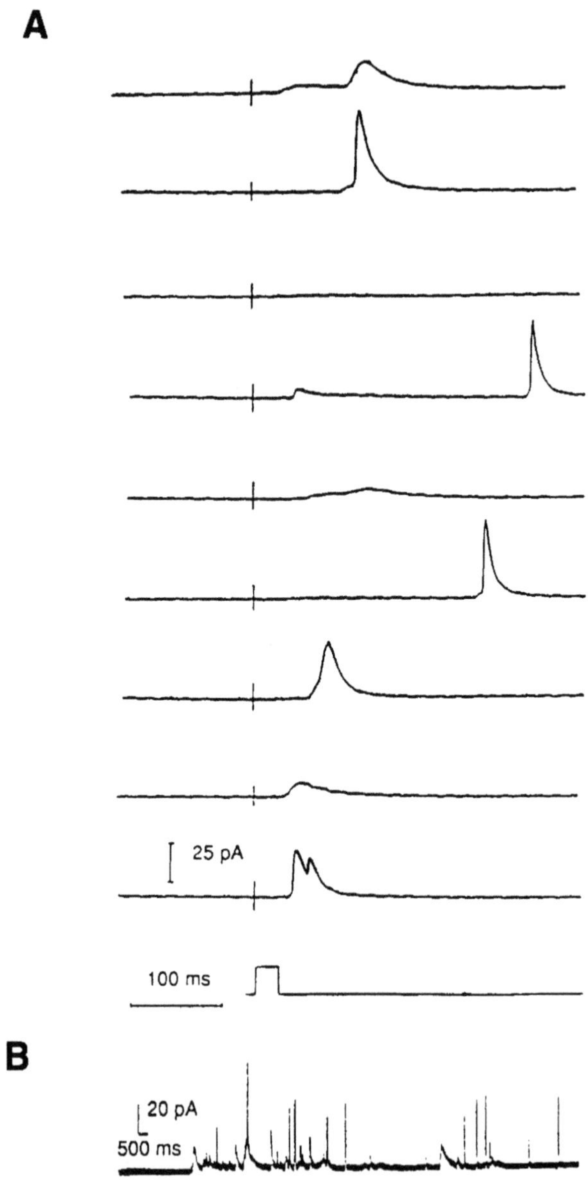

Fig. 12. Electrochemical detection of secretion from single vesicles. **(A)** Examples of signals recorded by the amperometric constant-voltage method. The voltage protocol is shown in the bottom trace, and representative amperometric current signals are shown above, aligned in time relative to the stimulating pulse. The signals vary in amplitude and duration, and occasionally there are "failures" (sweeps without a response). **(B)** Response to steady-state depolarization to 0 mV after the recording of (A) (note difference in time scales) (with permission from Chow et al., 1992).

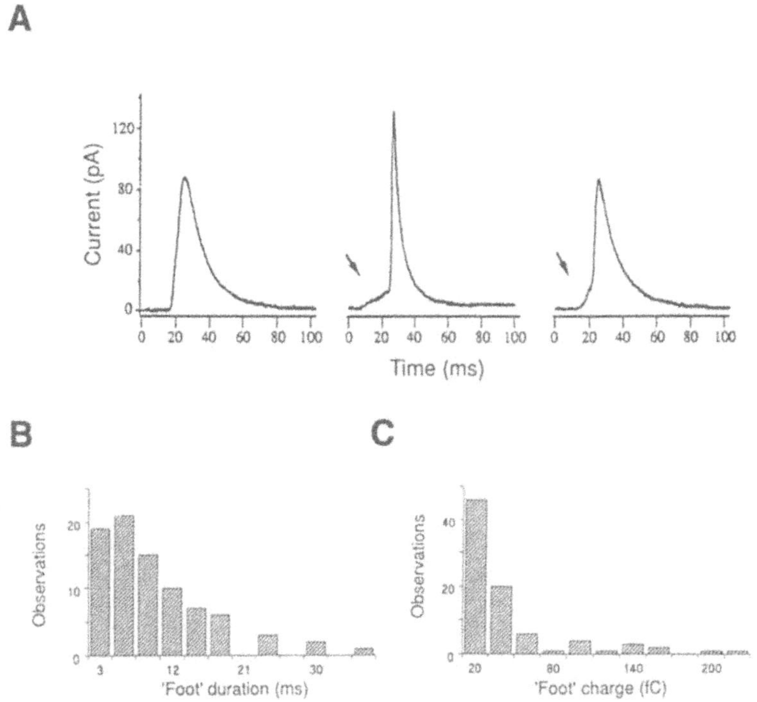

Fig. 13. Rising phase of amperometric current transients cannot be explained by instantaneous release of secretory vesicle contents. **(A)** Single amperometric current transients at high resolution. Note that the fast rising phase in two of the three examples is preceded by a small pedestal or "foot" (arrows). Such "foot" signals precede most of the transients with a fast rise time. The "foot" signals were typically much smaller than the average signal. **(B)** Histogram of duration of "foot" signals. Data from three cells are pooled, where 84 large (>20 pA) and fast (rise time <3 ms) signals were counted. Seventy-five of those were preceded by a discernible "foot." The number of occurrences is plotted against the duration of the "foot," as defined in (A). The mean of "foot" durations was 8.26 ms. **(C)** Histogram of "foot" charges. The same "foot" signals as those of (B) were integrated to calculate the charge involved. Mean charge: 34 fC or 1.05×10^5 molecules. The mean amplitude of the "foot" signal was 7.17 ± 5.28 pA (mean ± SD) (with permission from Chow et al., 1992).

the duration of this "foot" of the transients are shown in Fig. 13B,C. This is the first direct, real-time evidence for the existence of a fusion pore during exocytosis.

3.4. Temporal Characteristics of Exocytosis at Adrenal Cells

Catecholamine granules in adrenal medullary chromaffin cells are known to contain AA, ATP, and the protein chromogranin A in addition to catecholamine (Winkler and Westhead, 1980; O'Conner and Frigon, 1984). Dissociation of the intravesicular matrix of these cells has been described as a pH-dependent conformational change of the chromogranin A, which binds the catecholamines in an intravesicular matrix (Yoo and Lewis, 1992). Both amperometry and fast-scan cyclic voltammetry have been used to examine the diffusional profile of catecholamine released from individual exocytotic events as a function of distance between the electrode and the cell and of extracellular pH (Schroeder et al., 1992; Jankowski et al., 1993). Taken together, the data appear to indicate that catecholamine in adrenal cell granules is bound to chromogranin A and that this chromogranin A is bound to the vesicular membrane at the lower pH that exists in the granule prior to fusion with the plasma membrane and exposure to the pH 7.5 external medium. Thus, a conformational change of chromogranin A during exocytosis appears to be an important step defining the time-course of catecholamine release.

3.5. Application of Amperometry to Monitoring the Effects of Autoreceptors

The release of catecholamines to the extracellular synaptic cleft can be regulated by several factors, including presynaptic α_2-autoreceptor activation (Starke, 1981). The released extracellular catecholamines act on presynaptic α_2-autoreceptors to inhibit further catecholamine release (Starke, 1977; Westfall, 1977). Agonists of α_2-autoreceptors, such as clonidine or guanfacine, cause suppression of locus cell firing and decrease the release of catecholamines (Pelayo et al., 1980; Jackisch et al., 1992), whereas α_2-autoreceptor antagonists, such as piperoxan, yohimbine, and idazoxan, reverse these inhibitory effects and increase release (Dennis et al., 1987; Abercrombie et al., 1988).

Although some experiments have provided evidence about presynaptic mechanisms, it is not always possible to determine if the drugs affect the presynaptic terminals directly, or whether they act primarily on nearby cells that then emit a second signal to the presynaptic terminals. This is particularly difficult to decide in tissue studies. Basically, there are two methods to study the interaction of neurotransmitters or drugs with cells. The first procedure is to determine the biological response of an intact isolated organ, such as the guinea pig ileum, to applied agonists or antagonists. The disadvantage to this procedure lies in the difficulty in discriminating between a large number of processes taking place before the drug interacts with the receptor. The second approach to studying receptors is to measure ligand binding to a homogenate or slice preparation. This technique became feasible with the development of ligands of a high-specific radioactivity and a high affinity for the receptor. Neither of these techniques eliminates postsynaptic effects. The use of amperometric monitoring of catecholamine exocytosis, combined with the application of receptor agonists or antagonists is a new method that can functionally identify autoreceptors directly since there are no postsynaptic cells. The electrochemical measurements described here are spatially resolved to the single-cell level because of the size of the microelectrode sensor as well as the small volume of the applied chemical stimulus.

To demonstrate the effect of autoreceptors on exocytosis from adrenal cells, a protocol involving repeated nicotine stimulations of the same cell has been developed (Zhou et al., 1994). Individual cells were repeatedly stimulated with 3-s exposures to 100 μM nicotine at intervals of 100–400 s. The current responses for neurohormone secretion were integrated to obtain the total charge for all the current transient responses over a 200-s time window. For repetitive stimulations, the ratio of total secretion of the second to the first exposure is 0.32 ± 0.05 ($n = 7$), and the ratio of the third to the second exposure is 0.80 ± 0.02 ($n = 4$).

The effects of autoreceptors on single adrenal medullary chromaffin cells were examined by 5-s pressure applica-

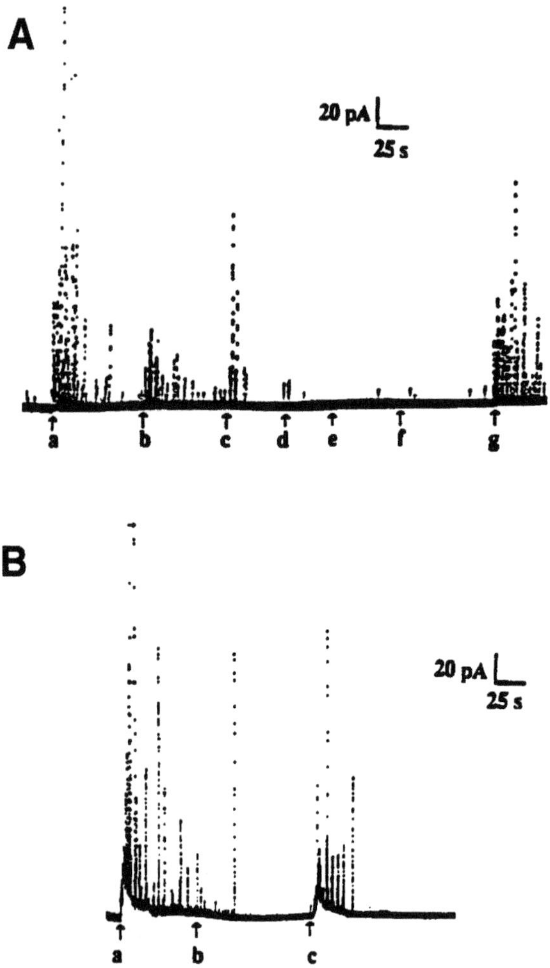

Fig. 14. Effect of piperoxan (20 μ*M*) on nicotine induced exocytosis of catecholamines. **(A)** a, g: 3-s application of 100 μ*M* nicotine, b, c, d, e, f: 5-s application of 20 μ*M* piperoxan. **(B)** a, c: 3-s application of 100 μ*M* nicotine. b: 5-s application of balanced salt solution. Electrode potential: 650 mV vs SSCE (with permission from Zhou et al., 1994).

tions of 20 μ*M* piperoxan approx 80 s after a 3-s exposure to nicotine. Figure 14 shows the response to piperoxan compared to a control experiment. To evaluate the effect of piperoxan on catecholamine exocytosis, the integrated current (large) for all the spikes (release events) following nicotine and piperoxan together has been compared to the total charge for release events after only nicotine. A 15% ($n = 3$) increase in catecholamine exocytosis is observed after the nicotine and piperoxan compared to a 46% ($n = 6$) decrease in release for the control experiment (identical protocol without application of piperoxan). These data strongly suggest that a negative feedback mechanism exists. Exposure to piperoxan enhances the release of catecholamines from nicotine-stimulated cells; however, piperoxan alone has no effect on release from cells not exposed to nicotine.

Perhaps the most widely known sympathomimetic drug is amphetamine. The actions of amphetamine appear to be manifested as an increased level of extracellular catecholamine in regions of the central nervous system (Zetterström et al., 1983; Hernandez et al., 1987; Sharp et al., 1987). A great deal of work has shown that amphetamine enhances release of catecholamines (Kuczenski, 1983), inhibits the catecholamine transporter (Wise and Bozarth, 1987), and accelerates exchange diffusion across the plasma membrane (Fischer and Cho, 1979; Kuczenski, 1983). In addition, recent evidence suggests that amphetamine acts as a weak base to disrupt catecholamine vesicularization and, therefore, enhances the cytoplasmic level of neurotransmitter for subsequent release via reverse transport (Sulzer and Rayport, 1990). All these actions appear to play a role in the neuronal response to amphetamine.

Amphetamine also appears to affect the exocytosis of catecholamine from adrenal cells (Zhou et al., 1994). A 5-s application of 20 μM amphetamine at approx 20 s after stimulation of an adrenal cell with nicotine results in a dramatic increase in catecholamine release events (Fig. 15). Integrating the total current under all the release events and comparing them to a control experiment reveal that amphetamine results in an 870% increase in exocytosis ($n = 3$). It appears that amphetamine might act as an antagonist of α_2-autoreceptors.

3.6. Amperometric Monitoring of Exocytosis from PC12 Cells

Mammalian neurons are difficult to work with in the laboratory since they stop proliferation once they have

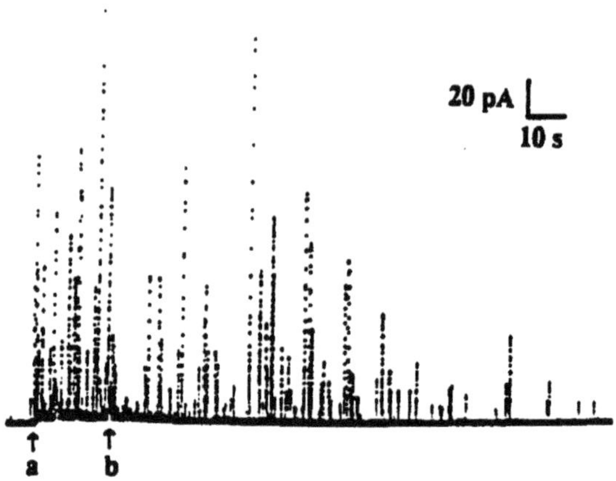

Fig. 15. Amperometric response of a carbon-fiber electrode placed at a single chromaffin cell following a 3-s application of 100 μ*M* nicotine **(A)**, and a 5-s application of 20 μ*M* amphetamine **(B)** (with permission from Zhou et al., 1994).

matured. Greene and Tischler established a nerve-growth-factor-responding clonal line of rat pheochromocytoma (PC12) cells in 1976. This cell line shares many physiological properties with primary cultures of sympathetic ganglion neurons (Green and Tischler, 1976; Schubert et al., 1980), and has been studied as a model for the developing sympathetic nerve (Wagner, 1985). PC12 cells can synthesize, store, and release catecholamines in a manner similar to sympathetic ganglion neurons (Green and Tischler, 1976; Schubert et al., 1980; Wagner, 1985). They are more similar to sympathetic ganglion neurons than to chromaffin cells, because they contain more DA than NE with no detectable level of epinephrine (Clift-O'Grady et al., 1990). Most importantly, PC12 cell vesicles are valid analogs of brain synaptic vesicles for four criteria: size, density, protein composition, and endocytotic origin (Wagner, 1985; Clift-O'Grady et al., 1990).

Carbon-fiber electrodes have been used to monitor single exocytotic events from individual PC12 cells (T. K. Chen et al., 1994). The electrode is placed against the top of a cell in culture, and the cell is stimulated in a manner similar to that described above for adrenal cells with KCl or nicotine. Figure 16 shows

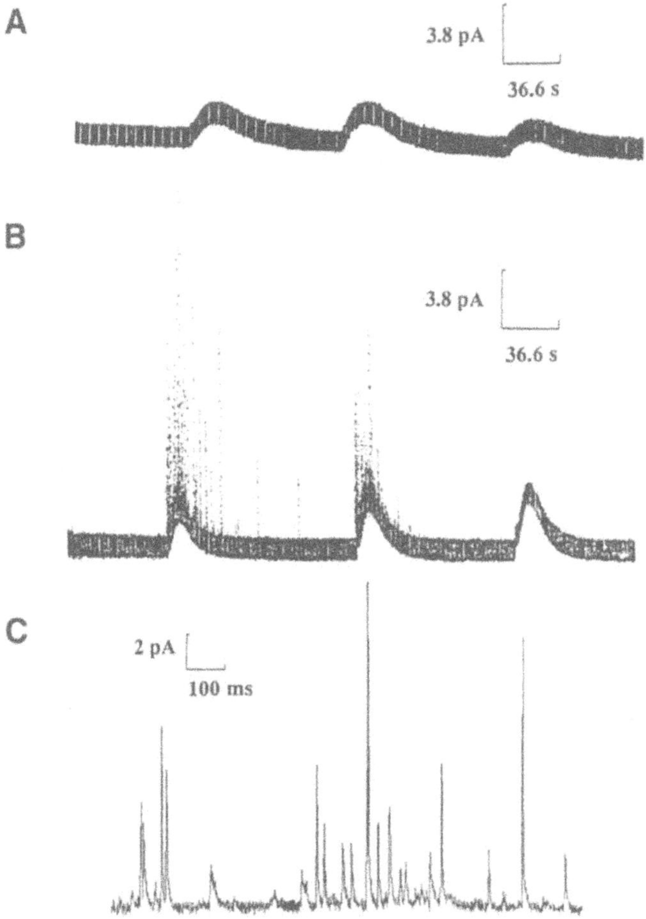

Fig. 16. Amperograms of single vesicle exocytosis from PC12 cells. **(A)** Amperogram of a control experiment with no cell near the tip of the working electrode. **(B)** Amperogram of exocytosis induced by 1 m*M* nicotine in 105 m*M* K^+ balanced salt solution. **(C)** Enlargement of a 1-s period of the amperogram from the first stimulation. Data displayed correspond to the time period near the middle of the first baseline disturbance in (B) (with permission from T. K. Chen et al., 1994).

the amperometric response following three successive stimulations with both 1 m*M* nicotine and 105 m*M* KCl. Stimulated release is not observed from these cells if calcium is omitted from the medium. In addition, preliminary data sug-

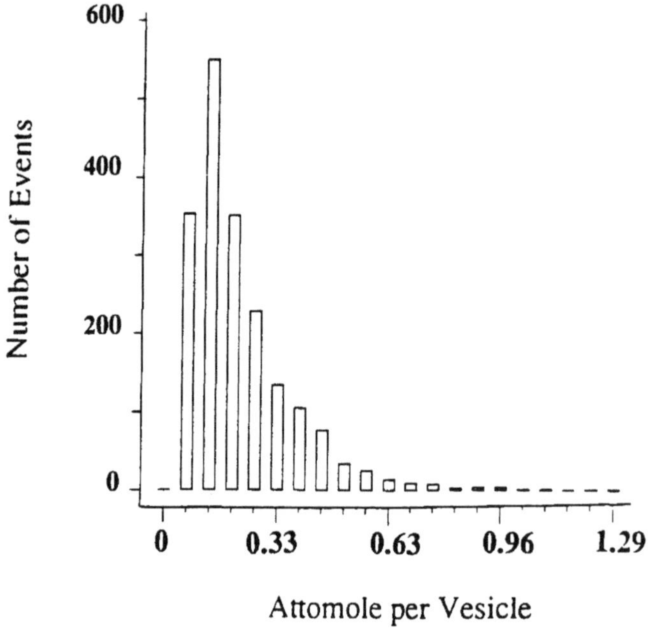

Fig. 17. Frequency distribution of quantal size from PC12 cells. The bins are divided into 66 zmol (40,000 molecules) intervals. Numbers on the x-axis represent the mean of each interval. Data shown here were obtained from the first stimulation of 13 individual cells and 1912 spikes are included (with permission from T. K. Chen et al., 1994).

gest that exposure to nicotine alone results in minimal exocytosis and that is only after a substantial delay. Exposure to elevated potassium chloride causes most of the exocytosis observed. Individual current spikes have an average half-width of 9.3 ± 0.1 ms ($n = 1912$ from 13 cells) and the average catecholamine level calculated with Faraday's law is 190 zmol (114,000 molecules)/vesicle. A histogram of the amount of catecholamine released per vesicle is shown in Fig. 17. This histogram has the same skewed profile observed for adrenal medullary chromaffin cells; however, the average vesicle content is much lower. The size distribution of vesicles is relatively narrow for PC12 cells with 85% of the release events corresponding to <330 zmol. This is an example of the tremedous sensitivity that can be obtained using amperometry to monitor catecholamine exocytosis.

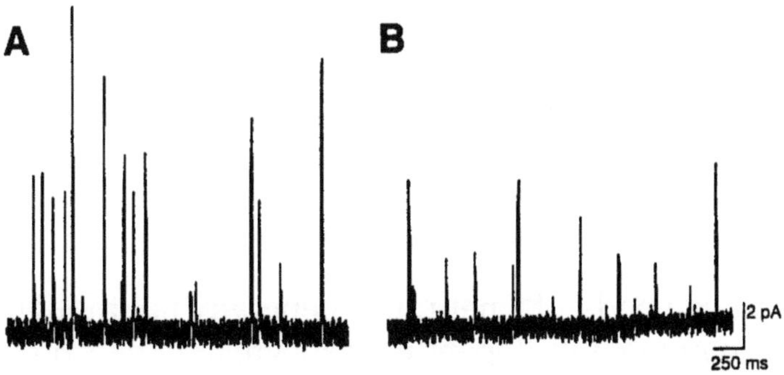

Fig. 18. Amphetamine exposure decreases the quantal amplitude of individual release events from PC12 cells. Representative electrochemical records of quantal release (selected randomly) for a control cell **(A)** and a cell that has been incubated in 10 μ*M* D-amphetamine for 10 min **(B)**. Cells were stimulated by local perfusion (30 nL of 1 m*M* nicotine in 105 m*M* KCl saline) to induce exocytosis. Amphetamine decreased the average quantal amplitude to 48 ± 5.4% of control (with permission from Sulzer et al., 1994).

The PC12 system has also been used to examine the effects of amphetamine (Sulzer et al., submitted). Since one of the proposed mechanisms of amphetamine is the redistribution of catecholamines from synaptic vesicles to the cytoplasm (Sulzer and Rayport, 1990; Sulzer et al., 1993), one can use voltammetric methods to examine the effects of amphetamine on quantal size. To measure the DA content of individual synaptic vesicles, carbon-fiber electrodes can be used to detect quantal packets containing as few as 31,000 molecules. Cells were stimulated with 1 m*M* nicotine in 105 m*M* KCl (30 nL). Incubation in 10 μ*M* amphetamine for 10 min prior to the stimulation reduced the stimulated catecholamine released per exocytotic event to 48 ± 5.4% of control ($n = 7$ cells) (Fig. 18). These results demonstrate that amphetamine can attenuate stimulus-dependent release by reducing quantal size. Quantal analysis studies generally assume that all vesicles contain a constant level of neurotransmitter. Data shown earlier in this chapter show that considerable variability in the size of release events exists. Here it is demonstrated that quantal size can be pharmacologically manipulated. These data combined suggest that both physiological and pharmacological alterations of quantal size might well play an important role in the modulation of catecholamine release.

3.7. Voltammetric Monitoring of Exocytosis from Cell Bodies

The cell body of a neuron is generally not thought to be involved in calcium-dependent exocytosis of neurotransmitters. Exocytosis of neurotransmitters is thought to occur at synaptic terminals. In fact, placement of a carbon-fiber microelectrode against the cell body of the identified DA neuron of *P. corneus* with subsequent stimulation by exposure to KCl results in a massive exocytotic response (Fig. 19). This release is calcium-dependent, and voltammograms obtained during the current transients are nearly identical to those obtained in authentic DA (Fig. 20). A histogram (Fig. 21) of the spike areas again shows the typical skewed shape characteristic of exocytosis of catecholamines shown above for adrenal and PC12 cells. The average vesicle contains 1.2 amol of DA with a broad distribution of sizes that is, perhaps, expected for a primitive neuron. Interestingly, the frequency of exocytotic events observed from the *Planorbis* cell does not diminish rapidly, as was observed for adrenal cells and PC12 cells. Although exocytosis from this cell does appear to show a bursting pattern, once a cell is stimulated, it can undergo rapid exocytosis for minutes to hours. As many as 89,000 release events have been observed from one spot on a cell after a single stimulation.

The DA cell used has been estimated to contain 5 pmol of DA based on immunofluorescence measurements (Powell and Cottrell, 1974). If virtually all the DA is estimated to be bound in vesicles (Chien et al., 1990), this indicates that there must be approx 4×10^6 vesicles in one cell (based on an average 1.2 amol of DA/vesicle). Hence 89,000 events represents only 2.1% of the total catecholamine in the cell. This represents the first direct measurement of exocytosis of neurotransmitter from the cell body of a processed neuron.

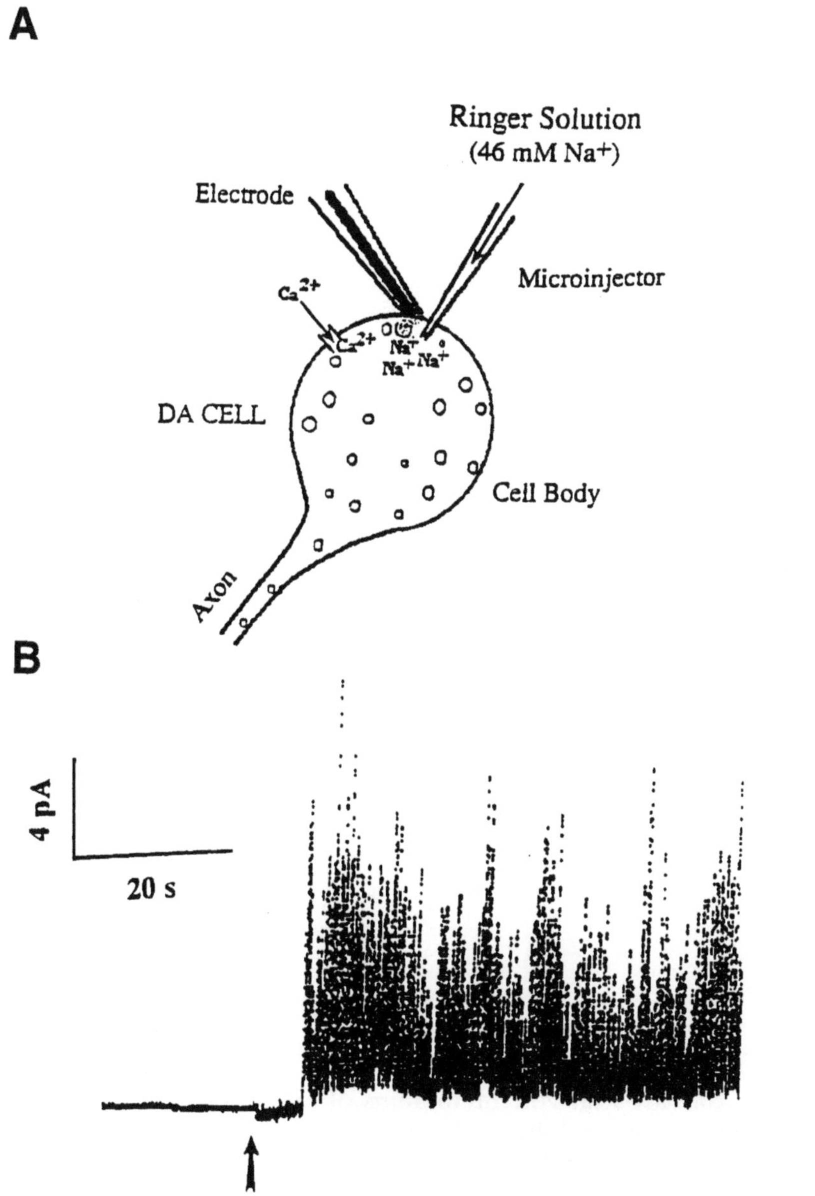

Fig. 19. Quantal release detected from the cell body of the DA neuron of *P. corneus.* **(A)** Schematic of the cell body, an intracellular pipet for injecting Ringer solution to the cell and a carbon-fiber electrode placed close to the penetrating site. **(B)** Typical response observed following intracellular stimulation with 30 pL of Ringer solution injected at the arrow. Transient quantal events were elicited about 5 s after injection.

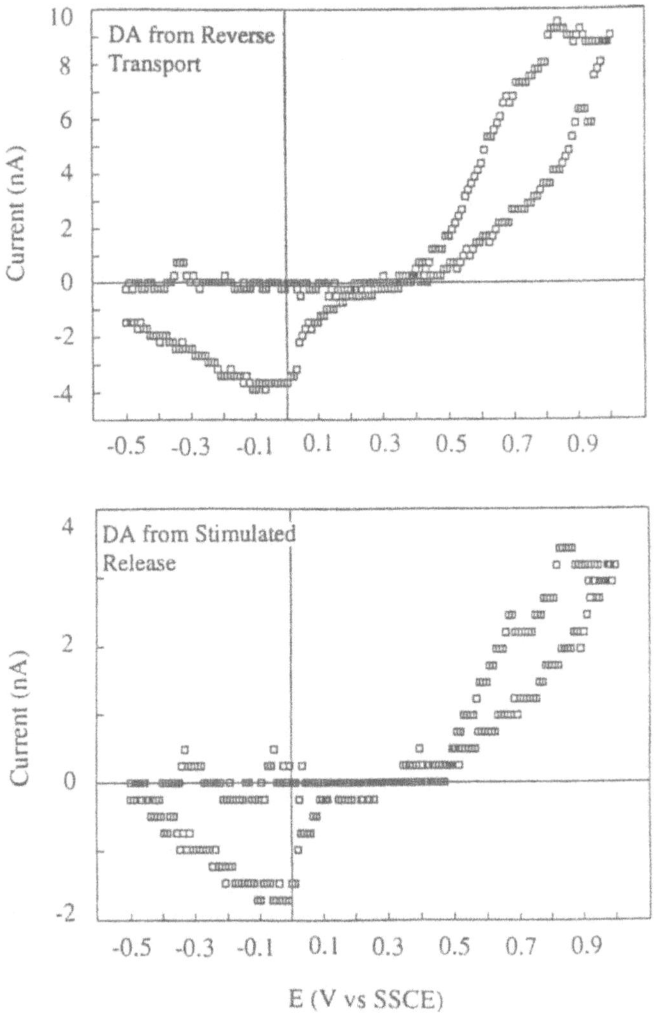

Fig. 20. Fast-scan voltammograms of DA. Background substrated voltammograms were obtained from DA ejection **(Top)** and during quantal release of DA at cell surface **(Bottom)** with a carbon-ring electrode (scan rate 300 V/s, 2-ms delay between each scan).

4. Summary and Future Outlook

The use of voltammetry and amperometry to monitor chemical dynamics at single nerve cells and nerve cell model systems has provided information about intracellular neurotransmitter dynamics and a means to monitor exocytosis

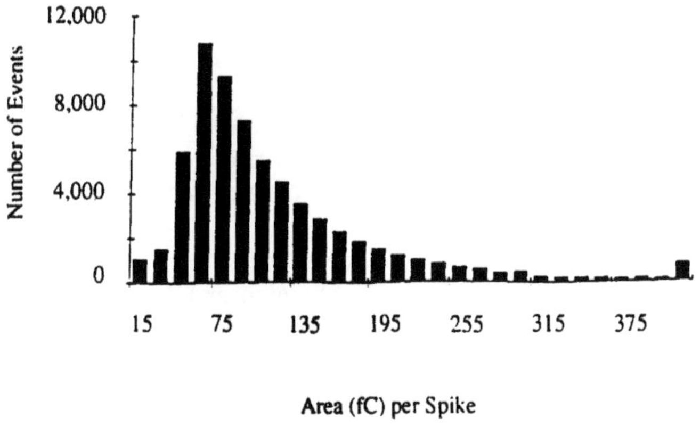

Fig. 21. Frequency distribution of spike areas from the DA neuron of *P. corneus*. The area of transient events is displayed in units of charge (fC). The histogram shown is for data from 5 cells. The average vesicle content for 17 cells has been estimated at 1.2 ± 0.5 amol (700,000 molecules)/vesicle.

directly. It is important, however, to emphasize that these techniques can as yet be used only for easily oxidized neurotransmitters, and concentration sensitivity is an issue. It is not yet possible to detect submicromolar levels of catecholamine in the cytoplasm of a cell voltammetrically. Finally, these techniques have as yet only been used on cells in culture and large identified cells in invertebrates. A large step in technology is needed to fabricate smaller electrode sizes and manipulation techniques before voltammetric measurements will be made in individual synapses. This will be one of the exciting experiments of the future.

Acknowledgments

We would like to thank our coworkers and colleagues who have contributed to the work in this chapter. The Office of Naval Research, National Science Foundation, and the National Institutes of Health supported the work in our laboratory. A. G. E. is a Camille and Henry Dreyfus Teacher-Scholar.

References

Abe T., Lau Y. Y., and Ewing A. G. (1991) Intracellular analysis with an immobilized-enzyme glucose electrode having a 2-μm diameter and subsecond response time. *J. Am. Chem. Soc.* **113**, 7421–7423.

Abe T., Lau Y. Y., and Ewing A. G. (1992) Characterization of glucose microsensors for intracellular measurements. *Anal. Chem.* **64,** 2160–2163.

Abercrombie E. D., Keller R. M., and Zigmond M. J. (1988) Characterization of hippocampal norepinephrine release as measured by microdialysis perfusion: pharmacological and behavioral studies. *Neuroscience* **27,** 897–904.

Amman D. (1986) *Ion Selective Microelectrodes.* Springer-Verlag, Berlin.

Betzig E., Trautman J. K., Harris T. D., Weiner J. S., and Kostelak R. L. (1991) Breaking the diffraction barrier: optical microscopy on a nanometric scale. *Science* **251,** 1468–1470.

Brownstein M. J., Saavedra J. M., Axelrod J., Zeman C. H., and Carpenter D. O. (1974) Coexistence of several putative neurotransmitters in single identified neurons of *Aplysia. Proc. Natl. Acad. Sci. USA* **71,** 4662–4665.

Chen G., Luo G. and Ewing A. G. (1994) Exocytosis of neuromessengers from nerve cell bodies revealed by electrochemical monitoring of single secretory events. submitted to *Nature.*

Chen T.-K., Luo G., and Ewing A. G. (1994) Amperometric monitoring of stimulated catecholamine quantal release from rat pheochromocytoma (PC12) cells at the zeptomole level. *Anal. Chem.* **66,** 3031–3035.

Chen T. K., Lau Y. Y., Wong D. K. Y., and Ewing A. G. (1992) Pulse voltammetry in single cells using platinum microelectrodes. *Anal. Chem.* **64,** 1264–1268.

Chien J. B., Saraceno R. A., and Ewing A. G. (1988) Intracellular voltammetry with ultrasmall carbon-ring electrodes, in *Redox Chemistry and Interfacial Behavior of Biological Molecules* (Dryhurst G. and Niki K., eds.), Plenum, New York, pp. 417–424.

Chien J. B., Wallingford R. A., and Ewing A. G. (1990) Estimation of free dopamine in the cytoplasm of the giant dopamine cell of *Planorbis corneus* by voltammetry and capillary electrophoresis. *J. Neurochem.* **54,** 633–638.

Chow R. H., von Rüden L., and Neher E. (1992) Delay in vesicle fusion revealed by electrochemical monitoring of single secretory events in adrenal chromaffin cells. *Nature* **356,** 60–63.

Ciolkowski E. L., Cooper B. R., Jankowski J. A., Jorgenson J. W., and Wightman R. M. (1992) Direct observation of epinephrine and norepinephrine cosecretion from individual adrenal medullary chromaffin cells. *J. Am. Chem. Soc.* **114,** 2815–2821.

Clift-O'Grady L., Linstedt A. D., Lowe A. W., Grote E., and Kelly R. B. (1990) Biogenesis of synaptic vesicle-like structures in a pheochromocytoma cell line. *J. Cell. Biol.* **110,** 1693–1703.

Cooper B. R., Jankowski J. A., Leszczyszyn D. J., Wightman R. M., and Jorgenson J. W. (1992) Quantitative determination of catecholamines in individual bovine adrenomedullary cells by reverse-phase microcolumn liquid chromatography with electrochemical detection. *Anal. Chem.* **64,** 691–694.

Dennis T., L'Heureux R., Carter C., and Scatton B. (1987) Presynaptic α_2-adrenoreceptors play a major role in the effects of idazoxan on cortical noradrenaline release (as measured by in vivo dialysis) in the rat. *J. Pharmacol. Exp. Ther.* **241,** 642–649.

Dunn L. A. and Holz R. W. (1983) Catecholamine secretion from digitonin-treated adrenal medullary chromaffin cells. *J. Biol. Chem.* **258,** 4989–4993.

Edwards F. A., Konnerth A., and Sakmann B. (1990) Quantal analysis of inhibitory synaptic transmission in the dentate gyrus of rat hippocampal slice: a patch-clamp study. *J. Physiol.* **430,** 213–249.

Ewing A. G., Strein T. G., and Lau Y. Y. (1992) Analytical chemistry in microenvironments: single nerve cells. *Accts. Chem. Res.* **25,** 440–447.

Fischer J. F. and Cho A. K. (1979) Chemical release of dopamine from striatal homogenates: evidence for an exchange diffusion model. *J. Pharm. Exp. Ther.* **208,** 203–209.

Giacobini E. (1968) Chemical studies on individual neurons (I) vertebrate nerves. *Neurosci. Res.* **1,** 1–71.

Giacobini E. (1987) Neurochemical analysis of single neurons. *J. Neurosci. Res.* **18,** 632–637.

Green L. A. and Tischler A. S. (1976) Establishment of a noradrenergic clonal line of rat adrenal pheochromocytoma cells which respond to nerve growth factor. *Proc. Natl. Acad. Sci. USA* **73,** 2424–2428.

Hawley M. D., Tatawawadi S. V., Piekarski S., and Adams R. N. (1967) Electrochemical studies of the oxidation pathway of catecholamines. *J. Am. Chem. Soc.* **89,** 447–450.

Hernandez L., Lee F., and Hoebel B. G. (1987) Simultaneous microdialysis and amphetamine infusion in the nucleus accumbens and striatum of freely moving rats: increase in extracellular dopamine and serotonin. *Brain Res. Bull.* **19,** 623–628.

Hogan B. L. and Yeung E. S. (1992) Determination of intracellular species at the level of a single erythrocyte via capillary electrophoresis with direct and indirect fluorescence detection. *Anal. Chem.* **64,** 2841–2845.

Ishikawa E., Hashida S., Kohno T., and Hirota K. (1990) Ultrasensitive enzyme immunoassay. *Clin. Chim. Acta* **194,** 51–72.

Jackisch R., Huang H. Y., Rensing H., Lauth D., Allgaier C., and Hertting G. (1992) α_2-adrenoceptor mediated inhibition of exocytotic noradrenaline release in the absence of extracellular calcium. *Eur. J. Pharm. Mol. Pharmacol. Sect.* **8(3),** 245–252.

Jankowski J. A., Schroeder T. J., Ciolkowski E. L., and Wightman R. M. (1993) Temporal characteristics of quantal secretion of catecholamines from adrenal medullary cells. *J. Biol. Chem.* **268,** 14,694–14,700.

Jankowski J. A., Schroeder T. J., Holz R. W., and Wightman R. M. (1992) Quantal secretion of catecholamines measured from individual bovine adrenal medullary cells permeabilized with digitonin. *J. Biol. Chem.* **267,** 18,329–18,335.

Kennedy R. T., Oates M. D., Cooper B. R., Nickerson B., and Jorgenson J. W. (1989) Microcolumn separations and the analysis of single cells. *Science* **246,** 57–63.

Kim Y.-T., Scarnulis D. M., and Ewing A. G. (1986) Carbon-ring electrodes with 1-μm tip diameter. *Anal. Chem.* **58,** 1782–1786.

Kuczenski R. (1983) Biochemical actions of amphetamine and other stimulants, in *Stimulants: Neurochemical, Behavioral, and Clinical Perspectives* (Creese I., ed.), Raven, New York, pp. 31–61.

Lau Y. Y., Abe T., and Ewing A. G. (1992) Voltammetric measurement of oxygen in single neurons using platinized carbon-ring electrodes. *Anal. Chem.* **64,** 1702–1705.

Lau Y. Y., Chien J. B., Wong D. K. Y., and Ewing A. G. (1991) Characterization of the voltammetric response at intracellular carbon-ring electrodes. *Electroanalysis* **3,** 87–95.

Leszczyszyn D. J., Jankowski J. A., Viveros O. H., Diliberto E. J. Jr., Near J. A., and Wightman R. M. (1990) Nicotinic receptor-mediated catecholamine secretion from individual chromaffin cells. *J. Biol. Chem.* **265,** 14,736–14,737.

Leszczyszyn D. J., Jankowski J. A., Viveros O. H., Diliberto E. J. Jr., Near J. A., and Wightman R. M. (1991) Secretion of catecholamines from individual adrenal medullary chromaffin cells. *J. Neurochem.* **56,** 1855–1863.

Llinas R., Steinberg I. Z., and Walton K. (1981) Relationship between presynaptic calcium current and postsynaptic potential in squid giant synapse. *Biophys. J.* **33,** 323–352.

Mantus D. S., Valaskovic G. A., and Morrison G. H. (1991) High mass resolution secondary ion mass spectrometry via simultaneous detection with a charge coupled device. *Anal. Chem.* **63,** 788–792.

Matioli G. T. and Niewisch H. B. (1965) Electrophoresis of hemoglobin in single erythrocytes. *Science* **150,** 1824–1826.

McCaman M. W., Weinreich D., and McCaman R. E. (1973) The determination of picomole levels of 5-hydroxytryptamine and dopamine in *Aplysia,* tritonia and leech nervous tissues. *Brain Res.* **53,** 129–137.

Meulemans A., Poulain B., Baux G., and Tauc L. (1987) Changes in serotonin concentration in a living neurone: a study by on-line intracellular voltammetry. *Brain Res.* **414,** 158–162.

Meulemans A., Poulain B., Baux G., Tauc L., and Henzel D. (1986) Micro carbon electrodes for intracellular voltammetry. *Anal. Chem.* **58,** 2088–2091.

Nicholson C. and Rice M. E. (1988) Use of ion-selective microelectrodes and voltammetric microsensors to study brain cell microenvironment, in *Neuromethods: Neuronal Microenvironment* (Boulton A. A., Baker G. B., and Walz W., eds.), Humana, Clifton, NJ, pp. 247–361.

O'Connor D. T. and Frigon R. P. (1984) Chromogranin A, the major catecholamine storage vesicle soluble protein. *J. Biol. Chem.* **259,** 3237–3247.

Osborne N. N., Priggemeier E., and Neuhoff V. (1975) Dopamine metabolism in characterized neurones of *Planorbis corneus. Brain Res.* **90,** 261–271.

Parsons D. W., Maat A. T., and Pinsker H. M. (1983) Selective recording and stimulation of individual identified neurons in freely behaving *Aplysia. Science* **221,** 1203–1206.

Pelayo F., Dubocovich M. L., and Langer S. Z. (1980) Inhibition of neuronal uptake reduces the presynaptic effects of clonidine but not of α_2-methylnoradrenaline on the stimulation-evoked release of ^{3}H-noradrenaline from rat occipital cortex slices. *Eur. J. Pharmacol.* **64,** 143–155.

Powell B. and Cottrell G. A. (1974) Dopamine in an identified neuron of *Planorbis corneus. J. Neurochem.* **22,** 605,606.

Schroeder T. J., Jankowski J. A., Kawagoe K. T., Wightman R. M., Lefrou C., and Amatore C. (1992) Analysis of diffusional broadening of vesicular packets of catecholamines released from biological cells during exocytosis. *Anal. Chem.* **64,** 3077–3083.

Schubert D., LaCorbiere M., Klier F. G., and Steinbach J. H. (1980) The modulation of neurotransmitter synthesis by steroid hormones and insulin. *Brain Res.* **190,** 67–79.

Sharp T., Zetterström T., Ljungbers T., and Ungerstedt U. (1987) A direct comparison of amphetamine-induced behaviours and regional brain dopamine release in the rat using intracerebral dialysis. *Brain Res.* **401,** 322–330.

Sossin W. S. and Scheller R. H. (1989) A bag cell neuron-specific antigen localizes to a subset of dense core vesicles in *Aplysia californica. Brain Res.* **494,** 205–214.

Starke K. (1977) Regulation of noradrenaline release by presynaptic receptor systems. *Rev. Physiol. Biochem. Pharmacol.* **77,** 1–124.

Starke K. (1981) Presynaptic receptors. *Annu. Rev. Pharmacol. Toxicol.* **21,** 7–30.

Sulzer D. and Rayport S. (1990) Amphetamine and other psychostimulants reduce pH gradients in midbrain dopaminergic neurons and chromaffin granules: a mechanism of action. *Neuron* **5,** 797–808.

Sulzer D., Maidment N. T., and Rayport S. (1993) Amphetamine and other weak bases act to promote reverse transport of dopamine in ventral midbrain neurons. *J. Neurochem.* **60,** 527–535.

Sulzer D., Chen T.-K., Lau Y. Y., St. Remy C., Kristensen H., Ewing A. G., and Rayport S. (1994) Amphetamine redistributes DA from synaptic vesicles to the cytosol and promotes reverse transport. submitted to *J. Neurosci.*

Tank D. W., Sugimori M., Connor J. A., and Llinas R. R. (1988) Spatially resolved calcium dynamics of mammalian Purkinje cells in cerebellar slice. *Science* **242,** 773–776.

Thomas L., Hartung K., Langosch D., Rehm H., Bamberg E., Franke W. W., and Betz H. (1988) Identification of synaptophysin as a hexameric channel protein of the synaptic vesicle membrane. *Science* **242,** 1050–1053.

Thomas R. C. and Moody W. J. (1980) Ion-sensitive microelectrodes for intracellular use. *Trends Biochem. Sci.* **5,** 86,87.

Trifarro J. M. and Lee R. W. H. (1980) Morphological characteristics and stimulus-secretion coupling in bovine adrenal chromaffin cell cultures. *Neuroscience* **5,** 1533–1546.

Wagner J. A. (1985) Structure of catecholamine secretory vesicles from PC12 cells. *J. Neurochem.* **45,** 1244–1253.

Westfall T. C. (1977) Local regulation of adrenergic neurotransmission. *Physiol. Rev.* **57,** 659–728.

Wightman R. M. (1981) Microvoltammetric electrodes. *Anal. Chem.* **53,** 1125A–1134A.

Wightman R. M. (1988) Voltammetry with microscopic electrodes in new domains. *Science* **240,** 415–420.

Wightman R. M. and Wipf D. O. (1988) Voltammetry at ultramicroelectrodes, in *Electroanalytical Chemistry* (Bard A. J., ed.), Marcel Dekker, New York, pp. 267–353.

Wightman R. M., Jankowski J. A., Kennedy R. T., Kawagoe K. T., Schroeder T. J., Leszczyszyn D. J., Near J. A., Diliberto E. J. Jr., and Viveros O. H. (1991) Temporally resolved catecholamine spikes correspond to single vesicles release from individual chromaffin cells. *Proc. Natl. Acad. Sci. USA* **88,** 10,754–10,758.

Winkler H. and Westhead E. (1980) The molecular organization of adrenal chromaffin granules. *Neuroscience* **5,** 1803–1823.

Wise R. A. and Bozarth M. A. (1987) Brain mechanisms of drug reward and euphoria. *Psych. Med.* **3,** 445–460.

Yoo S. H. and Lewis M. S. (1992) Effects of pH and Ca^{2+} on monomer-dimer and monomer-tetramer equilibria of chromogranin A. *J. Biol. Chem.* **267,** 11,236–11,241.

Zaczek R., Culp S., and DeSouza E. B. (1991) Interactions of [^{3}H]-amphetamine with rat brain synaptosomes. II. Active transport. *J. Pharmacol. Exp. Ther.* **257,** 830–835.

Zetterström T., Sharp T., Marsden C. A., and Ungerstedt U. (1983) In vivo measurement of dopamine and its metabolites by intracerebral dialysis: changes after D-amphetamine. *J. Neurochem.* **41,** 1769–1773.

Zhou R., Luo G., and Ewing A. G. (1994) Direct observation of the effect of autoreceptors on stimulated release of catecholamines from adrenal cells. *J. Neurosci.* **14,** 2402–2407.

Measurement of the Time-Resolved Kinetics of Biogenic Amine Release and Transporter Activity by Rotating Disk Electrode Voltammetry In Vitro

Susan M. Meiergerd and James O. Schenk

1. Introduction

Chemical neurotransmission is generally thought of as the fundamental process underlying communications between neurons of the central and peripheral nervous systems. Studies of this process are made by a variety of approaches and from different research perspectives, such as those of the molecular, genetic, molecular biological, cellular, and behavioral sciences. Chemical neurotransmission and its effects are transient events that in some cases are transformed into responses on longer time scales providing a unique challenge to the neuroscientist. The kinetic time scale of neurotransmission and its effects range from milliseconds in the case of synaptic transmission to minutes, hours, days, or perhaps a life time in the case of learning and memory. As a result, the design of experiments to elucidate kinetic events related to neurotransmission and its consequences is a challenging problem with a variety of "starting" places, each with a unique set of conditions and problems. Perhaps, the initial fundamental goal in research in neurotransmission is to elucidate the chemical events that control the initiation, duration, and termination of synaptic chemical signaling. These

From: *Neuromethods, Vol. 27: Voltammetric Methods in Brain Systems*
Eds: A. Boulton, G. Baker, and R. N. Adams

are membrane events, and a complete understanding will only be possible when all the intermediates involved are identified. Such a complicated task can only be accomplished through the efforts of many disciplines. Making time-resolved chemical measurements of the release and reuptake of neurotransmitters, the topic of this chapter, is just one of the goals of research in chemical neurotransmission.

The term "time-resolved" implies that different events can be separated from one another through their unique kinetic characteristics, for example, the resolution of the timing of neuronal fluxes of transmitters from timing defined by diffusion. Thus, in a time-resolved measurement, the biochemical events controlling release and reuptake are the only rate-limiting factors. The resolved time-course data is then said to be in the kinetic domain, a required condition if the goal is to study mechanism. Measurements made after equilibrium has been reached provide information on the state of a process after it has undergone a change, but nothing about the path that it took to get there. Data from the kinetic domain defines the relationships between each reactant in a process and can be used to write a rate law describing the rate constants, concentrations of reactants, and their power dependencies. From these results, one can eliminate mechanisms that are not described by the rate law and design experiments, such as trapping intermediates and so on, to confirm that a mechanism is consistent with the rate law.

The catecholamines and serotonin are easy to oxidize and detect quantitatively at solid electrodes. An electroanalytical technique capable of monitoring very rapid chemical events is rotating disk electrode (RDE) voltammetry. Thus, this technique was evaluated as a possible tool for monitoring the rapid rates of biogenic amine release and reuptake. The dopaminergic system was chosen as the system to work in because of this laboratory's interest in dopaminergic neurochemistry, physiology, and pharmacology. However, any of the approaches and principles illustrated in this chapter can be applied to the study of any neurochemically important electroactive compound.

The basic experiment consisted of placing dopaminergic tissue into a small volume of physiological buffer along with the RDE to detect extracellular dopamine (DA). The tissue was then stimulated by depolarization with K^+. The resulting DA concentration vs time profile illustrated in Fig. 1 (Schenk et al., 1990) shows the results of this study, which validate the use of the RDE technique for making rapid, time-resolved measurements. Because the components of release and reuptake are slower than the ca. 40 ms mixing time of the RDE, apparent time-resolved measurements of these events can be made by this technique. This suggests that the fluxes of DA observed in this experimental preparation are in the kinetic domain and are not influenced by a diffusive factor. Additional experiments were conducted to show that the system could monitor release with presynaptic control and that the uptake observed is mediated by the dopaminergic transporter. Blocking presynaptic D2 receptors for DA with haloperidol increased DA release, and adding cocaine inhibited reuptake as shown by the prolonged clearance time of released DA. These qualitative results are in accordance with those obtained in studies using more traditional methodologies.

The RDE technique has been used for additional experiments, including studies of the fundamental kinetic characteristics of the dopaminergic transporter, the chemistry of catechol transport, and the effects of cocaine on the function of the transporter after direct inhibition and withdrawal from repeated administration. The goal of this chapter is to explain the theoretical and practical basis for the experiment and show how the resulting data are analyzed. In addition, a brief overview of what new information can be gained by the technique and how it can be used to complement in vivo voltammetric data will be presented along with a summary of future experiments designed for the RDE system.

2. Experimental Technique

The RDE technique can be applied to the study of any biochemical system that can be detected by electroanalytical

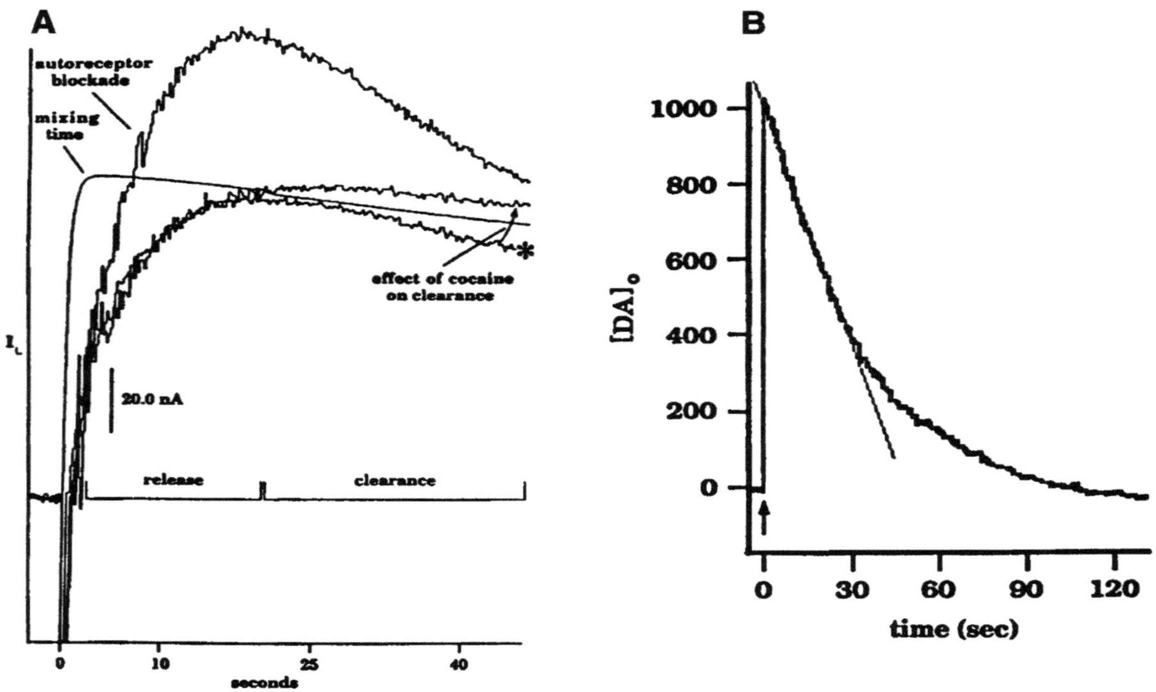

Fig. 1. Examples of a depolarization-induced DA release and reuptake concentration-time profile **(A)** from a striatal suspension and a transporter-mediated uptake concentration-time profile into resting suspensions **(B)**. In (A) the profile has two regions, labeled "release" and "clearance." The asterisk indicates a control profile. The maximum signal is 300 n*M* DA resulting from stimulation by 30 m*M* KCl. (B) shows the $[DA]_o$ vs *t* profile for the clearance of 1.0 μ*M* DA from the extracellular phase of a striatal suspension. The clearance has been shown to be mediated by the dopaminergic transporter (McElvain and Schenk, 1992b). The source of (A) is Schenk et al. (1990), used with kind permission from Elsevier Science Publishers BV.

methodology. In the neurosciences, this could include studies of DA, norepinephrine (NE), epinephrine, serotonin, uric acid, and ascorbic acid. Dryhurst et al. (1982) and Humphries and Dryhurst (1987) have reviewed the electroanalytical chemistry of many of these compounds. The remainder of this chapter focuses on DA, the neurotransmitter studied most often in this laboratory. However, similar studies with the compounds listed above can be easily envisaged.

2.1. Theoretical Basis

Dopamine is oxidized at solid electrode materials via the reaction (Dryhurst et al., 1982).

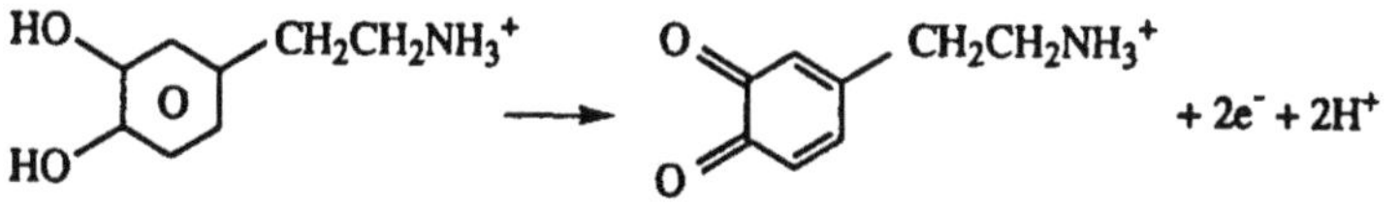

Rxn (1)

This reactivity is the basis of electrochemical detection in liquid chromatography and in vivo voltammetry (Adams and Marsden, 1982; Marsden, 1984; Justice, 1987). RDE voltammetry, a standard technique in the armamentarium of the electroanalytical chemist (Adams, 1969; Bruckenstein and Miller, 1977), has been used to study catechol and catecholamine chemistry (Adachi et al., 1990; Schenk et al., 1991) as well as reduction/oxidation (redox) chemistry (Malacheshy et al., 1966). Electron flow to the electrode from the oxidation of DA is detected as current. With an RDE, the electroactive species is brought to the detector (working) electrode by a stirring motion and the current detected, at potentials ≥100 mv more positive than the redox potential, can be described by the Levich equation,

$$i_L = 0.62nFAD^{2/3}C\upsilon^{-1/6}\omega^{1/2} \quad (1)$$

where i_L is the limiting current, n is the number of electrons transferred, F represents Faraday's constant, A is the area of the electrode in cm^2, D is the diffusion coefficient of the electroactive species in cm^2/s, C is the concentration of electroactive species in the bulk solution in mol/L, υ is a constant describing the resistance of a solution to motion (kinematic viscosity) in cm^2/s, ω is the angular velocity of rotation in rad/s where $\omega = 2\pi N$, and N is the number of rotations per second (Adams, 1969).

The stirring motion of the electrode is constant about its axis, thus setting up rapid and reproducible convective paths of solution (*see* dotted arrows in Fig. 2). As the disk rotates, it stirs a layer of solution close to the electrode surface in a circular motion, which is then spun out radially away from the center of the disk because of centrifugal force. The solution at the surface of the disk is replenished by an upward movement of new solution to the surface from a field perpendicular to the plane of the electrode. However, during detection, stirring is not the only factor bringing the electroactive species to the electrode. The electrolysis occurring at the electrode surface drives the concentration of detected species to zero, producing a thin diffusion layer at the electrode. Movement within this layer occurs by diffusion, which is driven by the concentration gradient between the bulk solution and the electrode surface. Thus, the current observed depends on the diffusion coefficient of the electroactive species as well as the movement of solution. Unlike the current profile measured at a stationary electrode, the current monitored at the RDE will remain constant, independent of time, since the concentration gradient across the diffusive layer is constant. A change in current occurs only by the appearance or disappearance of electroactive species into or from the bulk solution.

Fig. 2. *(opposite page)* Schematic diagram of the rotating disk electrode used for the measurement of the kinetics of biogenic amine release and reuptake. The combination incubation/electrochemical cell is shown in cross-section and consists of a glass, water-jacketed chamber capable or holding volumes of approx 800 µL. The contents of the cell (stippled) are stirred by the rotation of the glassy carbon rotating disk electrode attached to a precision rotator. The convection paths created by the stirring are indicated by the dotted line arrows. A pipet is positioned above and to one side of the chamber to direct a stream of 95% O_2/5% CO_2 across the solution to maintain pH and saturation of the buffer with O_2.

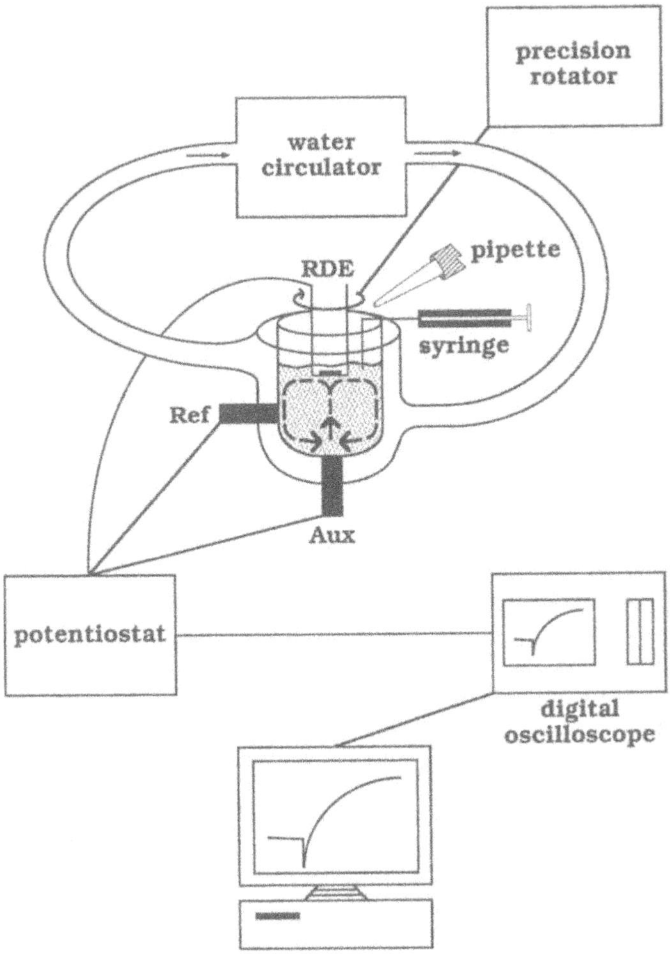

A constant-flow-rate syringe is used to add drugs or biogenic amines to the cell. The tip of the needle is bent and placed just below the surface. The potentiostat applies the potential to the RDE relative to the Ag/AgCl reference electrode. The Pt auxiliary electrode (AUX) completes the potential-controlling portion of the circuitry, and the RDE is the sink for electrons resulting from the oxidation of detected biogenic amines. The current flow is measured by the potentiostat via its current follower circuitry. Signals are recorded via a digital oscilloscope and are transferred to a computer for subsequent kinetic analyses. (Adapted from McElvain [1992])

These features make the RDE desirable for studies of chemical release and reuptake of DA or other electroactive biogenic amines. With the RDE, DA in solution can be detected oxidatively as i_L. In a tissue preparation its appearance and disappearance can be detected under two conditions, either as decreasing i_L as a result of uptake subsequent to additions from exogenous sources or as an initially increasing i_L followed by a decreasing i_L resulting from release of endogenous DA and its coupled reuptake. Since these two processes occur on a rapid time scale (subsecond to seconds), the forced convective property of the RDE system allows them to be measured under apparent kinetic control rather than diffusive control, making RDE voltammetry a powerful tool in the study of the kinetics and mechanisms of release and transport systems.

2.2. Tissue Preparation and General Experimental Protocol

The RDE technique for measuring release and reuptake has been performed in both striatal suspensions (McElvain and Schenk. 1992a) and synaptosomes (McElvain and Schenk, 1992b). The synaptosomes are prepared by the procedures of Booth and Clark (1978), and the striatal suspensions are prepared in the following manner. Male Sprague-Dawley rats, weighing 250–350 g, are rapidly decapitated. The striata (≈40 mg) is dissected from the brain, weighed, and placed on ice until experimentation. Just prior to experimentation, the tissue is finely chopped by hand with a razor blade on an ice-cold watch glass and transferred to the RDE cell containing 500 µL of physiological buffer at 37°C and pH 7.4. The composition of the buffer used is 124.0 m*M* NaCl, 1.80 m*M* KCl, 1.24 m*M* KH_2PO_4, 1.30 m*M* $MgSO_4$, 2.50 m*M* $CaCl_2$, 26.0 m*M* $NaHCO_3$, and 10.0 m*M* glucose. The buffer is saturated (McElvain, 1992) with a gas mixture of 95% O_2/5% CO_2 by gently directing a stream of the gas across the surface of the buffer (Fig. 2). The chopped tissue is then separated and further broken down by drawing it up into the tip of an Eppendorf pipet and releasing it back into the cell. This is performed about ten times. The tissue is washed by first allowing it to settle, then removing 250 µL of buffer, and replacing it with fresh buffer. This cycle is repeated six times. The washing steps can be automated easily with a Hamilton (Reno, NV) MicroLab M programmable autopipeter. Finally, the RDE is lowered into the cell and positioned just below the surface of the buffer (*see* Fig. 2). Rotation is begun at 2000 rotations/ min (rpm), and the working electrode is activated at an applied potential of +450 mV relative to the Ag/AgCl reference electrode. The rotation by the electrode disrupts the chopped tissue, producing a homogenous suspension of striatal tissue. The suspension is allowed to incubate under these conditions for 10 min. After this time, the experiment generally proceeds by either the addition of DA to the extracellular milieu with the observation of uptake under resting conditions or by addition of KCl to initiate depolarization-induced release of DA followed by its subsequent reuptake. These additions are made with a 20-µL spring-loaded constant-flow rate syringe (Hamilton, Part No. CR700-20). The tip of the needle is placed along the inside wall of the cell just below the surface of the solution. Signals are detected as voltage output from the potentiostat vs time (t), and are recorded onto a digital oscilloscope or a strip chart recorder depending on the time window under study. The voltages are converted to i_L vs t based on the sensitivity setting of the potentiostat. The profile is then converted to extracellular DA $[DA]_o$, vs t using a calibration factor. The calibration factor is the slope of a calibration curve obtained by measuring i_L as a function of $[DA]_o$ over a range of concentrations determined by the requirements of the experiment, generally from 0.1–3.0 µ*M*. The data can be analyzed by hand from the strip chart recorder or by computer; accomplished by transferring the data from the digital oscilloscope to a computer and a commercially available software, such as DADiSP Worksheet (DSP Development Corp., Cambridge, MA). Alternatively, the data can be transferred as an ASCII file into a spreadsheet, such as Quatro Pro (Borland International, Scotts Valley, CA). Excell (Microsoft Corp., Redmond, WA), and/or Sigma Plot (Jandel

Corp., Corta Madera, CA). If the effects of a drug on the events of release and reuptake or uptake alone are to be determined, the drug is added to the suspension 30 s, or some other constant preincubation time, prior to stimulation or addition of exogenous DA.

2.3. Required Equipment and Experimental Setup

The entire setup for the RDE system consists of an electrode, a glass cell, a precision rotator, a potentiostat, and an output device, such as an oscilloscope or strip chart recorder, and can be purchased, in 1994 prices, for around $10,000. A description of the setup depicted in Fig. 2 follows.

2.3.1. The Electrode

The working electrode, custom-built by Pine Instrument Company (Grove City, PA), Part No. AFMDO3GC, is comprised of a 3.0-mm diameter disk of glassy carbon encased in a 5.1-mm diameter Teflon™ shield. This assembly is attached to a stainless-steel shaft approx 6 in. in length. The upper part of the shaft is designed for a Pine Instrument AFMSR High-Performance Rotator capable of rotation speeds between 1 and 10,000 rpm. The rotator can operate either at a constant rotation speed or can be varied in any waveform. Contact to the electrode is made by silver carbon brushes (Pine Instruments, Grove City, PA).

2.3.2. The Electrochemical Cell and Circuitry

Experiments are carried out in a water-jacketed Pyrex™ glass cell with a maximum volume of 800 μL similar to that depicted in Fig. 2. The reference and auxillary electrodes are incorporated directly into the walls of the cell. The design of the two electrodes is similar. For each electrode, a platinum wire is heat-annealed through the glass wall of the cell. The wire runs through the water-jacket and extends approx 2 mm into the interior of the cell. The portion of the wire extending through the jacket is insulated with glass to prevent both leakage and electrical contact with the water used for heating and cooling. A bulb is made by heating the ends of the Pt wires. A drop of molten Ag is placed on the end of the Pt bulb that will serve as the reference electrode. The Ag button is then coated with AgCl by anodizing it in a solution of 0.1*M* HCl. Finally, copper tubing is soldered to the platinum wires extending outward from the cell and sealed to the external wall with Torr Seal epoxy (Varian Associates, Lexington, MA). The reinforcement of the tubing provides strong structural stability, and allows connections between cell and potentiostat to be made with alligator clips. The cell temperature is maintained at ±0.05°C by connecting the water jacket to any circulating, constant temperature water bath.

A constant potential is applied to the electrode with a potentiostat. The potentiostat currently in use in this laboratory is a Bioanalytical Systems (BAS, W. Lafayette, IN) LC-4A amperometric detector. Its response time was decreased to 20 ms by changing the capacitors in the feedback loop of the current-to-voltage converter (details on request). Any commercially available potentiostat with a current range of 0.1–100 nA/full-scale deflection, and an adjustable offset could be used. Currently, this laboratory is experimenting with a slightly different potentiostat. Its current-to-voltage converter, with variable gain, variable offset, and variable filtering, is designed specifically to enable us to make measurements of catechol uptake and release with the RDE system. The current sensitivity range was increased with graded selections from 1–1000 nA/full-scale deflection, the filtering is designed to be continuously selectable from 650 to 7 Hz, with a bandwidth of 9 KHz in the no-filtering position, and a probe for placing the initial amplification stage at the RDE connection is an added feature to reduce noise pickup. Details are available on request.

3. Signals and Signal Analyses

3.1. Selectivity of Chemical Signals

The measured current is the result of the oxidation of electroactive species at the surface of the electrode. Striatal tissue contains three major detectable species: DA, 3,4-dihydroxyphenylacetic acid (DOPAC), and ascorbic acid (AA). The selectivity of the RDE system for measuring these electroactive species has been shown to be 1:0.61:0.26, DA:DOPAC:AA (McElvain, 1992). However, DOPAC and AA do not contribute significantly to the signal measured in transport studies since: (1) 90% of the concentration of AA present in the suspension is washed out during tissue preparation (Schenk et al., 1982; McElvain, 1992), and (2) DOPAC does not change during the time-course of the RDE experiment (McElvain and Schenk, 1992a). The practical detection limit of DA in the absence of tissue as found to be 5.0 n*M* (McElvain, 1992) and in tissue is approx 34 n*M* (Meiergerd et al. 199b)

3.2. Timing of Signals

In the RDE experiment, once a biogenic amine is released from the neuron (or added to the tissue suspension), it must move from its site of release (or addition) to the working electrode in order to be measured. The relevant functional processes of release and uptake occur on subsecond time scales. Thus, time defined by diffusion must be minimized or eliminated, and the response time of the system must be defined. The stirring produced by the RDE minimizes the diffusion timing. To determine the whole system response time, a baseline is monitored in physiological buffer. Next, a small amount of DA or any detectable compound is added, and the timing of the resulting profile is measured. This experiment is then repeated in a striatal suspension either by addition of DA or K^+. The results of such an experiment, shown in Fig. 1, suggest that the kinetics of release and reuptake are slower than the mixing time of 40 ms at the experimental rotation rate of 2000 rpm and can therefore be resolved by this technique. By increasing the rotation rate to ca. 8000 rpm, the mixing time can be decreased to approx 20 ms (McElvain, 1992).

3.3. Mathematical Treatments of the Data

This section is written with the aim of describing the equations used previously to analyze the kinetic data obtained by the RDE system for characterizing DA release and uptake. Raw data are converted to $[DA]_o$ vs t as described, and from these plots, rates are determined and used in the following kinetic analyses.

3.3.1. Analysis of Stimulated Release and Reuptake by Apparent First-Order Kinetics

Profiles of DA release and reuptake follow an apparent first-order kinetic model (McElvain and Schenk, 1992a), which consists of two consecutive, irreversible reactions (Fersht, 1985):

$$A \xrightarrow{k_1} B \xrightarrow{k_2} C \qquad \text{Rxn (2)}$$

These reactions, expressed in terms of our measurements, are depicted as

$$[DA]_i \xrightarrow{k_r} [DA]_o \xrightarrow{k_u} [DA]_i' \qquad \text{Rxn (3)}$$

where $[DA]_o$, the transient intermediate monitored by the RDE, appears on the release of intracellular DA and disappears by reuptake. The variables, k_r and k_u, are the rate constants for release and reuptake, respectively. The variable $[DA]_i$ is distinguished from the variable $[DA]_i$ since the site of reuptake may differ from the site of release. In both cases, "*i*" signifies intracellular. The mathematical equation describing this model (Fersht, 1985) is:

$$[DA]_o = [DA]_i\, k_r/(k_u - k_r)[\exp(-k_r t) - \exp(-k_u t)] \qquad (2)$$

where all the variables have been described. Application of this model will be shown later. Please note that the original equation appearing in McElvain and Schenk (1992a) contained errors and that Eq. (2), above is the correct equation. The data in that paper were analyzed according to Eq. (2), above and its correct form is available in the literature cited in McElvain and Schenk (1992a) and above. The resulting $[DA]_o$ vs time data can be treated in the context of Eq. (2) in two ways: (1) by a graphical approach from which estimates of k_r and k_u can be made along with the value of $[DA]_i$; and

(2) by a computer-assisted, nonlinear curve fit to Eq. (2) (McElvain and Schenk, 1992a).

3.3.2. Analysis of Initial Uptake Velocities of Exogenously Added Dopamine Using Eadie-Hofstee Transformations of Michaelis-Menten Kinetics

The mechanism of dopaminergic uptake was characterized using the RDE system (McElvain and Schenk, 1992b) and applying the kinetic analyses of Fromm (1975). Dopamine uptake rates were measured for the first 25 s as initial apparent zero-order velocities. The criteria used to define zero-order kinetics are that a plot of $[DA]_o$ vs t is linear (by linear regression analyses) and has an intercept value in numerical agreement with and statistically indistinguishable from the initial $[DA]_o$ value. The velocities were analyzed by two approaches:

1. Graphically using the Eadie-Hofstee transformation of the Michaelis-Menten equation:

$$\nu = V_{max} - K_m \nu / [S]^x \qquad (3)$$

where n is the transport velocity of the substrate, S is the substrate (DA), V_{max} is the maximal transport velocity, K_m is the Michaelis-Menten constant defined as the $[DA]_o$ at which n = $1/2V_{max}$, and x is the reaction order; and
2. By nonlinear regression analysis of the velocity data to the Michaelis-Menten equation:

$$\nu = V_{max}[S]/(K_m + [S]) \qquad (4)$$

using a Quasi-Newton- or SIMPLEX-based fitting routine in the commercially available software, SYSTAT (Wilkinson, 1990). All variables have the same definition as given above. The values of n were obtained at concentrations of substrate extending well above and below the apparent K_m value.

The inward transport of DA is kinetically similar to that defined by Michaelis-Menten kinetics (Stein, 1986) and can be depicted as:

$$S + T \underset{k_{dis}}{\overset{k_{ass}}{\rightleftharpoons}} S - T \xrightarrow{k_{cat}} T + S \qquad \text{Rxn (4)}$$

where S is the substrate DA, T is the transporter, k_{ass} and k_{diss} are the rate constants of association and dissociation of the substrate with the transporter, and k_{cat} is the catalytic turnover that entails all the steps of substrate movement through the membrane following binding. The association and dissociation of intracellular substrate to the transporter is ignored, since the timing of these steps is assumed to be negligible as compared to the forward reaction (Stein, 1986). The uptake profiles are obtained by monitoring the disappearance of added DA under the steady-state condition where $d[\text{DA} - \text{T}]/dt = 0$. Consequently, the RDE technique appears to evaluate the entire forward reaction not distinguishing k_{cat} from k_{ass}. However, estimated values of k_{cat} and k_{ass} can be made by further treatment of the steady-state data. The details used to define these kinetic parameters are based on the work of Fersht and Stein, and are explained in Meiergerd and Schenk (1994a). In brief:

$$k_{cat} = V_{max}/T_d \qquad (5)$$

where the variables have been defined previously and T_d is the density of transporter sites, taken as 416 pmol/g wet wt, which is an average value obtained from from the literature for the B_{max} of ligand binding to the transporter (Meiergerd et al., 1994b). The value of k_{ass} is determined by:

$$k_{ass} = k_{cat}/K_m \qquad (6)$$

Although the value of k_{ass} obtained under steady-state conditions is a good approximation, it should be checked under presteady-state conditions where only the association of DA to the transporter is measured. Whereas other techniques are used to make steady-state kinetic measurements, to our knowledge there are no other laboratories using them to study presteady-state kinetics of transport systems. The RDE technique has been shown to be capable of such measurements and has been used to confirm the k_{ass} value estimated under steady-state conditions (Meiergerd et al., 1994b). Under presteady-state conditions, the binding reaction:

$$\mathrm{S} + \mathrm{T} \xrightarrow{k_{ass}} \mathrm{S-T} \qquad \text{Rxn (5)}$$

follows a second-order kinetic model in which the rate of the reaction is defined as (Capellos and Bielski, 1972),

$$dx/dt = k_{ass}\,([\mathrm{DA}]_{o,t=0} - [\mathrm{DA-T}])([\mathrm{T}]_{o,t=0} - [\mathrm{DA-T}]) \qquad (7)$$

where $[\mathrm{DA}]_{o,t=0}$ and $[\mathrm{T}]_{o,t=0}$ designate the concentrations of extracellular DA and transporter, respectively, at zero time. The terms, $[\mathrm{DA}]_{o,t=0} - [\mathrm{DA-T}]$ and $[\mathrm{T}]_{o,t=0} - [\mathrm{DA-T}]$, define the concentrations at any given t other than at $t = 0$. Equation (7) integrates to:

$$k_{ass}t = 1/\{[\mathrm{DA}]_{o,t=0} - [\mathrm{T}]_{o,t=0}\}\ \ln\{[\mathrm{T}]_{o,t=0}\,([\mathrm{DA}]_{o,t=0} - [\mathrm{DA-T}])/[\mathrm{DA}]_{o,t=0}\,([\mathrm{T}]_{o,t=0} - [\mathrm{DA-T}])\} \qquad (8)$$

where the boundary condition is $[\mathrm{DA}]_{o,t=0} \neq [\mathrm{T}]_{o,t=0}$ (Capellos and Bielski, 1972). Application of Eq. (8) is described later in Section 4.5.

3.3.3. Evaluation of Inhibition of Dopamine Uptake

Patterns observed in Eadie-Hofstee analyses (Fersht, 1985) are used to determine the apparent mechanism of inhibition of cocaine on the velocity of DA uptake. The concentrations of cocaine should be extended well below and above its K_i.

Another practical application of the RDE system is to assess the relationship between the sites of binding of two inhibitors of the transport process. Two inhibitors of the uptake process can act by binding to the transporter at the same site, at independent sites, or at sites that are separate yet interacting such that the binding of one inhibitor to the transporter negatively effects binding of the other. Mathematical descriptions for these three models have been developed (Devés and Krupka, 1980). The first, for a single binding site, is:

$$(v/v_{IL} - 1) = (v/v_L - 1) + (v/v_I - 1) \qquad (9)$$

where v is the velocity of DA transport in the absence of any inhibitor, v_I is the velocity at a given concentration of inhibitor I (the test inhibitor), v_L is the velocity of DA transport in the presence of inhibitor L (cocaine), and v_{IL} is the velocity of transport in the presence of both inhibitors.

The second, for two independent binding sites, is:

$$(v/v_{IL} - 1) = (v/v_I - 1) + (v/v_L - 1) + (v/v_I - 1)(v/v_L - 1)A \quad (10)$$

where the variables have been defined and A is $(1 + f_1/f_{-1})$ $(1 + [DA]_o/K_m)$, where f_1/f_{-1} is the equilibrium distribution of carrier between inward and outward-facing forms. When fitting data to this expression, the upper limit of f_1/f_{-1} is taken as unity (Devés and Krupka, 1980).

Finally, for two interacting sites, the expression is:

$$(v/v_{IL} - 1) = (v/v_I - 1) + (v/v_L - 1)(1 + [I]/K_{IL}) \quad (11)$$

where K_{IL} is the half saturation constant for the test inhibitor in the presence of saturating concentrations of cocaine, and the other variables represent the same conditions as given previously. From these models, velocities of DA transport in the presence of two test inhibitors can be predicted. Experimental velocities are then compared to the predicted values to obtain the model that best fits the data. In our work, the criteria for a fit are an experimental result that is statistically indistinguishable from a predicted value and numerical agreement within 5%.

4. Applications

This section illustrates how the RDE technique can be used to study DA release and reuptake, and includes summaries of characterization of DA release with coupled uptake, kinetic and chemical characterizations of the functioning of the dopaminergic transporter, and studies of the inhibition of the transport of DA. Combined, the results obtained add new information on dopaminergic chemical neutrotransmission.

4.1. Dopamine Release and Coupled Transporter-Mediated Reuptake

Having verified that the RDE technique is a suitable method for monitoring the rapid process of neurochemical transmission, experiments extending the observations depicted in Fig. 1 were conducted. The results of these studies indicate that the release process observed was Ca^{2+} and K^+-dependent with a steep temperature dependence (Q_{10} = 2.0 and no release at 0°C) (McElvain and Schenk, 1992a). The uptake proccss was shown to be capable of concentrating DA at ratios of 1–1000, buffer-to-tissue phase (McElvain and Schenk, 1992a,b). The complete concentration-time profiles for the release and reuptake of DA were kinetically analyzed and found to follow the apparent first-order kinetic model of Eq. (2) (*see* Fig. 3). Briefly, raw data were plotted as ln $[DA]_0$ vs t. The value of k_u was obtained from the slope of the reuptake portion of the curve and found to be 0.012/s. The rate constant of release, k_r, had a value of 0.132/s and was obtained from the slope of ln Δ vs t, where Δ is the difference between $[DA]_0$ measured at any given time, and $[DA]_i$. The $[DA]_i$ value, 715 n*M*, is shown in Fig. 3 as the *y*-intercept of the regression line. The values obtained graphically along with the independent variable, *t*, were used to calculate $[DA]_0$ at any given *t*. These values were then compared to the experimental $[DA]_0$ vs *t* data and found to be in agreement within 9% with some systematic error (Fig. 3B). A nonlinear curve-fitting approach eliminated the systematic error and improved agreement to within 2% (Fig. 3C) (McElvain and Schenk, 1992a).

Regulation of the release of DA is mediated by autoreceptors (reviewed in Wolf and Roth, 1987). Using the same experimental setup as described, the influence of haloperidol, a blocker of the DA autoreceptor, on the rate and magnitude of release of DA was determined (McElvain and Schenk, 1992a). Briefly, haloperidol was added just prior to the K^+ stimulus. Concentrations of haloperidol corresponding with those used in behavoral studies caused an increase in the release rate constant. There was no effect on the duration of release or on the reuptake rate constant. However, at higher concentrations, both constants were decreased. Thus, autoreceptors appear to mediate amplitude (as opposed to frequency) modulation of dopaminergic chemical signaling.

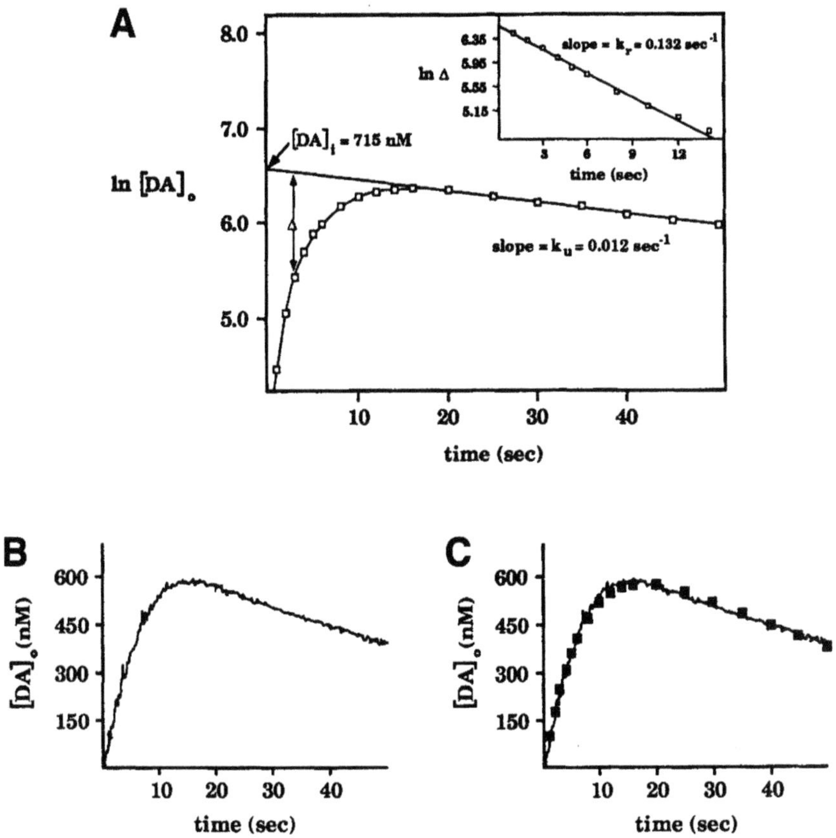

Fig. 3. Illustration of the results of kinetic analyses of depolarization-induced release of DA and its subsequent reuptake. **(A)** Results of the standard graphical solution to obtain the rate constants of Eq [2]. When the theoretical profile was evaluated by substituting the result back into Eq. (2), systematic error was observed **(B)**. However, nonlinear curve fitting was found to produce the best fits, within 2% of experimental values, with no apparent systematic errors **(C)**. Reprinted from McElvain and Schenk (1992b) and used with kind permission from from Pergamon Press, Headington Hill Hall, Oxford, UK.

4.2. Kinetics-Based Mechanistic Studies of the Dopaminergic Transporter

Another valuable application of the RDE has been to determine the mechanism by which exogenously added DA is taken up into striatal suspensions. The data were found to fit a model involving multiple substrates and have been described in detail (McElvain and Schenk, 1992b). Briefly, the mechanism of DA uptake is a partially random, sequential multisubstrate mechanism, wherein DA and Na^+ bind first to the transporter in random order followed by Cl^- binding prior to the movement of DA into or across the membrane.

In this study, the velocity data of DA uptake profiles were tested for adherence to Michaelis-Menten kinetics. Once confirmed, the Michaelis-Menten dependency of uptake on Na^+ and Cl^- was determined by substitution of each ion with choline or isethionate, respectively. Profiles of $[DA]_0$ vs t suggest that the clearance profile follows mixed-order kinetics, following apparent zero-order kinetics for the first 25 s and apparent first-order thereafter. Analyses in subsequent experiments were performed using data from the region of the profile defined by zero-order kinetics. The K_m and V_{max} values for DA were found to be 1.24 ± .06 μ*M* and 582 ± 11 pmol/s/g wet wt, respectively.

In the studies of ion dependencies, fitting the rate data to Eq. (4) identified the reaction order of the ion effect and provided estimates of the kinetic parameters. The apparent reaction orders in Na^+ and Cl^- were 2 and 1, respectively. Na^+ influenced DA uptake by a change in the V_{max}, but not the K_m value, an allosteric effect. Finally, the influence of Cl^- on DA transport was uncompetitive, since the V_{max}/K_m value was constant (McElvain and Schenk, 1992b). Graphs of the slopes of Eadie-Hofstee plots vs 1/[ion] and analysis of the results by the methods of Fromm (1975) gives evidence favoring a multisubstrate mechanism of the partially random, sequential type. Figure 4 illustrates the transport mechanism.

4.3. Resolution of k_{ass} and k_{cat} with RDE Techniques

Measurements were made only under the steady-state condition in the previously described studies of the transport process. It is possible to resolve k_{ass} from k_{cat} by analyses of steady-state data (Fersht, 1985), and we have applied these analyses to the study of the transport of DA (Meiergerd and Schenk, 1994a). Subsequently, the numerical value of k_{ass} was

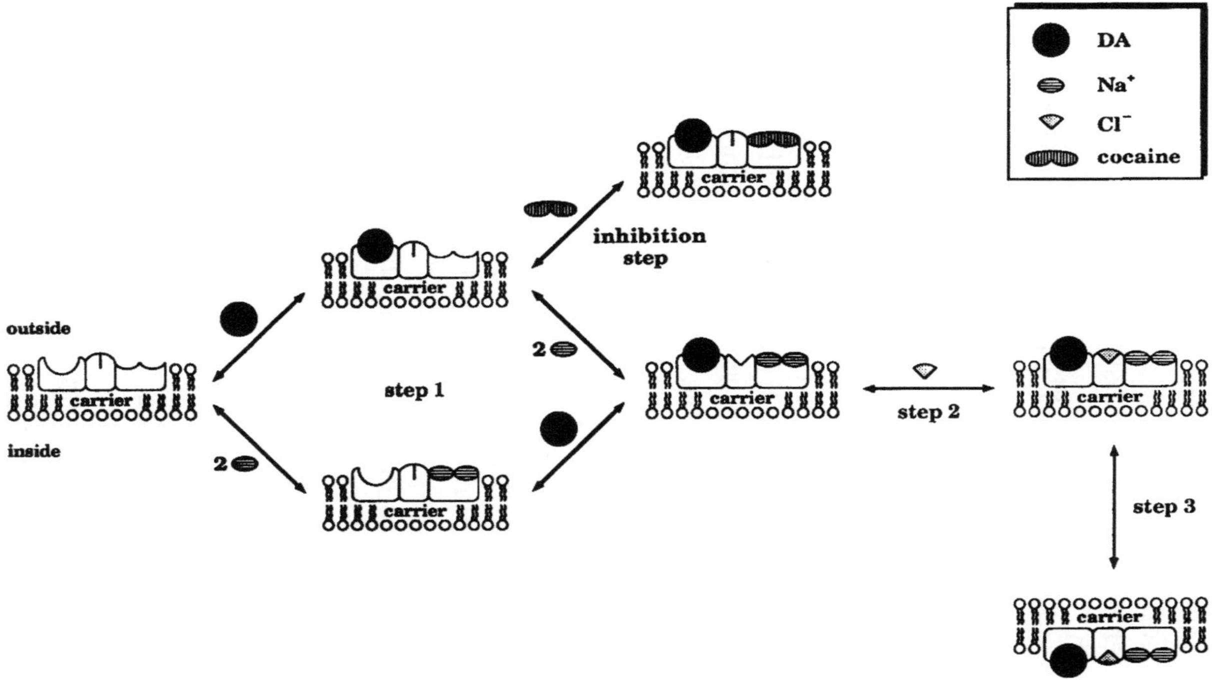

Fig. 4. Illustration of the kinetic model for transporter-mediated uptake of DA into resting suspensions. The mechanism is a partially random sequential process in which DA or Na^+ binds first to the transporter protein (Step 1) followed by the binding of Cl^- (Step 2) before movement (Step 3) of DA into or across the membrane. Cocaine was found to inhibit the process (indicated as the inhibition step) by competing with Na^+ for its binding site. Reprinted from McElvain and Schenk (1992a) and used with kind permission from Pergamon Press, Headington Hill Hall, Oxford, UK.

confirmed by apparent presteady-state kinetics (Meiergerd et al., 1994b). Briefly, the apparent zero-order velocity of DA uptake was measured following the addition of DA over the concentration range of 0.1–1.5 μ*M* DA (one concentration per striatal suspension). Values of V_{max} and K_m were obtained as described earlier with values indistinguishable from those previously given. The values of k_{cat} and k_{ass}, estimated using Eqs. (5) and (6), were found to be 1.4 ± 0.03/s and $1.2 \pm 0.10 \times 10^{-6}/M/s$, respectively.

The same experimental protocol used to make measurements under steady-state conditions was applied to presteady-state conditions where 50 n*M* DA was added to the suspension, and its disappearence was monitored until a value of 17 n*M* was reached. Justification of the concentration chosen was explained in detail (Meiergerd et al., 1994b). Based on the density of the transporter sites (416 pmol/g wet wt) and the sample weight (0.040 g wet wt), in a 500-μL vol, the tissue is capable of taking up approx 33 n*M* DA by the binding step alone. Values of k_{ass} predicted under steady-state conditions suggested that the binding event at this $[DA]_o$ should occur between 1.7 and 17 s, a range easily obtainable by the RDE. The data were fit to Eq. (8) and k_{ass} was obtained from the slope of the plot of the right-hand side of the equation vs *t*. The average presteady-state value of $1.4 \pm 0.2 \times 10^{-6}/M/s$ was found to be statistically indistinguishable from the steady-state value.

4.4. Structure-Activity Requirements for Catechol Transport at the Striatal Transporter for Dopamine

The RDE system has also been used to determine the structural requirements for substrate recognition and translocation at the striatal dopaminergic transporter (Meiergerd and Schenk, 1994a). The strategy used was to compare the rate of uptake of DA to the rate of uptake of a DA derivative to which alterations to the side-chain moiety or the catechol were made. In addition, the effect of the modified compound, if any, on DA uptake was monitored. Values of V_{max} and K_m

were calculated for the test compound. Two of the compounds tested were discussed as examples of how the information can be used. The rate of uptake of 4-ethylcatechol at 1.0 μ*M* was 79% less than the rate of DA uptake at the same concentration, but it inhibited the transport of DA and had a K_m value for the transporter equal to that of DA. However, its V_{max} was 77% less than that of DA. Thus, it appears that 4-ethylcatechol binds to the transporter, but is not transported as well as DA. 3-Methoxytyramine, however, has a K_m value seven times greater than that of DA and a V_{max} equivalent to that of DA. Taken together, these data suggest that the recognition of DA requires an intact catechol, whereas transport of dopamine is mediated by both the catechol and amine moieties (Fig. 5).

Chemical modifications of the transporter (Fig. 5) were found also to affect its function. As has been shown by others, external glycoslation as well as sulfhydryl groups are necessary (Meiergerd and Schenk, 1994a). In addition, the transporter may have a redox modulatory site since pyrroloquinoline quinone, a redox modulator of receptor function, also affects the functioning of the transporter for DA (Meiergerd and Schenk, 1994a).

4.5. Studies of Inhibition of Dopaminergic Transport

Cocaine inhibits the neuronal transport of DA (Horn, 1979), and this effect is believed to mediate the reinforcing properties of this drug of abuse (Ritz et al., 1987). Thus, a kinetic mechanistic investigation of the inhibition of DA transport by cocaine was undertaken using the RDE technique. Assuming other structurally dissimilar inhibitors of DA uptake could act in a different fashion, a kinetic evaluation of the sites of inhibition by various inhibitors, relative to cocaine, was made. This section is thus a brief overview of the results of these studies, as well as a discussion of the effects of inhibition on k_{ass} and k_{cat} in animals repeatedly treated with cocaine and in self-administering animals

Inhibitions of the transport of DA by cocaine were analyzed by Eadie-Hofstee analyses (McElvain and Schenk,

A Characteristics of the Substrate for Optimal Transport

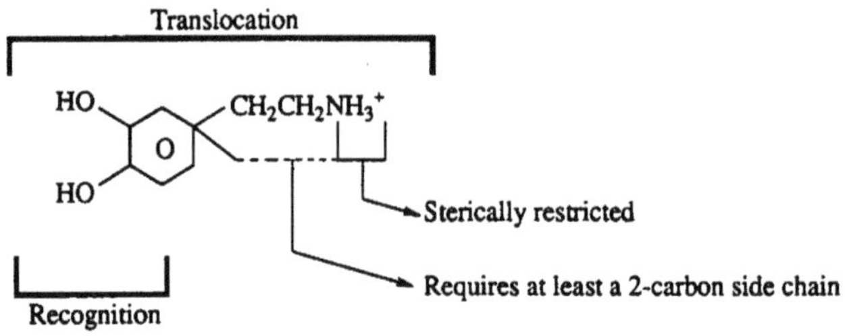

B Characteristics of the External Face of the Transporter for Optimal Function

Reduced by PQQ

Alkylated by chloroethylnorapomorphine

Not alkylated by EEDQ

Requires external glycosylation

Sulfhydryls required

Fig. 5. Summary of results of structure–reactivity studies of DA transport into striatal suspensions. **(A)** Illustration of the findings that the catechol moiety of DA appears to mediate recognition of catechol substrates by the transporter, and that both the catechol functionality and the ethylamine side chain are required for translocation. **(B)** List of the results of chemical modifications of the transporter and their effects on its kinetic function. NEM: *N*-ethylmaleimide; PQQ: pyrrolo-quinoline quinone.

1992b). Transport velocities were obtained from linear regression analysis of $[DA]_o$ vs t data over the zero-order region of the uptake curve. The values of K_m and V_{max} for DA uptake changed, yet the ratio of V_{max} to K_m remained constant at an average value of 462 ± 16 pmol/s/g wet wt/μ*M* (indistinguishable from the value obtained in the absence of cocaine, 469 pmol/s/g wet wt/μ*M*). Thus, cocaine inhibits DA uncompetitively. Estimates of V_{max}, K_m, and x were then used as first appproximations in nonlinear curve fitting to the Michaelis-Menten equation modified for uncompetitive inhibition to determine the value for the inhibition constant K_i. The K_i for cocaine was found to be 1.05 ± 0.03 μM (McElvain and Schenk, 1992b) The results from repeating this study in the presence of varying concentrations of either Na^+ or Cl^- suggested that cocaine inhibits the effects of both ions on DA uptake. In summary, cocaine inhibits the uptake of DA uncompetitively, and affects the Na^+ and Cl^- dependencies competitively.

To determine the effects of repeated cocaine on k_{cat} and k_{ass}, estimates of these constants were made in animals repeatedly treated with cocaine for 3 d followed by either a 24-h or 2-wk withdrawal period (Meiergerd et al., 1994b) and in animals trained to self-administer for 20 d with subsequent withdrawal for 3 wk (Meiergerd et al., 1994a). In both studies, the turnover of the transporter was increased relative to controls with no changes in the kinetics of substrate association. Under presteady-state conditions, k_{ass} was also unaffected by the addition of 1.0 μM cocaine to the striatal suspension. Taken together, these results suggest that the reinforcing action of cocaine may be the result of its inhibition of the k_{cat} of the dopaminergic transporter at an intramembrane site.

To support this hypothesis, studies of the inhibition of DA transport by mazindol were undertaken (Meiergerd and Schenk 1993, 1994b). Mazindol is another inhibitor of DA uptake, but has a lesser potential for abuse than cocaine (Chait et al., 1987) and has been used in the treatment of drug abuse in humans (Berger et al., 1989). A competitive relationship was found to exist between DA and mazindol with mazindol inhibiting k_{ass}, the kinetic constant of the binding step of inward transport of DA. Thus, there may be transporter-based mechanistic differences between DA transport inhibitors having different propensities for addiction. Further evaluations of the relationship between cocaine and mazindol were necessary to validate this hypothesis. If it were found that the two inhibitors compete for the same binding site on the transporter, the data obtained thus far would appear to make little sense. The relationship between cocaine and mazindol was determined by applying the kinetic models described by Devés and Krupka (1980) and given earlier as Eqs. (9)–(11). Briefly, the inhibition of the initial rate of the transport of DA was measured in the presence of each inhibitor acting alone and then acting in combination. The rates of DA uptake observed in the absence and presence of each inhibitor alone were used to provide a predicted value for the rate of DA transport in the presence of both inhibitors. The rate measured experimentally in the presence of the paired inhibitors was then compared to the predicted values for each of the models. It was found that cocaine and mazindol bind to the DA transporter at separate sites, but these sites are linked in such a way that the binding of one inhibitor negatively affects binding of the other. In addition, structurally similar analogs of cocaine, cocaethlene, and benztropine, were also tested in this model and found to be competitive with cocaine.

4.6. *Integration of Results Obtained with the RDE to Studies Conducted In Vivo*

Reuptake profiles conducted in vivo are qualitatively similar to those obtained by the RDE technique (compare profiles found in Patterson and Schenk [1991] to those of Fig. 1). In a study using the RDE technique designed to determine what effects presynaptic autoreceptors have on DA uptake, it was found that inward transport could be increased by D_2 receptor stimulation (Meiergerd et al., 1993). These findings caused us to evaluate the reuptake profiles observed in in vivo chronoamperometric studies following D_2 receptor antagonism, and resulted in the discovery that inwardly directed kinetic activity of the dopaminergic transporter is affected by D_2 receptors in vivo (Meiergerd et al., 1993).

In addition, results from the RDE system can be used to guide experimentation in vivo in the study of DA transport inhibitors. In our work with the RDE system, we find that cocaine inhibits the inwardly directed transport of DA by an uncompetitive mechanism, whereas mazindol inhibits transport competitively. Armed with this mechanistic information an in vivo voltammetric experiment would be expected to reveal more efficacious inhibitions of uptake by cocaine at the higher stimulations, assuming that the degree of stimulation is proportional to the amount of DA released, whereas mazindol inhibitions should be more efficacious at the lower levels of stimulation. Thus, we feel that an understanding of the apparent mechanisms of biogenic amine release and uptake observed in the RDE system can be used to guide the experiment in vivo, clearly the more work-intensive and expensive technique.

5. Limitations and Future Directions

The RDE appears to have relatively broad application to the study of biogenic amine release and reuptake with only a few limitations. The experiment described in this chapter for monitoring the uptake of DA approximates a zero trans entry transport experiment (Meiergerd and Schenk, 1994a,b) and has been shown to have kinetic characteristics of a transporter as opposed to a pore or channel (Meiergerd et al., 1994b). However, it should be noted that the zero trans entry experiment is only one of six types of experiments that can be designed to study the functioning of transporters (Stein, 1986, 1990; Devés and Krupka, 1989). Some of the other types of experiments can be conducted by appropriate changes in the experimental conditions used. However, obtaining information from the RDE system relevant to the equilibrium exchange experiment is difficult at the present time. Additional limitations of the RDE system are that it cannot be used to enhance time resolution of chemical measurements in intact tissues or measure release and reuptake of transporter activity for neurotransmitters, such as acetylcholine, the amino acids, or neuropeptides that are difficult to oxidize.

The RDE technique can be developed further to provide more details on the mechanism of biogenic amine neurotransmitter release and reuptake. Some developments are currently under way. One goal is to study the function of the transporter operating in reverse (outwardly directed activity). Atwell et al. (1993) have recently reviewed the importance of nonvesicular release, and the transporters are thought to mediate this event under both functional and

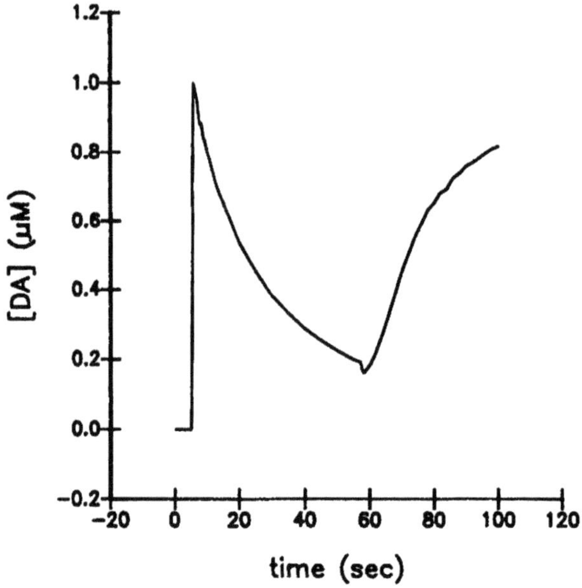

Fig. 6. Example of a veratridine-stimulated reversal of an inwardly directed $[DA]_o$ vs time profile. Prior to experimentation, the rat was treated with reserpine to destroy intracellular storage vesicles. The striatal suspension was then prepared and incubated in physiological buffer containing 100 μ*M* pargyline to inhibit monoamine oxidase. After a baseline was obtained, 1.0 μ*M* DA was added (at ca. 6 s). The signal instantaneously increased and subsequently decreased owing to inwardly directed transport of DA. Veratridine (10 μ*M*) was added after about 80% of the extracellular DA was cleared (at ca. 60 s), resulting in a release profile mediated by reversal of the transporter. Thus, the kinetics of the outwardly directed function of the transporter can be studied by the RDE technique.

pathological conditions. We have approached this problem by using veratridine to stimulate the dopaminergic transporter to run backward following the preloading of DA into striatal suspensions devoid of intracellular vesicles and monoamine oxidase activity (*see* Fig. 6). Another effort in this laboratory is to measure intracellular changes in inorganic ion concentrations along with $[DA]_o$ changes as a result of depolarization-induced release and/or uptake. Intracellular changes in inorganic ions are most often measured using spectroscopically active dyes. In such an experiment, RDE measurements would be conducted in a fluorometer so that the time-courses of DA and the target ions could be monitored simultaneously. At this time, we are still designing physical modifications to a Shimadzu fluorometer to accomplish this goal. If sucessful we should be able to study the electrogenic properties of the dopaminergic transporter by concomitantly monitoring spectroscopic properties of voltage-sensitive dyes and detect the transport of DA with the RDE system.

A more fundamental goal is to determine whether the temporal resolution of the measurements can be increased toward the low-millisecond time scale with good precision. It is expected that experiments conducted in different tissue preparations, such as synaptosomes, membrane preparations, or cells expressing cloned transporters, will require increases in signal-to-noise ratios. Such increases may be obtained by using hydrodynamic modulation. Before the method can be used effectively on these short time scales, theoretical developments related to understanding the effects of suspended particles on RDE responses will have to be developed. We feel that these developments will be forthcoming given that measurements in synaptosomes have been made (McElvain and Schenk, 1992b), procedures for increased signal-to-noise even on fast time scales using hydrodynamic modulation have been described (Miller et al., 1972), and background information on the effects of suspended particles, including biological cells, on RDE responses has been published (Anderson et al., 1989; Caprani et al., 1987, 1988; de Ficquelmont-Loizos, 1988).

6. Conclusions

The RDE technique has been shown to be a useful and relatively inexpensive tool for making measurements of the fluxes of DA in the kinetic domain. To our knowledge, it is the only chemical technique available that has been shown to be useful in the measurement of unidirectional fluxes of reuptake transport of a neurotransmitter subsequent to its stimulated release. The technique clearly has practical application to the study of the kinetic mechanisms of release and uptake transport and the effects of pharmacological agents on these processes.

Acknowledgments

The authors wish to thank James McElvain of Marion Merrill Dow (Kansas City, MO) for his continued interest in this work and discussions with us during the preparation of this chapter. The invaluable help of Bill Ryan and Fred Schuetze of Technical Services for their help in designing the incubation cell and instrumentation, respectively, is gratefully acknowledged. This work was supported by the National Institute on Drug Abuse (Grants DA07384 and DA00184 to JOS and Grant DA05568 to SMM), the National Institute of Mental Health (Grant MH42759 to JOS), the Washington Alcohol and Drug Abuse Program (Legislative Initiative 171), Washington State University, and the State of Washington.

References

Adachi D. K., Kalivas P. W., and Schenk J. O. (1990) Neurotensin binding to dopamine. *J. Neurochem.* **51,** 1321–1328.

Adams R. N. (1969) *Electrochemistry at Solid Electrodes,* Marcel Dekker, New York.

Adams R. N. and Marsden C. A. (1982) Electrochemical detection methods for monoamine measurements *in vitro* and *in vivo*, in *Handbook in Psychopharmacology,* vol. 15 (Iversen L. L., Iversen S. D., and Snyder S. H., eds.), Plenum, New York, pp. 1–74.

Anderson P. K., Muller R. H., and Tobias C. W. (1989) The effect of suspended particles on mass transfer to a rotating disk. *J. Electrochem.* **136,** 390–399.

Atwell D., Barbour B., and Szatkowski M. (1993) Non-vesicular release of neurotransmitter. *Neuron* **11,** 401–407.

Berger P., Gawin F., and Kosten T. R. (1989) Treatment of cocaine abuse with mazindol. *Lancet* **I,** 283.

Booth R. F. and Clark J. B (1978) A rapid method for the preparation of relatively pure metabolically competent synaptosomes from rat brain. *Biochem. J.* **176,** 365–370.

Bruckenstein S. and Miller B. (1977) Unraveling reactions with rotating electrodes. *Acc. Chem. Res.* **10,** 54–61.

Capellos C. and Bielski B. H. J. (1972) *Kinetic Systems: Mathematical Description of Chemical Kinetics in Solution,* Wiley-Interscience, New York.

Caprani A., Tamisier L., and de Ficquelmont-Loizos M. M. (1987) Novel tools for determining mechanically induced modifications of membrane structure: electrochemical and fluorescence methods. *Bioelectrochem. Bioenerg.* **17,** 303–315.

Caprani A. de Ficquelmont-Loizos M. M., Tamisier L., and Peronneau P. (1988) Mass transfer in laminar flow at a rotating disk electrode in suspensions of inert particles II. Theoretical evaluations of experimental results. *J. Electrochem. Soc.* **135,** 635–642.

Chait L. D., Uhlenhuth E. H., and Johanson C. E. (1987) Reinforcing and subjective effects of several anorectics in normal human volunteers. *J. Pharmacol. Exp. Thera.* **242,** 777–783.

de Ficquelmont-Loizos M. M., Tamisier L., and Caprani A. (1988) Mass transfer in laminar flow at a rotating disk electrode in suspensions of inert particles I. Experimental investigation on the influence of the electrode, the medium, and the particle. *J. Electrochem. Soc.* **135,** 626–634.

Devés R. and Krupka R. M. (1980) Testing transport systems for competition between pairs of reversible inhibitors, inhibition of erythrocyte glucose transport by cytochalasin B and steroids. *J. Biol. Chem.* **255,** 11,870–11,874.

Devés R. and Krupka R. M. (1989) Inhibition kinetics of carrier systems. *Meth. Enzymol.* **171,** 113–132.

Dryhurst G., Kadish K. M., Scheller F., and Renneberg R. (1982) Catecholamines, in *Biological Electochemistry,* vol. 1, Academic, New York, pp. 116–179.

Fersht A. (1985) *Enzyme Structure and Mechanism,* 2nd ed., Freeman, New York.

Fromm H. J. (1975) *Initial Rate Enzyme Kinetics,* Springer-Verlag, New York.

Humphries K. and Dryhurst G. (1987) Electrochemical oxidation of 5-hydroxytryptophan in acid solution. *J. Pharmaceut. Sci.* **76,** 839–846.

Horn A. S. (1979) Characteristics of DA uptake, in *The Neurobiology of Dopamine* (Horn A. S., Korf J., and Westernick B. H. C., eds.), Academic, London, pp. 217–235.

Justice J. B. (ed.) (1987) *Voltammetry in the Neurosciences: Principles, Methods, and Applications.* Humana, Clifton, NJ.

Malachesky P. A., Marcoux L. S., and Adams R. (1966) Homogeneous chemical kinetics with the rotating disk electrode. *J. Phys. Chem.* **70,** 4068–4070.

Marsden C. A. (ed.) (1984) *Measurement of Neurotransmitter Release In Vivo.* Wiley, Chichester, England.

McElvain J. S. (1992) In vitro studies of the kinetics of endogenous dopamine release and reuptake into rat striatal suspensions using rotating disk electrode voltammetry. Doctoral Dissertation, Washington State University, Pullman, WA.

McElvain J. S. and Schenk J. O. (1992a) Blockade of dopamine autoreceptors by haloperidol and the apparent dynamics of potassium-stimulated endogenous release of dopamine from and reuptake into striatal suspensions in the rat. *Neuropharmacology* **31,** 649–659.

McElvain J. S. and Schenk J. O. (1992b) A multisubstrate mechanism of striatal dopamine uptake and its inhibition by cocaine. *Biochem. Pharmacol.* **43,** 2189–2199.

Meiergerd S. M. and Schenk J. O. (1993) The functioning DA transporter: relationships between interactions with substrate and two structurally dissimilar inhibitors, cocaine and mazindol. *Soc. Neurosci. Abstr.* **19,** 745.

Meiergerd S. M. and Schenk J. O. (1994a) Striatal transporter for dopamine: catechol structure activity studies and susceptibility to chemical modification. *J. Neurochem.* **62,** 998–1008.

Meiergerd S. M. and Schenk J. O. (1994b) Kinetic evaluation of the commonality between the site(s) of action of cocaine and some other structurally similar and dissimilar inhibitors of the striatal transporter for dopamine. *J. Neurochem.* **63,** 1683–1692.

Meiergerd S. M., Hooks S., and Schenk J. O. (1994a) The striatal transporter for dopamine in the rat may be kinetically up-regulated following three weeks of withdrawal from cocaine self-administration. *J. Neurochem.* **63,** 1277–1281.

Meiergerd S. M., McElvain J. S., and Schenk J. O. (1994b) The effects of cocaine and repeated cocaine followed by withdrawal: alterations of dopaminergic transporter turnover with no changes in kinetics of substrate recognition. *Biochem. Pharmacol.* **47,** 1627–1634.

Meiergerd S. M., Patterson T. A., and Schenk J. O. (1993) D2 receptors may modulate the function of the striatal transporter for dopamine: kinetic evidence from studies in vitro and in vivo. *J. Neurochem.* **61,** 764–767.

Miller B., Bellavance M. I., and Bruckenstein S. (1972) Feasibility and applications of programmed speed control at rotating ring disk electrodes. *Analyt. Chem.* **44,** 1983–1992.

Patterson T. A. and Schenk J. O. (1991) Effects of acute and chronic systemic administration of some typical antipsychotic drugs on turnover of dopamine and potassium ion-induced release of dopamine in the striatum of the rat in vivo. *Neuropharmacology* **30,** 943–952.

Ritz M. C., Lamb R. T., Goldberg S. R., and Kuhar M. J. (1987) Cocaine receptors on dopamine transporters are related to self-administration of cocaine. *Science* **237,** 1219–1223.

Schenk J. O., Morocco M. T., and Ziemba V. A. (1991) Interactions between the argninyl moieties of neurotensin and the catechol protons of dopamine. *J. Neurochem.* **57,** 1787–1795.

Schenk J. O., Patterson T. A., and McElvain J. S. (1990) Rotating disk voltammetric measurements in neurobiology and neuropharmacology. *Trends Anal. Chem.* **9,** 325–330.

Schenk J. O., Miller E., Gaddis R., and Adams R. N. (1982) Homeostatic control of ascorbate concentration in CNS extracellular fluid. *Brian Res.* **253**, 353–356.

Stein W. D. (1986) *Transport and Diffusion Across Cell Membranes.* Academic, San Diego, CA.

Stein W. D. (1986) *Channels, Carriers, and Pumps: An Introduction to Membrane Transport.* Academic, San Diego, CA.

Wilkinson L. (1990) *SYSTAT: The System for Statistics.* SYSTAT Inc., Evanston, IL.

Wolf M. E. and Roth R. H. (1987) Dopamine autoreceptors, in *Dopamine Receptors* (Creese I. and Fraser C. M., eds.), Liss, New York, pp. 45–96.

Index

Index

GPSR Compliance
The European Union's (EU) General Product Safety Regulation (GPSR) is a set of rules that requires consumer products to be safe and our obligations to ensure this.

If you have any concerns about our products, you can contact us on

ProductSafety@springernature.com

In case Publisher is established outside the EU, the EU authorized representative is:

Springer Nature Customer Service Center GmbH
Europaplatz 3
69115 Heidelberg, Germany

www.ingramcontent.com/pod-product-compliance
Ingram Content Group UK Ltd.
Pitfield, Milton Keynes, MK11 3LW, UK
UKHW020134160926
475607UK00003B/44
9780896033122